Marine Artificial Reef Research and Development

Integrating Fisheries Management Objectives

Marine Artificial Reef Research and Development

Integrating Fisheries Management Objectives

Edited by

Stephen A. Bortone

Osprey Aquatic Sciences, LLC

47 West Shore Road, Windham, New Hampshire 03087, USA

American Fisheries Society Symposium 86

Proceedings of the Symposium "Marine Artificial Reef Research and Development: Integrating Fisheries Management Objectives"
Held in Tampa, Florida, USA
August 22–23, 2017

American Fisheries Society
Bethesda, Maryland, USA
2018

Suggested citation formats follow.

Entire book

Bortone, S. A., editor. 2018. Marine artificial reef research and development: integrating fisheries management objectives. American Fisheries Society, Symposium 86, Bethesda, Maryland.

Chapter in book

Jaxion-Harm, J., S. T. Szedlmayer, and P. A. Mudrak. 2018. A comparison of fish assemblages according to artificial reef attributes and seasons in the northern Gulf of Mexico. Pages 23–45 *in* S. A. Bortone, editor. Marine artificial reef research and development: integrating fisheries management objectives. American Fisheries Society, Symposium 86, Bethesda, Maryland.

Front cover: Red Snapper photograph provided by Carol and Robert Cox, Mexico Beach Artifical Reef Association; reef module photographs provided by Keith Mille, Florida Fish and Wildlife Conservation Commission.

Printed in the United States of America on acid-free paper.

Library of Congress Control Number 2018961018

ISBN 978-1-934874-51-6
ISSN 0892-2284

American Fisheries Society Web site: www.fisheries.org
American Fisheries Society
425 Barlow Place, Suite 110
Bethesda, Maryland 20814
USA

Contents

Preface

The subject of artificial reef research and development has expanded over the past several decades to address issues related to their function and utility relative to fishery management as well as ecological performance. Because of the potential that artificial reefs may have in the future of environmental management, a symposium and a conference were held in 2017 on the subject.

The symposium, entitled "Marine Artificial Reef Research and Development: Integrating Fisheries Management Objectives," was held August 22–23, 2017 in Tampa, Florida (USA) during the 147th annual meeting of the American Fisheries Society. The symposium built upon the interest in the subject created as a result of the 2016 National Artificial Reef Workshop (USA) that examined artificial reefs as a potential management tool to support and/or enhance sustainable fisheries. Topics presented in Tampa included the development and evaluation of standards for performance criteria, matching the design and siting of artificial reef networks to meet specific management objectives, novel monitoring approaches to assess reef-associated assemblages, and mechanisms for incorporating data from artificial reefs into stock assessments. A total of 26 oral presentations were offered during the symposium, with an additional 11 poster presentations.

The conference was the 11th International Conference on Artificial Reefs and Related Aquatic Habitats (CARAH) held September 11–13, 2017 at the Universiti Malaysia Terengganu, Kuala Nerus, Terengganu, Malaysia. This conference emphasized reducing conflicts and improving/developing integration between exploitation of marine living resources by fisheries and aquaculture and other human activities; monitoring and long-term studies; design of artificial reefs and man-made materials; multifunctional maritime infrastructures; marine conservation, fisheries, and aquaculture; and the environmental and socioeconomic aspects of artificial reefs. There were 40 oral presentations offered at the conference. Ten posters were offered as well.

This book includes chapters chiefly from papers presented (or modified from the original presentations) during the symposium and conference indicated above. Additional chapters, related to the general topic of artificial reefs but not presented at in either meeting, are also included here.

Note: Bortone, in his review of the international conferences on artificial reefs (i.e., CARAH; Bortone, S. A. 2015. CARAH [International Conference on Artificial Reefs and Related Aquatic Habitats]: an historical perspective of accomplishments. Journal of Applied Ichthyology 31[Supplement 2]:3–14.), indicated a lack of information regarding the second CARAH. His suspicion was that the literature references to "an artificial reef conference held in Brisbane in September 1977 called the Artificial Reef Symposium, 5th World Underwater Congress 'Downunder77', hosted by the Australian Underwater Federation of Brisbane" was a reference to the second CARAH. This suspicion has been confirmed. A colleague, Daniel

J. Sheehy (Aquabio, Inc.), was able to find a page from the agenda of the second international conference on artificial reefs. While incomplete, the information on the agenda page indicates that the meeting was held at the Crest Hotel in Brisbane, Australia, beginning on September 11, 1977. Thus, the papers described in Bortone (2015) were in fact part of the second CARAH. Thanks to Dan Sheehy for his assistance on verifying this missing piece of information.

Acknowledgments

Since the majority of chapters offered here are the result of two recent professional meetings on the subject of artificial reefs, it is important to acknowledge and recognize the contribution of the individuals and agencies that contributed to the success of these meetings.

I thank Keith Mille and Sean Keenan, both with the Florida Fish and Wildlife Conservation Commission, who served as chairs of the American Fisheries Society symposium and were instrumental in organizing the symposium. I also thank Dr. Jarina Mohd Jani and Professor Mazlan Ghaffar, from the School of Fisheries and Aquaculture at the Universiti Malaysia Terengganu, who were chiefly responsible for organizing the 11th International Conference on Artificial Reefs and Related Aquatic Habitats (CARAH). Dr. Jarina Mohd Jani was especially helpful in encouraging CARAH participants to submit chapters included here. Providing support for the 11th CARAH were the Universiti Malaysia Terengganu and the Coral Triangle Initiative on Coral Reefs, Food Security and Fisheries. Institutional support for the conference was provided by the Ministry of Higher Education, Malaysia and the Department of Fisheries Malaysia.

No effort like this is possible without the expert advice and critical comments provided by reviewers of the chapters. I thank Sean Keenan (Florida Fish and Wildlife Conservation Commission), Matthew Streich (Texas A&M University, Corpus Christi), Rob Andrews (U.S. National Oceanic and Atmospheric Administration), Donna Schroeder (U.S. Bureau of Ocean Energy Management), and Stephen T. Szedlmayer (Auburn University) for their help in reviewing early drafts of manuscripts.

I also thank Aaron Lerner and Debby Lehman(publications director and book production editor, respectively) of the American Fisheries Society for their expert help in seeing the book through to completion. Photographs for the cover were gratefully provided by Carol and Robert Cox from the Mexico Beach Artificial Reef Association (Red Snapper) and Keith Mille from the Florida Fish and Wildlife Conservation Commission (reef modules).

Last, I thank the contributors to this volume, as well as the participants and attendees of both meetings, for their earnest contributions to the discussion that has helped continue the exploration of artificial reefs as an option in the management of fisheries and associated aquatic resources.

—Stephen A. Bortone

Symbols and Abbreviations

The following symbols and abbreviations may be found in this book without definition. Also undefined are standard mathematical and statistical symbols given in most dictionaries.

A	ampere
AC	alternating current
Bq	becquerel
C	coulomb
°C	degrees Celsius
cal	calorie
cd	candela
cm	centimeter
Co.	Company
Corp.	Corporation
cov	covariance
DC	direct current; District of Columbia
D	dextro (as a prefix)
d	day
d	dextrorotatory
df	degrees of freedom
dL	deciliter
E	east
E	expected value
e	base of natural logarithm (2.71828…)
e.g.	(exempli gratia) for example
eq	equivalent
et al.	(et alii) and others
etc.	et cetera
eV	electron volt
F	filial generation; Farad
°F	degrees Fahrenheit
fc	footcandle (0.0929 lx)
ft	foot (30.5 cm)
ft^3/s	cubic feet per second ($0.0283\ m^3/s$)
g	gram
G	giga (10^9, as a prefix)
gal	gallon (3.79 L)
Gy	gray
h	hour
ha	hectare (2.47 acres)
hp	horsepower (746 W)
Hz	hertz
in	inch (2.54 cm)
Inc.	Incorporated
i.e.	(id est) that is
IU	international unit
J	joule
K	Kelvin (degrees above absolute zero)
k	kilo (10^3, as a prefix)
kg	kilogram
km	kilometer
l	levorotatory
L	levo (as a prefix)
L	liter (0.264 gal, 1.06 qt)
lb	pound (0.454 kg, 454g)
lm	lumen
log	logarithm
Ltd.	Limited
M	mega (10^6, as a prefix); molar (as a suffix or by itself)
m	meter (as a suffix or by itself); milli (10^{-3}, as a prefix)
mi	mile (1.61 km)
min	minute
mol	mole
N	normal (for chemistry); north (for geography); newton
N	sample size
NS	not significant
n	ploidy; nanno (10^{-9}, as a prefix)
o	ortho (as a chemical prefix)
oz	ounce (28.4 g)

P	probability
p	para (as a chemical prefix)
p	pico (10^{-12}, as a prefix)
Pa	pascal
pH	negative log of hydrogen ion activity
ppm	parts per million
qt	quart (0.946 L)
R	multiple correlation or regression coefficient
r	simple correlation or regression coefficient
rad	radian
S	siemens (for electrical conductance); south (for geography)
SD	standard deviation
SE	standard error
s	second
T	tesla
tris	tris(hydroxymethyl)-aminomethane (a buffer)
UK	United Kingdom
U.S.	United States (adjective)
USA	United States of America (noun)
V	volt
V, Var	variance (population)
var	variance (sample)
W	watt (for power); west (for geography)
Wb	weber
yd	yard (0.914 m, 91.4 cm)
α	probability of type I error (false rejection of null hypothesis)
β	probability of type II error (false acceptance of null hypothesis)
Ω	ohm
μ	micro (10^{-6}, as a prefix)
′	minute (angular)
″	second (angular)
°	degree (temperature as a prefix, angular as a suffix)
%	per cent (per hundred)
‰	per mille (per thousand)

American Fisheries Society Symposium 86:1–5, 2018

An Introduction to Artificial Reef Research and Development

Stephen A. Bortone*
Osprey Aquatic Sciences, LLC
47 West Shore Road, Windham, New Hampshire 03087, USA

The deployment of artificial reefs to improve fisheries and fishing has been occurring at least since the earliest records of human civilization (Seaman and Sprague 1991). Artificial reefs, especially as a topic of interest to the general and fishing public, environmental resource managers, and scientists, have received considerable interest since 1974 when the first international conference on artificial reefs was held (Clark et al. 1974). While there are many objectives being served by their use in the aquatic environment, the unabashed purpose of artificial reef deployment has been to explore "the potential use of artificial reefs in fishery management" (Clark et al. 1974; see Bortone [2006, 2015] for a more complete review).

It is clear from the literature that the interest in deploying artificial reefs has focused on the presumption that they function to increase the attraction (i.e., aggregation), increase the production (increase in fish biomass), or simultaneously serve in both the attraction and production of living aquatic resources in a given area (Polovina 1991; Bortone 2008). Concomitantly, with few exceptions, there has been little actual application of artificial reefs in the prescribed management of fisheries (Bortone 2011). Hence, research regarding artificial reefs continues to develop and expand to provide much needed answers aimed at resolving questions that managers continue to raise.

* Corresponding author: steve.bortone@gmail.com

Artificial Reef: A Definition

There have been many definitions proposed for artificial reefs over the past four decades. One can look to Wikipedia (June 2013) for a definition: "An artificial reef is a human-made underwater structure, typically built to promote marine life in areas with a generally featureless bottom, control erosion, block ship passage, or improve surfing." The definition from Seaman and Jensen (2000:5) is a more thoughtful consideration: "An artificial reef is one or more objects of natural or human origin deployed purposefully on the seafloor to influence physical, biological, or socioeconomic processes related to living marine resources."

The word "artificial" obviously implies something that should not normally be present in a location. It is not in opposition to the word "natural," however. Many artificial reefs are in fact made of natural materials, but when used in the context of artificial reefs, "artificial" refers to materials in the environment that are artificial to their normal location or composition. The word "allochthonous" comes to mind as a possible alternative to the word "artificial." "Allochthonous" is used by ecologists to describe items that are out of their normal place, but it is not generally used to describe deployed reef materials. The term "artificial reef" is chosen throughout this discourse to indicate a structure (either human-made or natural) that has been placed (or sometimes inadvertently deposited or dropped) into the aquatic environment. Here, artificial reef is meant to denote structure. Thus, any structure when placed into the aquatic environment (either intentionally or inadvertently) can be construed as being an artificial reef. The word "reef" in most contexts refers to a structure, large or small, that occurs in the aquatic environment. Several other clarifications are in order. It should be noted that Seaman and Jensen included "physical, biological, or socioeconomic processes." Clearly, each of these processes are not mutually exclusive, so an "and" should be inserted to be more accurate. Last, artificial reefs are not exclusively marine as there are many applications in fresh and estuarine waters around the world.

Bortone (2016) suggested a more refined definition, building on that of Seaman and Jensen (2000): an artificial reef is one or more objects of natural or human origin deployed purposefully or inadvertently on the seafloor to influence physical, biological, and/or socioeconomic processes related to living marine, estuarine, and freshwater resources.

Summary of Themes

As indicated in the preface, this book is the result of presentations offered at two recent symposia/conferences on the subject of artificial reefs. Herein are offered a series of chapters that help enlighten our understanding of artificial reef function, especially when artificial reefs are deployed in pursuit of improving fisheries management.

Seven chapters (Caddy 2018; Froehlich et al. 2018; Gaetani et al. 2018; Jaxion-Harm et al. 2018; Kohno et al. 2018; Lee et al. 2018; Szedlmayer and Miller 2018; all this volume) pertain directly to the interaction of organisms and the artificial reef as habitat. It is clear that adding inanimate objects to a water body can do little to improve the overall energy-input/biomass-output ratio relative to environmental productivity. However, it is potentially possible to channel energy in a way that enhances or improves productivity or sustainability among target species. These chapters examine assemblage and species specific interactions to uncover the relationships between biological limiting factors and habitat characteristics.

The next four chapters (Keenan et al. 2018; Martore and Bell 2018; Pondella et al. 2018; Schroeter et al. 2018; all this volume) explore the larger ecosystem interrelationship that artificial reefs have with the environment. Artificial reefs have been considered as mitigation for modified natural environments, and these chapters address this issue from a variety of perspectives. Moreover, artificial reefs can have a proper role in the conservation of our natural habitat as human communities continue to expand and develop.

The next set of papers (Cross et al. 2018; Delgadillo and Toro 2018; Mohd Jani et al. 2018; Mohd Salleh et al. 2018; all this volume) offer insight to facilitating the development of artificial reefs through a combination of community and agency interactions. Fishing communities stand as the biggest beneficiaries of successful artificial reef deployment and development. It is only fitting that all members of the user-group community, especially fishers, should be included in the decision-making process regarding artificial reefs, as the reefs are deployed in the user group's neighboring waters. Buy-in and community support are, by and large, thought to be key features in the future application of artificial reefs in aquatic resource management.

The last set of papers (Ali et al. 2018; Bortone 2018; Zakariya and Sakae 2018; all this volume) are directly related to the future design aspects of artificial reefs. Future aspects include both technological advances in reef assessments, but also include refinement of design, either through evaluations of a plethora of design concepts or consideration of the development of artificial reef plan objectives.

No summary is offered here as the efforts into the research and development of artificial reefs continues steadily, perceptively, and purposefully.

References

Ali, A., N. I. Tajudin, and R. Hazizi. 2018. Research and development on the design and construction of artificial reefs in Malaysia (1975–2017). Pages 289–313 *in* S. A. Bortone, editor. Marine artificial reef research and development: integrating fisheries management objectives. American Fisheries Society, Symposium 86, Bethesda, Maryland.

Bortone, S. A. 2006. A perspective of artificial reef research: the past, present, and future. Bulletin of Marine Science 37:11–39.

Bortone, S. A. 2008. Coupling fisheries with ecology through marine artificial reef deployments. Pages 917–924 *in* J. Nielsen, J. J. Dodson, K. Friedland, T. R. Hamon, J. Musick, and E. Verspoor, editors. Reconciling fisheries with conservation: proceedings of the fourth world fisheries congress, volume 1. American Fisheries Society, Symposium 49, Bethesda, Maryland.

Bortone, S. A. 2011. A pathway to resolving an old dilemma, lack of artificial reefs in fisheries management. Pages 311–321 *in* S. A. Bortone, F. P. Brandini, G. Fabi, and S. Otake, editors. Artificial reefs in fisheries management. CRC Press, Boca Raton, Florida.

Bortone, S. A. 2015. CARAH (International Conference on Artificial Reefs and Related Aquatic Habitats): an historical perspective of accomplishments. Journal of Applied Ichthyology 31(Supplement 2):3–14.

Bortone, S. A. 2016. Artificial reef. Pages 37–41 *in* M. J. Kennish, editor. Encyclopedia of estuaries. Springer, Dordrecht, Netherlands.

Bortone, S. A. 2018. An alternative approach to designing artificial reefs to manage marine fisheries. Pages 315–321 *in* S. A. Bortone, editor. Marine artificial reef research and development: integrating fisheries management objectives. American Fisheries Society, Symposium 86, Bethesda, Maryland.

Caddy, J. F. 2018. Reef fish depend on corals and natural outcrops: are predation and migration linked to the fractal characteristics of habitats? Pages 117–140 *in* S. A. Bortone, editor. Marine artificial reef research and development: integrating fisheries management objectives. American Fisheries Society, Symposium 86, Bethesda, Maryland.

Clark, W., J. Mosely, and R. Stone. 1974. Foreword. Page 3 *in* L. Colunga and R. Stone, editors. Proceedings of the international conference on artificial reefs. Texas A&M University Sea Grant College, TAMU-SG-74-103, College Station.

Cross, T. A., B. Sauls, R. Germeroth, and K. Mille. 2018. Methods to quantify recreational angling effort on artificial reefs off Florida's Gulf of Mexico coast. Pages 265–277 *in* S. A. Bortone, editor. Marine artificial reef research and development: integrating fisheries management objectives. American Fisheries Society, Symposium 86, Bethesda, Maryland.

Delgadillo-G, O., and N. N. Toro. 2018. An integrative artificial reef project for conservation and coastal management in the Columbian Caribbean. Pages 215–234 *in* S. A. Bortone, editor. Marine artificial reef research and development: integrating fisheries management objectives. American Fisheries Society, Symposium 86, Bethesda, Maryland.

Froehlich, C. Y. M., A. M. Lee, R. Oquita, C. E. Cintra-Buenrostro, J. D. Shively, and J. B. Shipley. 2018. A comparison of population dynamics from Red Snapper associated with inshore and offshore artificial reefs in the northwestern Gulf of Mexico. Pages 81–95 *in* S. A. Bortone, editor. Marine artificial reef research and development: integrating fisheries management objectives. American Fisheries Society, Symposium 86, Bethesda, Maryland.

Gaetani, A., A. N. Tassetti, S. Guicciardi, C. Ferrà, and G. Fabi. 2018. How does fish behavior change during the day around gas platforms? Pages 7–21 *in* S. A. Bortone, editor. Marine artificial reef research and development: integrating fisheries management objectives. American Fisheries Society, Symposium 86, Bethesda, Maryland.

Jaxion-Harm, J., S. T. Szedlmayer, and P. A. Mudrak. 2018. A comparison of fish assemblages according to artificial reef attributes and seasons in the northern Gulf of Mexico. Pages 23–45 *in* S. A. Bortone, editor. Marine artificial reef research and development: integrating fisheries management objectives. American Fisheries Society, Symposium 86, Bethesda, Maryland.

Keenan, S. F., T. S. Switzer, K. A. Thompson, A. J. Tyler-Jedlund, and A. R. Knapp. 2018. Comparison of reef-fish assemblages between artificial and geological habitats in the northeastern Gulf of Mexico: implications for fishery-independent surveys. Pages 141–163 *in* S. A. Bortone, editor. Marine artificial reef research and development: integrating fisheries management objectives. American Fisheries Society, Symposium 86, Bethesda, Maryland.

Kohno, N., S. Okubo, S. Otake, W. Fujie, T. Kaneko, T. Miyamukai, and K. Imai. 2018. Environmental factor analysis of snow crab associated with artificial reefs. Pages 69–80 *in* S. A. Bortone, editor. Marine artificial reef research and development: integrating fisheries management objectives. American Fisheries Society, Symposium 86, Bethesda, Maryland.

Lee, A. M., C. E. Cintra-Buenrostro, and J. D. Shively. 2018. Investigating reproductive characteristics of Gray Triggerfish on three artificial reefs in the northwest Gulf of Mexico. Pages 97–116 *in* S. A. Bortone, editor. Marine artificial reef research and development: integrating fisheries management objectives. American Fisheries Society, Symposium 86, Bethesda, Maryland.

Martore, R. M., and M. Bell. 2018. The Charleston Deep Reef: creating an artificial reef marine protected area to enhance fisheries resources. Pages 187–196 *in* S. A. Bortone, editor. Marine artificial reef research and development: integrating fisheries management objectives. American Fisheries Society, Symposium 86, Bethesda, Maryland.

Mohd Jani, J., E. Olson, and G. Patenaude. 2018. Re-exploring the application of artificial reefs for community-based fishery management in Malaysia. Pages 235–249 *in* S. A. Bortone, editor. Marine artificial reef research and development: integrating fisheries management objectives. American Fisheries Society, Symposium 86, Bethesda, Maryland.

Mohd Salleh, N. H., A. A. Aman, and S. A. Hamid. 2018. Selection of artificial reef deployment sites by using evidential reasoning. Pages 251–264 *in* S. A. Bortone, editor. Marine artificial reef research and development: integrating fisheries management objectives. American Fisheries Society, Symposium 86, Bethesda, Maryland.

Polovina, J. J. 1991. Fisheries applications and biological impacts of artificial habitats. Pages 153–176 *in* W. Seaman, Jr. and L. M. Sprague. Artificial habitats for marine and freshwater fisheries. Academic Press, San Diego, California.

Pondella, D. J., II, J. P. Williams, C. M. Williams, J. T. Claisse, and D. Witting. 2018. Restoring a nearshore rocky reef ecosystem in the challenge of an urban setting. Pages 165–186 *in* S. A. Bortone, editor. Marine artificial reef research and development: integrating fisheries management objectives. American Fisheries Society, Symposium 86, Bethesda, Maryland.

Schroeter, S. C., D. C. Reed, and P. Raimondi. 2018. Artificial reefs to mitigate human impacts in the marine environment: the Wheeler North Reef as a test case. Pages 197–213 *in* S. A. Bortone, editor. Marine artificial reef research and development: integrating fisheries management objectives. American Fisheries Society, Symposium 86, Bethesda, Maryland.

Seaman, W., Jr., and A. Jensen. 2000. Purposes and practices of artificial reef evaluation. Pages 1–9 *in* W. Seaman, Jr., editor. Artificial reef evaluation with application to natural marine habitats. CRC Press, Boca Raton, Florida.

Seaman, W., Jr., and L. M. Sprague. 1991. Artificial habitat practices in aquatic systems. Pages 1–29 *in* W. Seaman, Jr. and L. M. Sprague. Artificial habitats for marine and freshwater fisheries. Academic Press, San Diego, California.

Szedlmayer, S. T., and D. R. Miller. 2018. A comparison of fish and epibenthic assemblages on artificial reefs with and without copper-based, anti-fouling paint. Pages 47–67 *in* S. A. Bortone, editor. Marine artificial reef research and development: integrating fisheries management objectives. American Fisheries Society, Symposium 86, Bethesda, Maryland.

Zakariya, R., and L. S. Sakai. 2018. Evaluating a *teranas* natural reef off Malaysia using a multibeam echosounder: a method to support artificial reefs. Pages 279–288 *in* S. A. Bortone, editor. Marine artificial reef research and development: integrating fisheries management objectives. American Fisheries Society, Symposium 86, Bethesda, Maryland.

American Fisheries Society Symposium 86:7–21, 2018

How Does Fish Behavior Change during the Day around Gas Platforms?

Annalisa Gaetani, Anna Nora Tassetti,* Stefano Guicciardi, Carmen Ferrà, and Gianna Fabi
Institute of Marine Science of the National Research Council, Largo Fiera della Pesca 1, Ancona 60125, Italy

Abstract.—It is well known that offshore gas and oil extraction platforms act as artificial reefs that create a haven for finfish species. Being attracted to hard substrates, fish schools usually aggregate around the jackets or legs of the platforms where they are afforded some protection from predators and benefit from high concentrations of prey and smaller organisms foraging in the area. However, even though higher fish densities are reported near the structures, behavior and movements of fish schools around the platforms have not yet been properly investigated.

In such context, this chapter investigates the diurnal spatial movements of schooling fish around three gas extraction platforms placed in the north-central Adriatic Sea to improve knowledge on how these structures influence the surrounding fish assemblage.

Hydroacoustic multibeam echosounder surveys were conducted for 1 year at dawn, noon, and dusk. Results highlighted the variation in fish biomass around the platforms related to the time of the day and distance from the structure. Recognizing a constant attraction of fish to the platforms during daylight hours (i.e., dawn to dusk), fish biomass was greater near the gas platforms, especially at dusk. Oppositely, variation in the morphometric features of fish schools occurred during the daytime only for the school volume associated with platforms at deeper depths. Complementary information on the fish assemblages associated with the gas platforms derived from fishing surveys allowed a better understanding of the diurnal behavior of fish schools around the structures.

Introduction

Offshore gas and oil extraction platforms improve and/or increase structural diversity in already-existing marine habitats. Moreover, these structures act as artificial reefs, creating havens for finfish species (Fabi et al. 2004; Love and York 2005; Wilhelmsson et al. 2006). Being attracted to hard substrates, fish schools usually aggregate around the platform legs where they are provided some protection from predators and benefit from a high concentration of prey (Gallaway and Lewbel 1982; Pitcher 1993; Stachowitsch et al. 2002).

* Corresponding author: nora.tassetti@an.ismar.cnr.it

Scientific investigations on fish assemblages associated with offshore petroleum platforms began in the late 1970s in regions such as Southern California (USA), the Gulf of Mexico, the western coast of Africa, and the North Sea. These investigations used a variety of methods to survey these assemblages, including visual surveys by scuba divers, remotely operated underwater vehicles, and fixed underwater cameras (Hastings et al. 1976; Gallaway 1980; Putt 1982; Stanley and Wilson 1991; Castriota et al. 2011). These surveys were mostly of short duration (even as brief as a single survey) and were often affected by limited visibility, diver avoidance by fishes, and a lack of a standardized survey methodology. Consequently, describing the abundance and assemblage composition of fishes associated with offshore platforms was chiefly based on a single site or during a few points in time.

In response to the above-mentioned limits, other multiseasonal surveys have been conducted with gill nets and/or midwater traps (Valdemarsen 1979; Fabi et al. 2002, 2004; Løkkeborg at al. 2002; Fujii 2015), as well as long-term observations conducted using stationary acoustic transducers (i.e., split-beam transducers attached to the platform; Gerlotto et al. 1989; Jørgensen et al. 2002; Soldal et al. 2002; Stanley and Wilson 1990, 1996, 1997, 2000; Wilson et al. 2003). Some surveys employed a combination of active sampling/fishing gears and biological assessments using vessel-based hydroacoustics (Soldal et al. 2002).

These and other studies provided evidence of fish aggregations close to the platforms, quantitative data on the extent of fish attraction by these offshore structures (Gerlotto et al. 1989; Stanley and Wilson 1996, 1997; Wilson et al. 2003; Scarcella et al. 2011; Punzo et al. 2015), and information on the horizontal and vertical migrations of platform-associated fish schools (Stanley and Wilson 1997; Wilson et al. 2003). The results were often inconclusive regarding species-specific, temporal, and spatial patterns in the fish schools, as there was only information about diurnal and/or seasonal behavior of fishes associated with the platforms. Moreover, previous studies often indicated complex relationships between fish movements and the offshore platforms, confirming the need for long-term monitoring at multiple locations over a wide spatial area.

Information on fish school behavior around platforms is vital for the assessment of any negative and/or positive effects that gas and oil production might have on fisheries. Additionally, more information will allow a better understanding of the overall effect of the presence of several offshore structures in the same area and their subsequent removal at the end of the platform's production life (Fowler et al. 2014; Smyth et al. 2014).

The aim of the present research was to investigate the diurnal movements of fish schools around offshore gas extraction platforms to increase our knowledge of the influence these structures on fish behavior.

With an established data acquisition/sampling protocol, a pilot study was conducted at three different gas platforms situated in the north-central Adriatic Sea using hydroacoustic multibeam echosounder (MBES) surveys to (1) measure and compare the density and distribution of fish schools associated with the structures at different times of day (TODs), (2) determine the diurnal interactions between fish population biomass and distance from the platform relative to water depth, and (3) define the influence of the platform on the dimensional and acoustic characteristics of fish schools.

Methods

Study Site

Two three-legged gas platforms (E and F) placed in the north-central Adriatic Sea, 50 km (78 m deep) and 45 km (70 m deep) off the Italian coast, were investigated 1 year after deployment. A four-legged platform (B), deployed 87 m deep and 60 km from the coast, was monitored shortly after installation (Figure 1). The study location was off Ancona, Italy, in the Adriatic Sea (Figure 1). Surveys were conducted each season from summer 2015 to spring 2016, three times a day (dawn, noon, and dusk), during calm or nearly calm sea for a total of 36 surveys (3 platforms × 3 times per day × 4 seasons).

Acoustic data acquisition and processing

Acoustic surveys were conducted using a high-resolution Kongsberg EM3002D (dual-head) MBES covering a total receiving beam angle of 120° with range of 254–508 beams. The EM 3002D system uses frequencies in the 300 kHz band, an ideal frequency for shallow-water applications that assures a high maximum range capability and robustness in water high in solid particles. The 508 beams were updated simultaneously at least 40 times per second with a pulse length of 150 μs. In addition to bathymetric soundings, this MBES can provide georeferenced water column data. This feature offers the potential of detecting the fine-scale distribution of fish schools. The instrument's technical specifications, together with a specific survey protocol, allowed us to monitor the

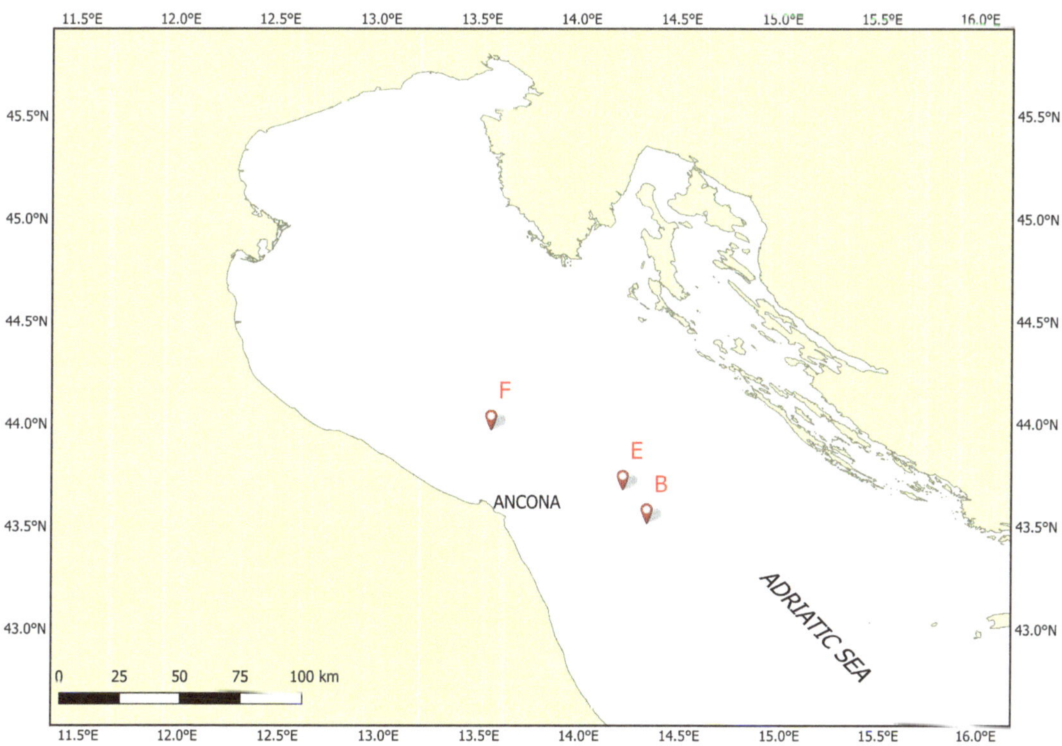

Figure 1. Geographical position of the three gas platforms (B, E, and F) in the Adriatic Sea.

habitat around the platforms. At each platform, a 1.5-km^2 area was investigated with 13 transects. Each transect was 125 m apart and conducted in a north–south direction with a small overlap, thus ensuring almost total coverage of the water column (Figure 2) at each platform. A profiling sound velocimeter (Smart SV & Pressure 4609) was used to record sound velocity profiles and optimize the detection performance during each survey.

Raw data were processed using the Echoview Software, version 8.0 (EV). All echograms were initially examined to identify and remove bad (i.e., unreliable or questionable) data regions due to electronic noise, cavitation, and bottom intrusion within the sampling area. After employing the bottom detection algorithm to identify the three-dimensional object surface that better represents the seafloor, a school-detection module allowed us

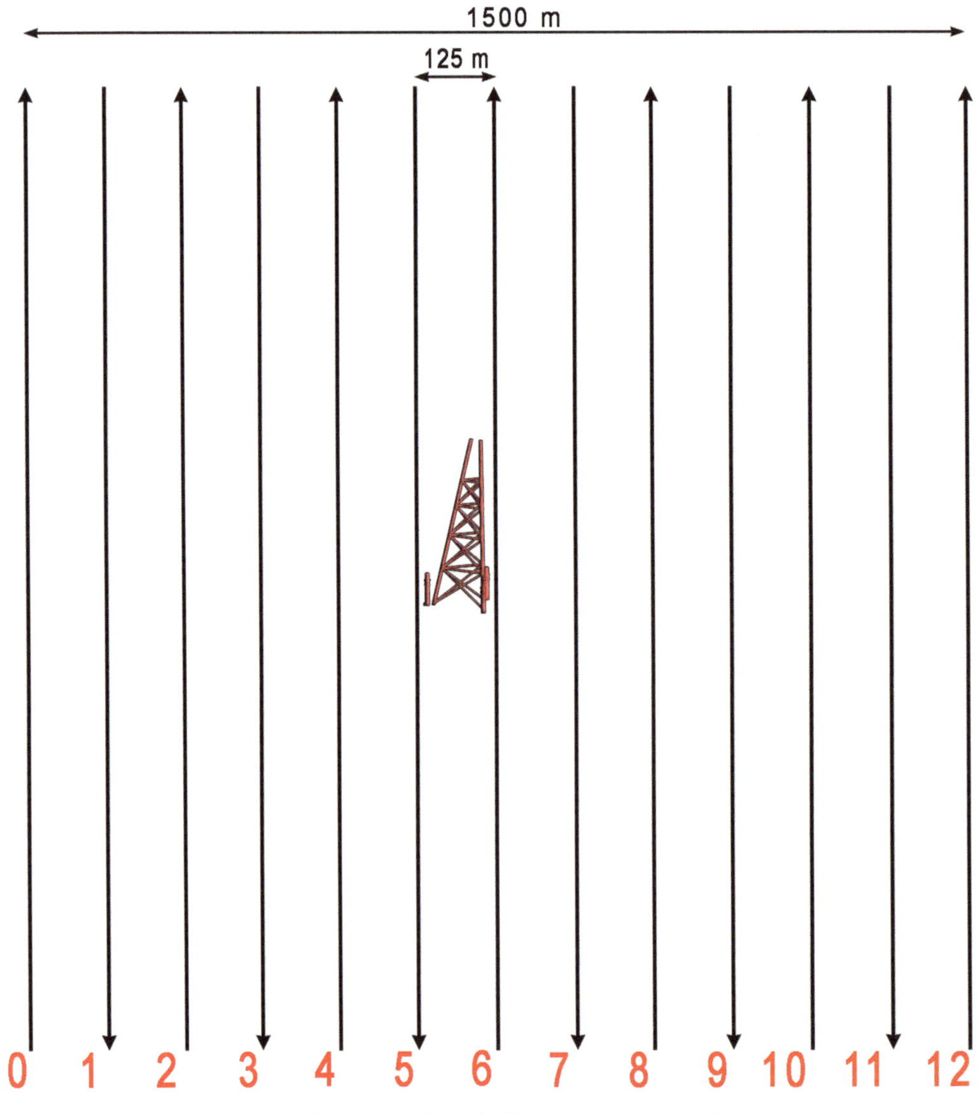

Figure 2. Acoustic survey plan around a platform (center image).

to individuate and analyze fish schools in the water column. All data within 5 m of the surface were excluded from the analysis to minimize in influence of noise from surface waves and bubbles. Also, data obtained within 0.5 m from the seafloor were excluded to avoid confusing fish schools with bottom noise. Prior to detecting a school, a volume backscattering strength (Sv) range was set. Numerous approaches have been taken into account to determine which Sv thresholds should be used for school detection (Richards et al. 1991; MacLennan and Simmonds 1992; Barange 1994; Swartzman et al. 1994; Scalabrin et al. 1996) as different thresholds can lead to significant differences in estimating school shape and energetic properties. The corrections for beam width effects used here performed best at processing thresholds between –30 and –54 decibels. After estimating the Sv, fish schools were identified using the three-dimensional cruise-scanning algorithm, setting the three dimensions for each aggregation at 80 cm (minimum longest, minimum middle, and minimum shortest dimensions in EV) that takes into account the mean size and aggregation features of local schooling fish species. Targets smaller than these dimensions were discarded during the three-dimensional school detection process. Detected fish schools were then characterized by their schooling metrics (i.e., the school's geographical position, depth, volume, and density).

Spatial and Statistical Analysis

Spatial analyses were conducted to investigate the diurnal position of fish schools in relation to the gas platforms. The study area was divided into five concentric circular buffer areas surrounding the platforms. The radius for each buffer area was 150, 300, 450, 600, and 750 m, respectively, from the platforms. For each buffer, the value for fish biomass was standardized by the relative water volume (kg/m^3) within each buffer area and was computed for each dawn, noon, and dusk (TODs) survey period using QGIS. Between 52% and 74% of schools were detected within 10 m of the seabed. The water column was divided into two layers: a deeper layer (the first 10 m from the bottom) and an upper layer (the remaining water column). The effect of depth, using the two water column layers, on fish school biomass (kg/m^3) standardized by the relative water volume was statistically compared between the two depth layers using the nonparametric Wilcox test (Rosner 2016). These data were not normally distributed due to the high number of zero (i.e., total absence) responses.

Subsequently, a nonparametric three-way analysis of variance (ANOVA) was performed that considered fish biomass as a dependent variable and TOD, distance from the platform (in terms of the previously defined buffers), and depth layer (hereinafter referred to as layer) as factors.

When the ANOVA level of significance (P-value) was lower than 0.05, semiparametric analyses by generalized additive model (GAM) were performed to account for differences in fish distribution in relation to TOD and the buffer. These GAMs were initially performed to investigate the entire water column and then repeated for each of the two layers separately. Further GAMs were used to investigate differences between fish school features (i.e., density, [kg/m^3] and volume [m^3]) for each buffer area as a distance from the platforms throughout the water column during each TOD. All statistical analyses were conducted using R software (R Core Team 2017).

Results

In total, 142, 136, and 181 fish schools were detected at platforms B, E, and F, respectively. The initial spatial analysis indicated a qualitative trend in fish schools in relation to the structures. In general, fish schools were concentrated near the gas platforms (≤150 m) at dawn and dusk while they were more scattered during the hours of maximum light (i.e., noon; Figure 3). Evidence for this observation was weaker at platform E where the fish biomass was similar at dawn and noon (1.012×10^{-5} kg/m^3 and 1.295×10^{-5} kg/m^3, respectively) and became considerably greater at dusk (3.095×10^{-5} kg/m^3). The average fish biomasses associated with the three platforms during the sampling period are reported in Table 1.

Statistical comparisons between the two depth layers indicated significantly higher fish biomass in the deeper layer for all sites and during all TODs ($p \leq 0.005$). According to the results of the three-way nonparametric ANOVA, fish biomass varied relative to buffer size. Additionally, fish biomass varied relative to TOD and depth layer with significant interactions among the factors (i.e., buffer : TOD, buffer : layer, TOD : layer, and buffer : TOD : layer; Table 2). In-

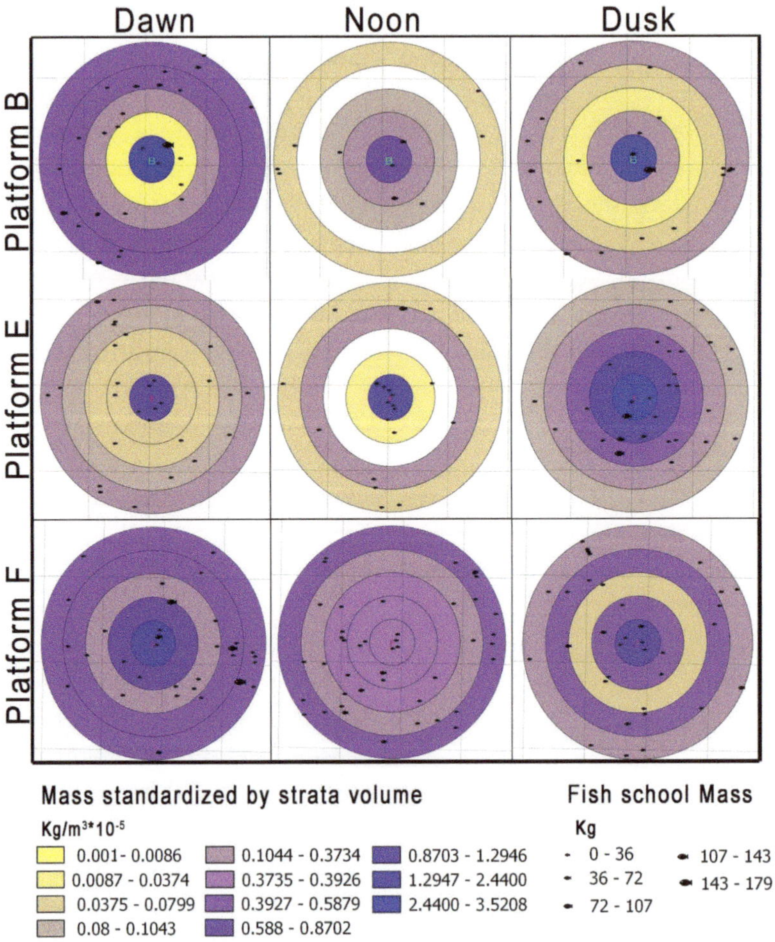

Figure 3. Spatial analysis using acoustic data according to time of day: fish biomass (kg/m^3), standardized per water column layer volume and calculated for each buffer area surrounding a platform.

Table 1. Mean values of fish biomass recorded at platforms B, E, and F according to buffer distance, depth layers, and times of day. Down = deeper layer (0.5–10 m off the bottom); Up = upper layer (0.5 m from the sea surface to 10 m from the bottom).

	Buffer radius	Dawn		Noon		Dusk	
	(m)	Up	Down	Up	Down	Up	Down
Platform B	150	–	0.372±0.228	–	0.096±0.096	0.793±0.616	–
	300	0.261±0.261	0.046±0.034	–	0.963±0.605	–	0.310±0.298
	450	–	0.373±0.134	0.002±0.002	0.167±0.167	–	0.078±0.063
	600	0.043±0.020	0.172±0.120	–	–	–	0.044±0.023
	750	0.027±0.024	0.424±0.201	0.011±0.007	0.033±0.023	0.021±0.011	0.365±0.160
Platform E	150	–	0.889±0.409	–	0.477±0.319	0.005±0.005	4.877±3.741
	300	0.006±0.006	0.043±0.043	–	0.447±0.396	0.171±0.168	0.752±0.705
	450	0.017±0.017	0.084±0.047	–	0.135±0.131	0.013±0.005	0.635±0.508
	600	–	0.077±0.027	–	0.119±0.094	0.023±0.013	0.068±0.040
	750	–	0.185±0.120	0.001±0.001	0.214±0.159	0.016±0.013	0.035±0.013
Platform F	150	–	2.801±1.705	–	0.379±0.207	0.007±0.007	1.534±1.310
	300	0.006±0.006	0.775±0.720	0.084±0.071	0.117±0.067	0.003±0.003	0.520±0.395
	450	0.005±0.004	0.224±0.090	0.007±0.007	0.280±0.168	0.003±0.003	0.043±0.031
	600	0.116±0.091	0.089±0.065	0.009±0.009	0.210±0.073	0.018±0.010	0.467±0.177
	750	0.090±0.089	0.157±0.073	0.031±0.026	0.116±0.047	0.011±0.006	0.211±0.062

Table 2. Results of the three-way robust analysis of variance performed on the mean values of fish biomass recorded at the three gas platforms, considering the factors buffer, time of day (TOD), and depth layer, both individually and combined. * significant; ** highly significant.

Factor	df	Platform B	Platform E	Platform F
Buffer	4	0.01368*	0.00000**	0.00007**
TOD	2	0.00045**	0.00000**	0.01595*
Layer	1	0.23752	0.00072**	0.00000**
Buffer * TOD	8	0.00423*	0.00000**	0.00168*
Buffer * layer	4	0.00037**	0.00000**	0.00001**
TOD * layer	2	0.00164*	0.04654*	0.01374*
Buffer * TOD * layer	8	0.00250*	0.00000**	0.00033**

terestingly, layer was not a significant factor at platform B.

Generalized additive model analyses, performed to investigate the influence of the buffer on fish distribution, indicated different trends for each platform at specific TOD. The biomass decreased with increasing distance from the structure at dusk, noon, and dawn at platform F, at dawn and dusk at platform E, and only at dusk at platform B (Figure 4).

Fish biomass throughout the entire water column was compared with fish biomass in the deeper layer during the TODs. The GAM analyses indicated similar trends and levels of significance at both three-legged platforms, but not at platform B. This was due to the absence of fish schools within

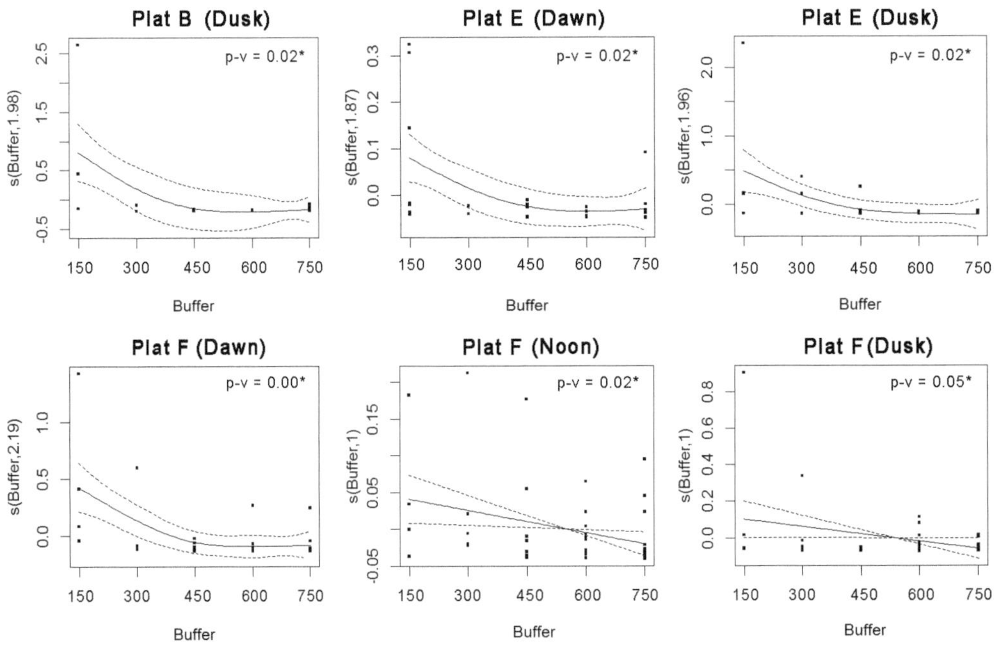

Figure 4. Generalized additive model smoothed for fish biomass (kg/m^3) recorded throughout the water column at platforms B, E, and F as a function of buffer distance.

150 m and an increase of fish mass at 750 m from platform B (Figures 4 and 5). Conversely, there was no variation in fish biomass associated with buffer considering only the fish schools detected in the upper layer.

Regarding the morphological features of fish schools (Table 3), a few differences were recorded only at platform B during dusk. Here, the school volume was significantly small within the 150-m buffer area in comparison to the more distant buffers (p = 0.04).

Discussion

Previous hydroacoustic studies conducted around gas and oil platforms indicated that there were significantly higher fish densities in the immediate vicinity of the structures, especially when compared with fish biomass over the surrounding, open soft-bottom areas. Particularly in the Gulf of Guinea and in the Gulf of Mexico, Gerlotto et al. (1989) and Stanley and Wilson (1996, 1997) reported relatively greater fish density around shallow-water platforms (5 versus 50 times and 3 versus 25 times, respectively). Moreover, they reported comparable near-field areas of influence (30 and 50 m, respectively), beyond which fish abundances were comparable to background levels in open waters. Punzo et al. (2015) recorded a predominance of fish schools, both in terms of number of individuals and biomass, within a 300-m radius associated with a four-legged gas platform in the northern Adriatic Sea compared to an open-sea area located 900–1,200 m from a platform. In addition, long-term studies have reported that fish populations associated with petroleum platforms were annually highly variable. Putt (1982) observed that fish density could vary by a factor of

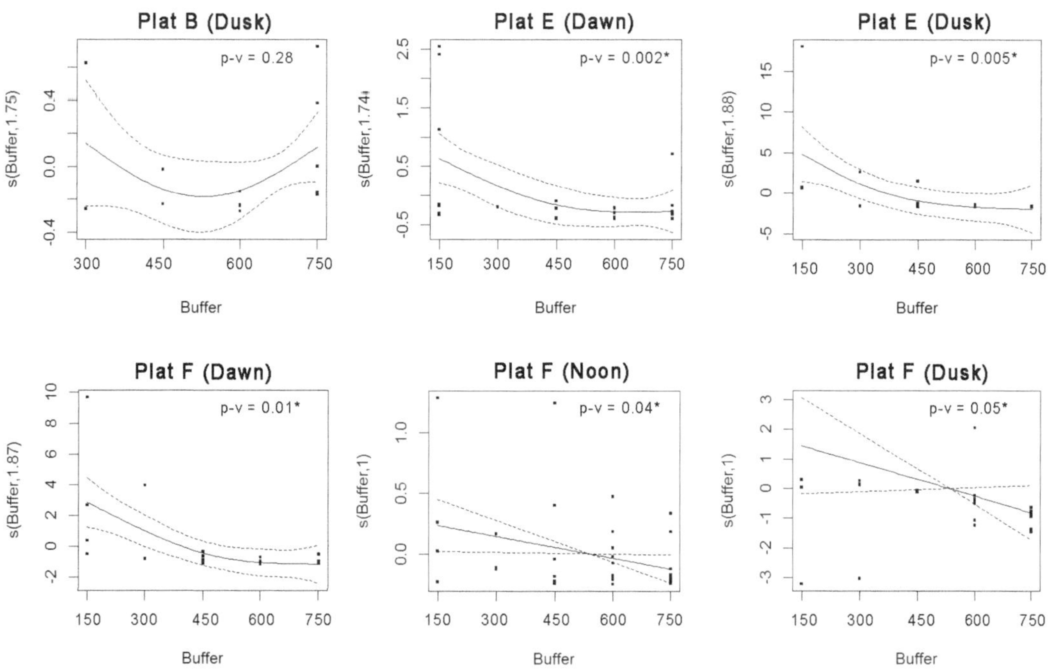

Figure 5. Generalized additive model smoothed for fish biomass (kg/m^3) recorded in the deeper layer (0.5–10 m from the bottom) at platforms B, E, and F as a function of buffer distance.

Table. Mean volume (m^3) and density (kg/m^3) of fish schools detected at different times of day (TODs) in the different buffers around platforms B, E, and F.

TOD	Buffer radius (m)	Platform B		Platform E		Platform F	
		Volume	Density	Volume	Density	Volume	Density
Dawn	150	54.28±7.85	0.12±0.01	61.19±22.45	0.10±0.01	235.56±119.19	0.13±0.03
	300	465.39±456.45	0.10±0.01	25.06±12.96	0.13±0.01	420.63±375.72	0.11±0.02
	450	76.37±23.05	0.17±0.04	39.40±15.39	0.12±0.01	39.40±15.39	0.12±0.01
	600	190.62±56.94	0.13±0.01	48.99±15.51	0.12±0.02	179.82±76.02	0.11±0.02
	750	177.83±53.02	0.15±0.03	90.52±46.15	0.12±0.02	175.88±89.29	0.10±0.01
Noon	150	31.95[a]	0.08[a]	42.52±26.03	0.09±0.01	49.34±22.91	0.08±0.00
	300	155.53±17.67	0.33±0.02	110.24±88.16	0.12±0.02	121.46±87.16	0.14±0.03
	450	147.94±108.74	0.07±0.02	86.82±76.52	0.10±0.04	86.82±76.52	0.1±0.04
	600	–	–	86.38±70.45	0.11±0.01	92.39±27.37	0.12±0.01
	750	46.16±16.56	0.14±0.03	144.12±96.77	0.08±0.01	99.26±40.46	0.10±0.01
Dusk	150	560.46±382.61	0.11±0.01	297.40±189.42	0.18±0.06	42.65±24.62	0.17±0.06
	300	70.17±65.24	0.03±0.02	188.21±108.12	0.08±0.02	153.59±107.99	0.12±0.03
	450	44.03±20.01	0.11±0.04	140.46±74.91	0.10±0.02	147.94±108.74	0.07±0.02
	600	25.02±7.12	0.09±0.02	80.76±27.00	0.09±0.01	150.32±43.98	0.14±0.03
	750	162.29±45.98	0.13±0.02	78.94±55.52	0.10±0.01	109.82±28.42	0.11±0.01

[a] School number = 1.

two from month to month while Stanley and Wilson (1996, 1997) reported that there were monthly and seasonal changes in density by a factor of five. However, the results of our evaluation regarding fish density relative to TOD were inconclusive. Stanley and Wilson (1997, 2000) documented vertical migrations of fish related to TOD, but the mixed outcomes here did not indicate a distinct trend, while Wilson et al. (2003) recorded higher fish density at dusk close to an eight-legged platform in the Gulf of Mexico. Our study, conducted during daylight, indicated diurnal spatial and dimensional variations of fish schools associated with three gas platforms located in the Adriatic Sea by recording the position of the schools and measuring some of their physical features using a Kongsberg EM3002D MBES.

Our results indicated that there was a large concentration of fish schooling close to the seabed during all TODs (dawn, noon, and dusk). Such behavior, also reported by Punzo et al. (2015), is attributable to the species' gregarious behavior as the schools tended to aggregate near the bottom during daytime and moved up in the water column, above the thermocline, nocturnally. These vertical migrations are most likely related to their feeding activity, plausibly intertwined with predation avoidance and bioenergetic optimization (Pitcher 1993; Cardinale et al. 2003). The initial qualitative spatial analyses conducted using QGIS indicated a decrease in fish biomass at increasing distances from the platforms at every TOD. Fish biomass recorded throughout the water column was lower at distances from 150 to 750 m from the platforms. Such near-field attraction to the structure was similarly reported by Scarcella et al. (2011) at another extraction platform situated in the north-central Adriatic Sea. This "halo" of attraction is broader than the attraction effect reported by Gerlotto et al. (1989) and Stanley and Wilson (1996, 1997). However, our first hypothesis was not entirely supported by statistical test, which found evidence of a decrease in density at increasing buffer distances only for certain structures during certain TODs. This tendency occurred during all TODs at platform F, during dawn and dusk at platform E, and only during dusk at platform B. The decreasing trend was mainly due to the behavior of fish schools on the bottom at platforms E and F, where the occurrence of surficial aggregations was not observed during all TODs. Oppositely, at platform B, there was a tendency for a few large fish schools to aggregate in the upper water column layer close to the platform legs at dusk. This was also confirmed by our report of larger fish-school dimensions detected within the 150-m buffer. Indeed, the analysis of these parameters indicated a decrease in school size (i.e., volume) at increasing distances from the structures at dusk, underlining a tendency for fish school to shoal near this structure at dusk.

At platform E, the greater school volume observed at dawn and dusk was associated with changes in measured metrical characteristics of the fish schools. This factor indicated that schooling tendency was considerable near the structures during low-light hours. An even more static situation was recorded at platform F, where fish did not display any spatial migration or different degrees of aggregation at increasing distances from the platform legs during any TOD. The variability in the school density, either at different buffer distances from the platforms or during different times of the day, is likely due to the different fish assemblage composition at the three platforms. Indeed, catch data of pelagic and nektobenthic fish (benthic fish were not considered here as they occurred too close to the seabed to be detected by MBES) from experimental fishing surveys conducted in the same year at the same platforms revealed that 74% of fish caught at platform B were pelagic fishes. Most of these were European Anchovy *Engraulis encrasicolus*, which usually aggregate in large schools during daytime (Iglesias et al. 2003; Tsagarakis et al. 2012). Other abundant pelagic species were Bogue *Boops boops*, mackerels (Atlantic Mackerel *Scomber scombrus* and Pacific Chub Mackerel *S. japonicus*) and horse mackerels (European Horse Mackerel *Trachurus trachurus* and Mediterranean Scad *T. mediterraneus*). In the open sea, the fish assemblage was composed of nektobenthic species and subspecies, mainly European Hake *Merluccius merluccius* and Poor Cod *Trisopterus minutus capelanus*, which tended to aggregate in smaller schools than the pelagic species (Institute of Marine Sciences of the Italian National Research Council, unpublished data). Catch data relative to platform E showed a lower frequency of occurrence of pelagic fish (69%) near the legs than observed at platform B. At platform E, the main species were Poor Cod, Bogue, Blackspot Seabream (also known as Red Seabream) *Pagellus bogaraveo*, and horse mackerels. The latter three species groups often formed mixed-species schools. Finally, pelagic fish occurred at an even lower percentage of the catch around platform F compared to the other platforms (66%). The fish assemblage at platform was dominated by the same species recorded at platform E. In addition to the greater depth, the diverse fish assemblage composition at platform B (when compared to assemblages associated with platforms E and F) is likely related to the different dimensions and complexity of its jacket. Indeed, besides having a larger volume of water associated with it, the four-legged jacket of platform B was characterized by higher structural complexity than the three-legged platforms. Therefore, it is reasonable to assume that platform B acts as a more suitable refuge for small finfish species as well.

The outcomes of this study do not differ from that which was previously documented by Howell (2008) and Punzo et al. (2015) concerning the utility of the most recent MBES surveying technology. Indeed, MBES is able to simultaneously survey the seafloor and the water column. Moreover, it can collect quantitative information on the spatial distribution and abundance of fish in wide areas over time. It can also help gather detailed information on the fish schooling behavior in complex habitats with a relative low effort in comparison to other sampling methods (e.g., fishing surveys, remotely operated underwater vehicles). However, according to Punzo et al. (2015), the main limitation of this system occurs when mapping schools swim within a radius of 10 m from a platform. This is because of the noise generated by the extraction activity on the platforms, which interferes with MBES frequencies. The effect of this problem can be reduced by using a multifrequency MBES and setting the instrument to a frequency that is interfered with less from the vibration related to platform activity.

Acknowledgments

The authors acknowledge the staff of the Marine Environment Management Unit of the Institute of Marine Sciences of the Italian National Research Council in Ancona and the crew of R/V Tencopesca II for their active contribution to the data collection.

References

Barange, M. 1994. Acoustic identification, classification and structure of biological patchiness on the edge of the Agulhas Bank and its relation to frontal features. South African Journal of Marine Science 14:333–347.

Cardinale, M., M. Casini, F. Arrhenius, and N. Håkansson. 2003. Diel spatial distribution and feeding activity of Herring *Clupea harengus* and Sprat *Sprattus sprattus* in the Baltic Sea. Aquatic Living Resources 16:283–292.

Castriota, L., M. Falautano, M.G. Finoia, P. Consoli, C. Pedà, V. Esposito, P. Battaglia, and F. Andaloro. 2011. Trophic relationships among scorpaeniform fishes associated with gas platforms. Helgoland Marine Research 66:401–411.

Fabi, G., F. Grati, A. Lucchetti, and L. Trovarelli. 2002. Evolution of the fish assemblage around a gas platform in the northern Adriatic Sea. ICES Journal of Marine Science 59:309–315.

Fabi, G., F. Grati, M. Puletti, and G. Scarcella. 2004. Effects on fish community induced by installation of two gas platforms in the Adriatic Sea. Marine Ecology Progress Series 273:187–197.

Fowler, A. M., P. I. Macreadie, D. O. B. Jones, and D. J. Booth. 2014. A multi-criteria decision approach to decommissioning of offshore oil and gas infrastructure. Ocean and Coastal Management 87:20–29.

Fujii, T. 2015. Temporal variation in environmental conditions and the structure of fish assemblages around an offshore oil platform in the North Sea. Marine Environmental Research 108:69–82.

Gallaway, B. J. 1980. Pelagic, reef and demersal fishes and macrocrustacean/biofouling communities. *In* W. B. Jackson and E. F. Wilkens, editors. Environmental assessment of Buccaneer gas and oil field in northwestern Gulf of Mexico, 1975–1980, volume 2. NOAA Technical Memorandum NMFS-SEFC 48.

Gallaway, B. J., and G. S. Lewbel. 1982. The ecology of petroleum platforms in the northwest Gulf of Mexico: a community profile. U.S. Fish and Wildlife Service, Office of Biological Sciences, FWS/OBS-82/27, Washington D.C.

Gerlotto, F., C. Bercy, and B. Bordeau. 1989. Echo integration survey around offshore oil extraction platforms off Cameroon: observations of the repulsive effect on fish of some artificially emitted sounds. Proceedings of the Institute of Acoustics 19:79–88.

Hastings, R. W., L. H. Ogren, and M. T. Mabry. 1976. Observations on the fish fauna associated with offshore platforms in the northeastern Gulf of Mexico. U.S. National Marine Fisheries Service Fishery Bulletin 74:387–402.

Howell, W. H. 2008. Development of multibeam sonar as a fisheries tool for stock assessment and essential fish habitat identification of groundfish in the Western Gulf of Maine. University of New Hampshire, Final Report, Durham.

Iglesias, M., P. Carrera, and R. Muiño. 2003. Spatio-temporal patterns and morphological characterisation of multispecies pelagic fish schools in the north-western Mediterranean Sea. Aquatic Living Resources 16:541–548.

Jørgensen, T., S. Løkkeborg, and A. V. Soldal. 2002. Residence of fish in the vicinity of a decommissioned oil platform in the North Sea. ICES Journal of Marine Science 59:S266–S271.

Løkkeborg, S., O.-B. Humborstad, T. Jørgensen, and A. V. Soldal. 2002. Spatio-temporal variations in gillnet catch rates in the vicinity of North Sea oil platforms. ICES Journal of Marine Science 59:S294–S299.

Love, M. S., and A. York. 2005. A comparison of the fish assemblages associated with an oil/gas pipeline and adjacent seafloor in the Santa Barbara Channel, Southern California Bight. Bulletin of Marine Science 77:101–117.

MacLennan, D. N., and E. J. Simmonds. 1992. Fisheries acoustics. Chapman and Hall, London.

Pitcher, T. J. 1993. Behaviour of teleost fishes, 2nd edition. Fish and Fisheries Series 7. Champman and Hall, London.

Putt, R. E., Jr. 1982. A quantitative study of fish populations associated with a platform within Buccaneer Oil Field, northwestern Gulf of Mexico. Master's thesis. Texas A&M University, College Station.

Punzo, E., S. Malaspina, F. Domenichetti, P. Polidori, G. Scarcella, and G. Fabi. 2015. Fish detection around offshore artificial structures: preliminary results from hydroacoustics and fishing surveys. Journal of Applied Ichthyology 31(S3):48–59.

R Core Team. 2017. R: a language and environment for statistical computing. R Foundation for Statistical Computing, Vienna, Austria. Available: www.R-project.org/. (April 2018).

Richards, L. J., R. Kieser, T. J. Mulligan, and J. R. Candy. 1991. Classification of fish assemblages based on echo integration surveys. Canadian Journal of Fisheries and Aquatic Sciences 48:1264–1272.

Rosner, B. 2016. Fundamentals of biostatistics, 8th edition, Cengage Learning, Boston.

Scalabrin, C., N. Diner, A. Weill, A. Hillion, and M. C. Mouchot. 1996. Narrowband acoustic identification of monospecific fish shoals. ICES Journal of Marine Science 53:181–188.

Scarcella, G., F. Grati, and G. Fabi. 2011. Temporal and spatial variation of the fish assemblage around a gas platform in the northern Adriatic Sea, Italy. Turkish Journal of Fisheries and Aquatic Sciences 11:433–444.

Smyth, K., N. Christie, D. Burdon, J. P. Atkins, R. Barnes, and M. Elliott. 2014. Renewables-to-reefs: response to Fowler et al. Marine Pollution Bulletin 98:372–374.

Soldal, A. V., I. Svellingen, T. Jorgensen, and S. Lokkeborg. 2002. Platforms-to-reefs in the North Sea: hydroacoustic quantification of fish in the vicinity of a "semi-cold" platform. ICES Journal of Marine Science 59:281–287.

Stachowitsch, M., R. Kikinger, J. Herler, P. Zolda, and E. Geutebrück. 2002. Offshore oil platforms and fouling communities in the southern Arabian Gulf (Abu Dhabi). Marine Pollution Bulletin 44:853–860.

Stanley, D. R., and C. A. Wilson. 1990. A fishery-dependent based study of fish species composition and associated catch rates around oil and gas structures off Louisiana. U.S. National Marine Fisheries Sdrvice Fishery Bulletin 88:719–730.

Stanley, D. R., and C. A. Wilson. 1991. Factors affecting the abundance of selected fishes near oil and gas platforms in the northern Gulf of Mexico. U.S. National Marine Fisheries Service Fishery Bulletin 89:149–159.

Stanley, D. R., and C. A. Wilson. 1995. Detection of the effect of scuba divers on fish density and target strength utilizing dual-beam hydroacoustics. Transactions of the American Fisheries Society 124:946–949.

Stanley, D. R., and C. A. Wilson. 1996. Abundance of fishes associated with a petroleum platform as measured with dual-beam hydroacoustics. ICES Journal of Marine Science 53:473–475.

Stanley, D. R., and C. A. Wilson. 1997. Seasonal and spatial variation in the abundance and size distribution of fishes associated with a petroleum platform in the northern Gulf of Mexico. Canadian Journal of Fisheries and Aquatic Sciences 54:1166–1176.

Stanley, D. R., and C. A. Wilson. 2000. Variation in the density and species composition of fishes associated with three petroleum platforms using dual beam hydroacoustics. Fisheries Research 47:161–172.

Swartzman, G., W. Stuetzle, K. Kulman, and N. Wen, N. 1994. Modeling the distribution of fish schools in the Bering Sea: morphological school identification. Natural Resource Modeling 8:177–194.

Tsagarakis, K., M. Giannoulaki, S. Somarakis, and A. Machias. 2012. Variability in positional, energetic and morphometric descriptors of European Anchovy *Engraulis encrasicolus* schools related to patterns of diurnal vertical migration. Marine Ecology Progress Series 446:243–258.

Valdemarsen, J. W. 1979. Behaviour aspects of fish in relation to oil platforms in the North Sea. International Council for the Exploration of the Sea, CM 1979/B:27, Copenhagen, Denmark.

Wilhelmsson, D., T. Malm, and M. C. Ohman. 2006. The influence of offshore windpower on demersal fish. ICES Journal of Marine Science 63:775–784.

Wilson, C. A., A. Pierce, and M. W. Miller. 2003. Rigs and reefs: a comparison of the fish communities at two artificial reefs, a production platform, and a natural reef in the northern Gulf of Mexico. U.S. Department of the Interior, Minerals Management Service, New Orleans, Louisiana.

American Fisheries Society Symposium 86:23–45, 2018

A Comparison of Fish Assemblages According to Artificial Reef Attributes and Seasons in the Northern Gulf of Mexico

Jessica Jaxion-Harm, Stephen T. Szedlmayer*, and Peter A. Mudrak
School of Fisheries, Aquaculture and Aquatic Sciences, Auburn University 8300 State Highway 104, Fairhope, Alabama 36532, USA

Abstract.—Visual census scuba surveys (n = 87) were used to compare fish assemblages among three artificial reef types: big reefs (e.g., ships), tank reefs (i.e., U.S. Army tanks) and small reefs (e.g., metal cages and concrete pyramids), over three locations on the continental shelf (inner shelf, 18–26-m depths; mid-shelf, 26–34-m depths; outer shelf, 34–41-m depths) from April 20, 2012 to November 30, 2015 in the northeast Gulf of Mexico. These surveys identified 66 fish taxa (lowest taxon: 58 species, five genera, three families), and 65 taxa were used in community comparisons. Artificial reefs were dominated by Red Snapper *Lutjanus campechanus* (35.3% of total fish observed), Tomtate *Haemulon aurolineatum* (22.4%), Vermilion Snapper *Rhomboplites aurorubens* (19.5%), Atlantic Spadefish *Chaetodipterus faber* (7.0%), Greater Amberjack *Seriola dumerili* (3.0%), and Gray Triggerfish *Balistes capriscus* (1.8%). These six most abundant species comprised 89% of the total number of individuals observed. Red Snapper and Greater Amberjack mean sizes (total length mm) were larger at big reefs, Vermilion Snapper and Atlantic Spadefish were larger at tank reefs, and Tomtates were larger at small reefs. Red Snapper, Atlantic Spadefish, and Greater Amberjacks were larger at reefs on the outer shelf, and Red Snapper, Tomtates, Vermilion Snapper, Atlantic Spadefish and Greater Amberjacks were larger in the spring. Richness and Shannon–Wiener diversity indices were higher on big reefs and tank reefs compared to small reefs. Evenness, richness, and Shannon–Wiener diversity were lower in winter compared to other seasons. Fish assemblages, based on Bray–Curtis similarities, were different among reef type, location, and season, but no interactions effects were identified. In the present study, fish assemblages on big reefs were more similar to assemblages on tank reefs in comparison to small reefs. The larger size, longer life span, and relative stability of the big reefs and tank reefs were the reef attributes most likely responsible for these assemblage associations. Similarly, more stable conditions at deeper depths (less affected by tropical storms) and proximity to deepwater reef fish communities (e.g., pinnacle reefs) most likely influenced

* Corresponding author: szedlst@auburn.edu

the increased assemblage diversity on the artificial reefs at outer-shelf locations. Diversities and densities were highest during the fall. This was most likely due to increased recruitment of tropical species and new age-0 recruits that were spawned during the same year. The attributes of all artificial reefs are not identical; consequently, it is important for managers to consider how reef type, shelf location, and season affect each species' affinity and association with artificial reefs.

Introduction

Artificial reefs are deployed to achieve various objectives (e.g., facilitate habitat mitigation, prevent trawling, and enhance fisheries), but ultimately, they provide hard substrate that can potentially increase the stock size of fish species (Baine 2001; Gallaway et al. 2009; Shipp and Bortone 2009; Smith et al. 2015). Hard-structure habitats can provide protection from predators and increase food resources from the growth of epibenthic communities (Hixon and Beets 1993; Redman and Szedlmayer 2009). Historically, deploying artificial reefs for enhancing fish production has focused on individual fish species (Lindberg et al. 2006; Shipp and Bortone 2009), but deploying artificial reefs to achieve the objectives of ecosystem-based management has recently become more prevalent (Crowder et al. 2008; deReynier et al. 2010). For example, the loss of biodiversity in marine systems has negatively affected ecosystem stability and production (Worm et al. 2006). Additionally, Pickering et al. (1998) suggested that one potential approach to help increase biodiversity is to deploy artificial reefs.

In the northeast Gulf of Mexico, there are few natural reefs (Dufrene 2005; Jenkins et al. 2008; Williams et al. 2012). Thus, the deployment of artificial reefs can increase available reef-like habitat and possibly increase faunal diversity of species that are reef-associated (Lingo and Szedlmayer 2006; Gallaway et al. 2009). Historical evidence suggests that reef species such as Red Snapper *Lutjanus campechanus* have become more abundant subsequent to artificial reef deployments on the sand-mud substrate of the continental shelf in the northern Gulf of Mexico (Gallaway et al. 2009; Shipp and Bortone 2009). There has been extensive research depicting artificial reefs as fish attractors and potential producers (Bohnsack and Sutherland 1985; Pickering and Whitmarsh 1997; Bortone et al. 2011), yet few studies have examined fish assemblages on substantially different types of artificial reefs in the northern Gulf of Mexico (Stanley and Wilson 1997; Lingo and Szedlmayer 2006; Dupont 2008; Dance et al. 2011; Ajemian et al. 2015). For example, studies have examined concrete pyramids and reef balls (Dance et al. 2011), small 1-m^3 concrete blocks (Lingo and Szedlmayer 2006), and oil platforms (Stanley and Wilson 1997; Ajemian et al. 2015). Yet, a variety of artificial reefs have been deployed in the northeast Gulf of Mexico that differ in size and complexity, from small concrete pyramids and metal cages to large ships and oil-gas platforms.

As the deployment of reefs increases (McGurrin et al. 1989; Bohnsack et al. 1994), it is necessary to consider both reef type and location in relation to fish assemblages that recruit to these artificial structures. Thus, an important new aspect of the present study was the examination of fish assemblages over diverse artificial reef types. In addition, the present study examined artificial reefs over much of the continental shelf as distributions of many fish species are depth-related (Gul et al. 2011; Sherman et al. 1999). Specifically, the present study examined different artificial reef types over three shelf areas: inner shelf (18–26-m depths), mid-shelf (26–34-m depths), and outer-shelf (34–41-m depths), to provide a better understanding of fish assemblages and densities of juvenile and adult fishes that associate with these artificial reef habitats. Such information will help managers with the difficult decisions to employ artificial reefs in fishery resource management and give direction to future reef research efforts.

Methods

Study Area

The study area was located 16–64 km south of Mobile Bay, Alabama (USA), in the northern Gulf of Mexico. All reefs were from three locations on the continental shelf over three depth zones (inner shelf, 18–26 m; mid-shelf, 26–34 m; and outer shelf, 34–41 m). Artificial reef types that were surveyed included concrete pyramids, army tanks, concrete rubble, metal cages, barges, gas platforms, ships, and several other miscellaneous structures. All reefs were separated into three categories: big reefs (2,500–3,800 m^3; e.g., ships, gas platforms), mid-size tank reefs (volume = 51 m^3; i.e., U.S. Army tanks), and small reefs (volume = 4–15 m^3; e.g., pyramids). Both public reefs (i.e., reefs with published locations made available to fishers by the Alabama Department of Conservation and Natural Resources) and private reefs (locations usually known only to the builder) were investigated in the present study. Unpublished or private reefs were located with side-scan sonar surveys (Edgetech 4125-Dual 400/900 Hz). Private reefs consisted of a wide variety of structures, from metal cages to concrete rubble. Stratified random sampling was used to select reefs among the three shelf locations and three artificial reef types (big reefs, tank reefs, and small reefs). Sample sites were selected using a random numbers table from a total of 835 identified reef structures (268 private reefs located with the side-scan surveys and 567 public reef sites). A total of 87 artificial reef sites were surveyed from 16 to 64 km offshore of coastal Alabama (Figure 1). These included 23 big reefs (8 inner shelf, 8 mid-shelf, and 7 outer shelf), 24 tank reefs (5 inner shelf, 11 mid-shelf, and 8 outer shelf), and 40 small reefs (14 inner shelf, 11 mid-shelf, and 15 outer shelf). To evaluate seasonal differences in the reef-associated fish assemblages, surveys were conducted during the spring (22 surveys), winter (10), summer (26), and fall (29), pooled over reef types and locations from April 20, 2012 to November 30, 2015 (Figure 2).

Fish Surveys

Divers (using scuba) estimated fish length to the nearest 25 mm total length (TL) using a stationary point-count method during the day (from 2 h before to 4 h after 1200 hours) to calculate fish densities (Greene and Alevizon 1989). Water clarity was determined using horizontal measures of a Secchi disk during each survey. The composition material and size of each reef were also recorded at each site. A YSI 6920 environmental meter was used to measure temperature (°C), salinity (‰), and dissolved oxygen (mg/L) within 2 m of the bottom and less than 20 m of each reef site.

The area surveyed was calculated by using a maximum radius of 6 m for density estimations when horizontal water clarity exceeded 6 m from the Secchi disk measure-

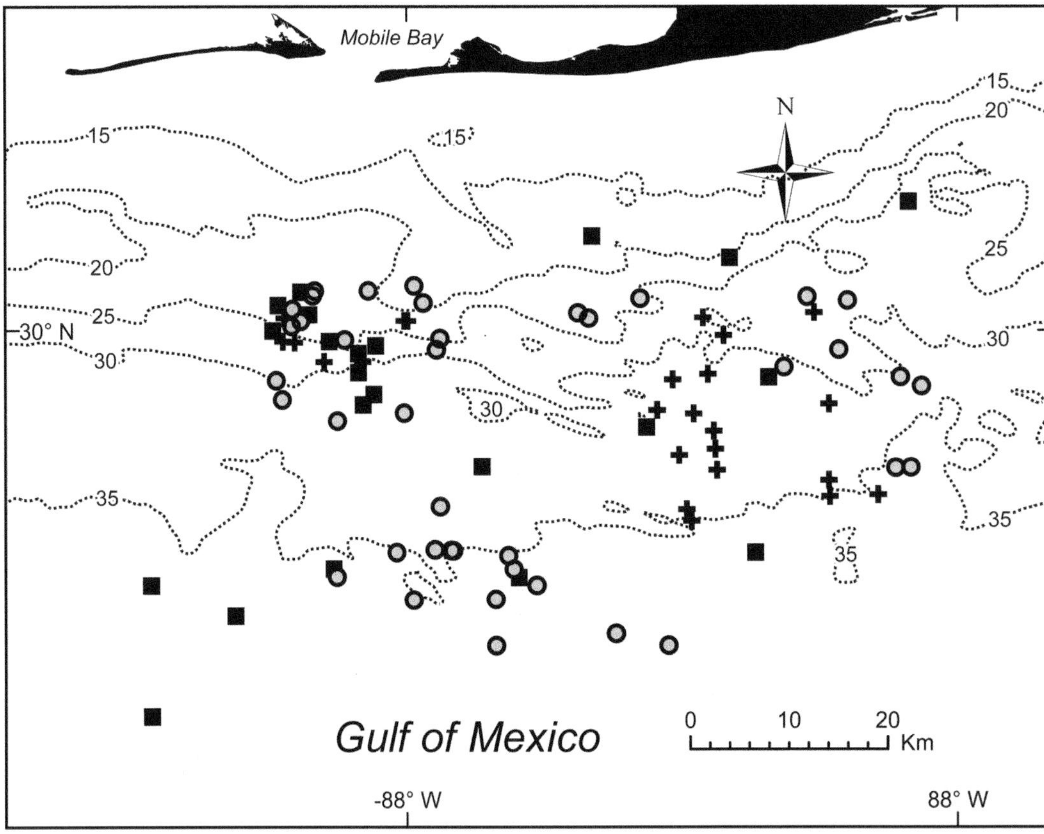

Figure 1. Artificial reef study sites in the northern Gulf of Mexico. Big reefs = squares, tank reefs = crosses, and small reefs = gray dots. Dotted lines represent depth contours at 5-m intervals.

ment. The maximum 6-m radius was selected as the distance that divers could accurately identify fish species and estimate fish lengths. When visibility was between 3 and 6 m, the Secchi disk visibility distance was used as the radius for area-surveyed calculations. If visibility was less than 3 m, the survey data were not included in the analyses (10 sites were removed out of 97 sites = 87 survey sites). Fish density (number of fish/100 m^2) was calculated for each site surveyed.

Statistical Analyses

Diversity indices were compared relative to the independent variables: reef type, shelf location, and season (winter = December–February, spring = March–May, summer = June–August, and fall = September–November) using analyses of variance (ANOVAs; SAS version 9.4; Zar 2010). Densities of fish were square-root-transformed prior to calculations of diversity measures (Field et al. 1982). Fish communities were compared using the Shannon–Wiener diversity index (H'), richness (S), and evenness ($J = H'/H'$max; Magurran 1988).

Fish densities (number of fish/100 m^2) were compared using generalized linear models (PROC GENMOD, SAS version 9.4) with negative binomial distributions and log link functions (Huelsenbeck and Crandall 1997; Seavy et al. 2005; Bolker et al. 2009). The statistical model examined all main effects and interactions, but after analyses,

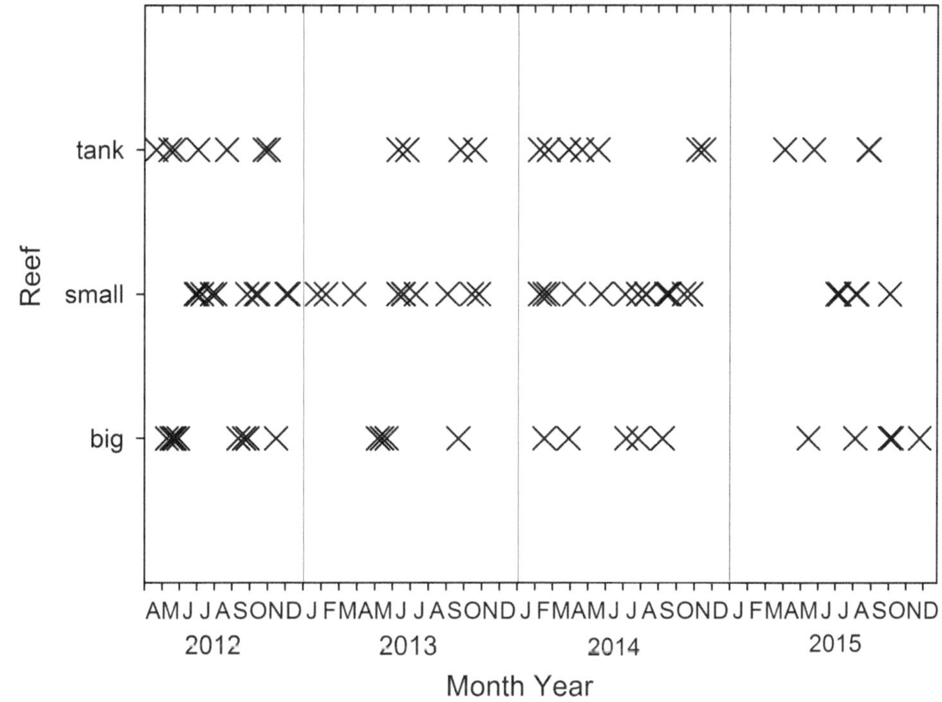

Figure 2. Sampling frequency by month and year for fish assemblage surveys on artificial reefs in the northern Gulf of Mexico.

most three-way interactions were excluded because of the small sample sizes in three factor cells.

Densities of fish species were square-root-transformed prior to calculations of Bray–Curtis similarities (Field et al. 1982). These similarity indices were compared among all surveys and visually examined with nonmetric multidimensional scaling (MDS) ordination plots (Bray and Curtis 1957; Clarke and Green 1988; Szedlmayer and Able 1996; Lingo and Szedlmayer 2006). The MDS plots spatially indicates the relative similarity (or dissimilarity) of reef fish assemblages. The fish assemblage similarities among all individual reef surveys were examined statistically among reef types, shelf locations, and seasons with three way permutational multivariate analysis of variance (PERMANOVA; Anderson 2001; Anderson et al. 2008).

Environmental variables (temperature, salinity, dissolved oxygen, and depth) similarities (based on Euclidian distance) were compared to fish species similarities among surveys with the BEST procedure in the Primer 7 statistical program (Clarke et al. 2014). This procedure compares the species rank order similarities with the environmental rank order similarities and tests for significance (Spearman's rho, $P \leq 0.05$) from a statistical probability distribution derived from permutations of surveys (Clarke et al. 2014).

Mean lengths (TL mm) of the dominant species were compared among reef types, locations, and seasons with three-way ANOVA. All differences were considered significant at $P \leq 0.05$. If significant differences were detected with ANOVA or generalized linear models, specific differences were assessed with a Tukey post-hoc test (Zar 2010).

Results

Total Fish Densities

The density of the fish assemblage (all species combined) was higher on mid-shelf reefs compared to inner- or outer-shelf reefs (χ^2 = 9.2, P = 0.010) and higher in the fall compared to other seasons (χ^2 = 10.8, P = 0.013; Table 1). Fish density was also associated with an interaction of reef type and location, with higher densities on big reefs at mid-shelf locations compared to small reefs at outer-shelf locations and big reefs at inner-shelf locations (χ^2 = 15.5, P = 0.004; Table 2).

Individual Species

We identified 66 fish taxa (lowest taxa: 58 species, five genera, three families) from visual surveys on artificial reefs in the northern Gulf of Mexico. When examining reef attributes and affinities of the associated species, we removed Round Scad *Decapterus punctatus*. Even though it was the most abundant species (40.2%), it is generally considered a transient, pelagic species (McBride et al. 2002). After removal of Round Scad, the artificial reefs examined in the present study were dominated by Red Snapper (35.3%), Tomtate *Haemulon aurolineatum* (22.4%), Vermilion Snapper *Rhomboplites aurorubens* (19.5%), Atlantic Spadefish *Chaetodipterus faber* (7.0%), Greater Amberjack *Seriola dumerili* (3.0%) and Gray Triggerfish *Balistes capriscus* (1.8%). These six most abundant species comprised 89% of the total number of individuals (Table 1). Other common species (1.0–1.4%) on reefs included Cubbyu *Pareques umbrosus*, Gray Snapper *Lutjanus griseus*, lionfish *Pterois* spp., and Whitespotted Soapfish *Rypticus maculatus*.

There were significant differences in density associated with reef type, location, season, and interactions that depended on species. Tomtates had higher densities on tank and small reefs, at inner- and mid-shelf locations, and during summer and fall seasons. Vermilion Snapper had higher densities at mid- and outer-shelf locations, and during the fall. Atlantic Spadefish had higher densities at mid-shelf locations, and during the spring. Greater Amberjacks had higher densities during the fall (Table 1).

Vermilion Snapper had higher densities associated with interactions of big and tank reefs at inner-shelf locations. Greater Amberjacks had higher densities on big reefs at outer-shelf locations, compared to big reefs on inner-shelf locations (Table 2). Tomtates had higher densities associated with interactions of big reefs during the fall and with mid-shelf locations during the fall, and Atlantic Spadefish had higher densities associated with mid-shelf locations during the winter (Table 3).

Tomtates had higher densities associated with three-way interactions at big reefs at mid-shelf locations during the fall (χ^2 = 13.5, P = 0.019). Vermilion Snapper had higher densities associated with three-way interactions of big reefs at mid-shelf locations during the fall (χ^2 = 11.9, P = 0.036). Atlantic Spadefish had higher densities associated with three-way interactions for small reefs at mid-shelf locations during the winter (χ^2 = 17.6, P = 0.004).

Size Comparisons

Larger mean sizes (TL mm) were detected for Red Snapper and Greater Amberjack on big reefs, for Atlantic Spadefish on big and tank reefs, for Vermilion Snapper on tank reefs, and for Tomtate on small reefs (Table 4). Larger sizes were detected for Red Snapper, Atlantic Spadefish, and Greater Amberjack on outer-shelf locations; for Vermilion Snapper on outer- and mid-shelf locations; and for Tomtate on inner-shelf locations

Table 1. Mean and percent densities (number/100 m^2) for all fish and dominant fish species ($\geq 1.3\%$ of the total density) for all surveys; and least-square mean densities by reef type, shelf location and season, on artificial habitats in the northern Gulf of Mexico for 2012–2015. Wi = winter, Sp = spring, Su = summer and Fa = fall. Different letters indicate significant ($P \leq 0.05$) differences within reef type, shelf location, and season.

			Reef type			Shelf location			Season			
Species	Mean	%	Big	Tank	Small	Inner	Mid	Outer	Wi	Sp	Su	Fa
All species			237	278	225	224 y	313 z	201 y	226 y	177 y	212 y	352 z
Red Snapper *Lutjanus campechanus*	98	35.3	85	105	34	97	95	79	64	91	97	92
Tomtate *Haemulon aurolineatum*	62	22.4	14 y	30 z	28 z	28 z	38 z	13 y	5 y	8 y	41 z	53 z
Vermilion Snapper *Rhomboplites aurorubens*	54	19.5	11	19	19	8 y	26 z	20 z	9 yx	3 x	13 y	82 z
Atlantic Spacefish *Chaetodipterus faber*	19	7.0	9	4	1	3 zy	5 z	2 y	3 xy	13 z	<1 x	4 y
Greater Amberjack *Seriola dumerili*	8	3.0	6	6	4	2	8	6		2 y	5 y	13 z
Gray Triggerfish *Balistes capriscus*	5	1.8	3	5	4	3	5	4	2	4	3	6

Table 2. Least-square mean density (number/100 m^2) for all species, Vermilion Snapper, and Greater Amberjack associated with an interaction of reef type and shelf location on artificial habitats in the northern Gulf of Mexico for 2012–2015. Different letters indicate significant differences ($P \leq 0.05$) within species.

		Reef type		
Species	Location	Big	Tank	Small
All species	Inner	167 yx	243 zx	274 zx
	Mid	522 z	253 zx	350 zy
	Outer	216 zx	397 zy	156 x
Vermilion Snapper	Inner	<1 w	1 xw	89 z
	Mid	119 z	38 zy	41 zy
	Outer	77 zy	121 z	10 yx
Greater Amberjack	Inner	3 yx		8 zy
	Mid	12 zy	12 zy	7 zx
	Outer	9 zx	21 z	3 yx

(Table 4). Seasonally, larger sizes were detected for Red Snapper in the winter and spring; for Tomtate, Vermilion Snapper, and Greater Amberjack in the spring; and for Atlantic Spadefish in the spring and summer (Table 4).

Several species were larger when associated with an interaction of reef type and shelf location. Red Snapper were larger on big reefs at outer-shelf locations, Tomtates on tank reefs at inner-shelf locations and small reefs at mid-shelf locations, Vermilion Snapper on tank reefs at inner- and mid-shelf locations and small reefs at outer-shelf locations, Atlantic Spadefish on big and tank reefs at mid-shelf locations, and Greater

Table 3. Least-square mean density (number/100 m^2) for Tomtate associated with an interaction of reef type and season, and for Tomtate and Atlantic Spadefish associated with an interaction of shelf location and season on artificial habitats in the northern Gulf of Mexico for 2012–2015. Different letters indicate significant differences ($P \leq 0.05$) within species.

		Season			
Species	Reef type	Winter	Spring	Summer	Fall
Tomtate	Big		12 xw	7 yxw	196 z
	Tank	1 w	11 xw	170 zy	82 zx
	Small	33 zw	89 zw	43 zw	53 zw
	Location	Winter	Spring	Summer	Fall
Tomtate	Inner	43 zx	27 yxw	43 zx	65 zx
	Mid	8 zw	6 xw	53 zx	213 z
	Outer		36 zw	82 zy	24 yxw
Atlantic Spadefish	Inner		13 yx	1 xv	6 yv
	Mid	531 z	22 y	1 wv	11 yxw
	Outer		29 yx	<1 v	6 yv

Table 4. Comparison of least square mean sizes ± SD (total length mm) within reef type, shelf location, and season for dominant species (>1.3% of the total density) on artificial habitats in the northern Gulf of Mexico for 2012–2015. Different letters indicate significant differences ($P \leq 0.05$) within species.

	Reef type			Shelf location			Season			
Species	Big	Tank	Small	Inner	Mid	Outer	Winter	Spring	Summer	Fall
Red Snapper	433	416	376	374	388	452	427	435	396	385
	±141 z	121 y	±157 x	±145 x	±132 y	145 z	±181 z	±142 z	±139 y	±134 x
Tomtate	121	62	158	145	121	71	117	208	91	115
	±78 y	±60 x	±70 z	±69 z	±83 y	±63 x	±38 y	±40 z	±76 x	±80 y
Vermilion Snapper	145	166	91	60	157	152	98	315	99	134
	±95 y	±124 z	±83 x	±47 y	±103 z	±115 z	±76 x	±37 z	±92 x	±105 y
Atlantic Spadefish	287	290	139	209	173	241	138	287	278	252
	±63 z	±46 z	±24 y	±67 y	±75 x	±63 z	±20 x	±52 z	±72 z	±84 y
Greater Amberjack	617	495	410	400	477	560		622	490	469
	±275 z	±157 y	±159 x	±179 x	±193 y	±218 z		±144 z	±172 y	±228 y

Amberjacks on big reefs at outer-shelf locations (Table 5).

Several species were larger when associated with an interaction of reef type and season. Red Snapper were larger on big reefs during the fall and on big and tank reefs during the winter, Vermilion Snapper on tank reefs during the winter and spring, Atlantic Spadefish on big and tank reefs during the spring and tank and small reefs during the summer, and Greater Amberjacks on big reefs during the spring and fall (Table 6).

Several species were larger when associated with an interaction of shelf location and season. Red Snapper were larger at outer-shelf locations during the winter and spring, Tomtates at mid-shelf locations during the spring, Vermilion Snapper at outer-shelf locations during the winter and inner- and at mid-shelf locations during the spring, Atlantic Spadefish at outer- and inner-shelf locations during the summer and at mid-shelf during the spring and fall, and Greater Amberjacks at inner- and mid-shelf reefs during the spring (Table 7).

Also, there were larger fish associated with three-way interactions. Red Snapper were larger on big reefs at outer-shelf locations during the winter ($F_{5,\,8334} = 27.5$, $P < 0.001$). Tomtates were larger on big reefs at mid-shelf locations during the summer ($F_{2,\,5585} = 10.2$, $P < 0.001$). Vermilion Snapper were larger on tank reefs at outer-shelf locations during the winter ($F_{1,\,5279} = 34.8$, $P < 0.001$). Atlantic Spadefish were larger on small reefs at inner-shelf locations during the spring and on tank reefs at outer-shelf locations during the summer ($F_{1,\,1753} = 148$, $P < 0.001$). Greater Amberjacks were larger on big reefs at outer-shelf locations during the fall ($F_{3,\,745} = 25.9$, $P < 0.001$).

Table 5. Least square mean sizes (Total length mm) for dominant species (>1.3% of the total density) associated with an interaction of reef type and shelf location on artificial habitats in the northern Gulf of Mexico for 2012–2015. Different letters indicate significant differences ($P \leq 0.05$) within species.

		Reef type		
Species	Location	Big	Tank	Small
Red Snapper	Inner	396 x	446 y	308 u
	Mid	435 y	374 w	353 v
	Outer	504 z	448 y	438 y
Tomtate	Inner	174 y	205 zy	136 x
	Mid	122 w	62 v	203 z
	Outer	46 v	54 x	145 x
Vermilion Snapper	Inner	241 zywv	267 zyx	59 v
	Mid	127 w	253 z	133 xw
	Outer	176 y	128 w	231 z
Atlantic Spadefish	Inner	260 yx	246 xw	156 u
	Mid	307 z	303 z	138 t
	Outer	232 w	281 y	185 v
Greater Amberjack	Inner	274 v		435 w
	Mid	550 y	532 y	300 v
	Outer	813 z	468 xw	542 yx

Table 6. Least square mean sizes (total length mm) for dominant species (>1.3% of the total density) associated with an interaction of reef type and season on artificial habitats in the northern Gulf of Mexico for 2012–2015. Different letters indicate significant differences ($P \leq 0.05$) within species.

		Season			
Species	Reef type	Winter	Spring	Summer	Fall
Red Snapper	Big	452 zy	471 z	338 u	453 z
	Tank	470 z	415 yxw	422 yx	389 v
	Small	390 wv	378 wvu	406 xv	339 u
Tomtate	Big		210 zy	237 zy	114 v
	Tank	241 zxwvut	228 z	43 t	64 u
	Small	117 v	190 yx	155 w	176 x
Vermilion Snapper	Big		306 y	289 zy	136 w
	Tank	379 z	323 zy	79 u	172 x
	Small	79 u		108 v	87 u
Atlantic Spadefish	Big		293 z		280 y
	Tank		304 z	278 zy	271 y
	Small	138 w	173 x	280 zy	159 x
Greater Amberjack	Big		698 z	391 wv	631 zy
	Tank		565 yx	513 xw	461 w
	Small			505 xw	342 v

Table 7. Least square mean sizes (total length mm) for dominant species (>1.3% of the total density) associated with an interaction of shelf location and season on artificial habitats in the northern Gulf of Mexico for 2012–2015. Different letters indicate significant differences ($P \leq 0.05$) within species.

		Season			
Species	Reef type	Winter	Spring	Summer	Fall
Red Snapper	Inner	391 w	437 yx	318 v	333 v
	Mid	241 u	420 x	376 w	379 w
	Outer	517 z	488 z	449 y	426 x
Tomtate	Inner	116 v	211 y	130 wv	137 w
	Mid	134 xwv	252 z	138 w	116 v
	Outer		170 x	63 u	77 u
Vermilion Snapper	Inner	77 u	267 zyxw	43 t	50 t
	Mid	140 xwvu	335 zy	111 v	163 x
	Outer	379 z	305 y	101 v	149 w
Atlantic Spadefish	Inner		258 y	280 zyx	147 w
	Mid	138 w	311 z	265 y	303 z
	Outer		247 yx	343 z	232 x
Greater Amberjack	Inner		597 zyxw	504 yx	326 w
	Mid		670 z	386 xw	430 x
	Outer		564 y	539 y	571 y

Richness, Evenness, and Diversity

Richness ($F_{2,58} = 6.1, P = 0.004$) and Shannon–Wiener diversity ($F_{2,58} = 7.0, P = 0.002$) indices were higher on tank and big reefs (Figure 3). Richness was lower during the winter ($F_{3,58} = 5.4, P = 0.003$), evenness was lower during the winter ($F_{3,58} = 5.7, P = 0.002$), and Shannon–Wiener diversity was higher during the fall and lower during the winter ($F_{3,58} = 9.5, P < 0.001$; Figure 4). No significant interactions were associated with reef type, shelf-location, or season for richness, evenness, or diversity indices.

Fish Communities

Three-way PERMANOVA analysis indicated differences in fish assemblages relative to reef type (pseudo-$F_{2,58} = 2.8, P = 0.001$), location (pseudo-$F_{2,58} = 3.3, P = 0.001$), and season (pseudo-$F_{2,58} = 2.0, P = 0.007$). Pairwise tests showed specific differences (Table 8). Visual examination of the MDS plot indicates differentiation of small reefs from big and tank reefs (Figure 5) but little separation for shelf locations or seasons (not shown).

Temperature, salinity, and dissolved oxygen were similar among shelf locations (Figure 6). Salinity was also similar over seasons, temperatures were lowest in winter, and dissolved oxygen had minimum levels in late summer but never dropped below 3 mg/L (Figure 6). Abiotic factors (temperature, salinity, dissolved oxygen, and depth) had no significant effect on the fish assemblage for any single or combinations of environmental variables (Spearman's Rho = 0.117, $P = 0.076$).

Discussion

Diversity Measures

Reef fish diversity measures in the present study were significantly associated with all three main reef attributes or factors: reef type, shelf location, and season. It was not surprising that higher assemblage diversity was observed at big and tank reefs in comparison to small reefs as these diversity to habitat size relations reflect classical species area patterns well documented in the literature (Molles 1978; Sale and Douglas 1984; Schroeder 1987; Bohnsack et al. 1994; Chittaro 2002; Jordan et al. 2005). Clearly, larger reefs (e.g., ships or army tanks) provide greater vertical relief that can attract more upper water column species, more stable structure that allows for species that tend to borrow under structures (many Serranidae and Lutjanidae), and perhaps more shelter habitat for smaller species to avoid predators. Further, assemblage diversity relations documented here align with the time-stability hypothesis where the big and tank artificial reefs were older and served as a reef for longer time periods to attract and retain more species and individuals (Hessler and Sanders 1967; Sanders 1968; McClain and Schlacher 2015).

When compared to open habitat, it is generally accepted that artificial reefs will increase numbers of individuals and species diversity compared to natural sand-mud habitat in marine shelf habitats (Bohnsack 1989; Milon 1989; Fabi and Fiorentini 1994; Mills et al. 2017). Although not tested in the present study (open habitats were not sampled), previous comparisons of reef complexity among artificial habitats confirm such patterns in the northern Gulf of Mexico (Lingo and Szedlmayer 2006). In addition to structural complexity, these patterns may also be related to increased epifaunal communities that provide increased forage base associated with artificial reefs (Hueckel and Buckley 1987; Redman and Szedlmayer 2009).

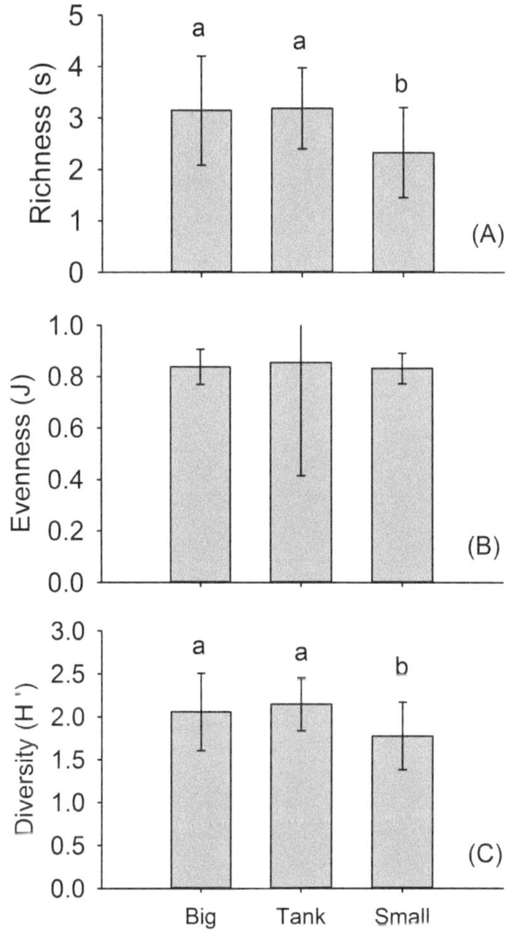

Figure 3. Comparison of diversity indices by reef type. **(A)** Species richness, **(B)** evenness, and **(C)** Shannon–Wiener diversity. Significant differences ($P < 0.05$) are indicated by different letters.

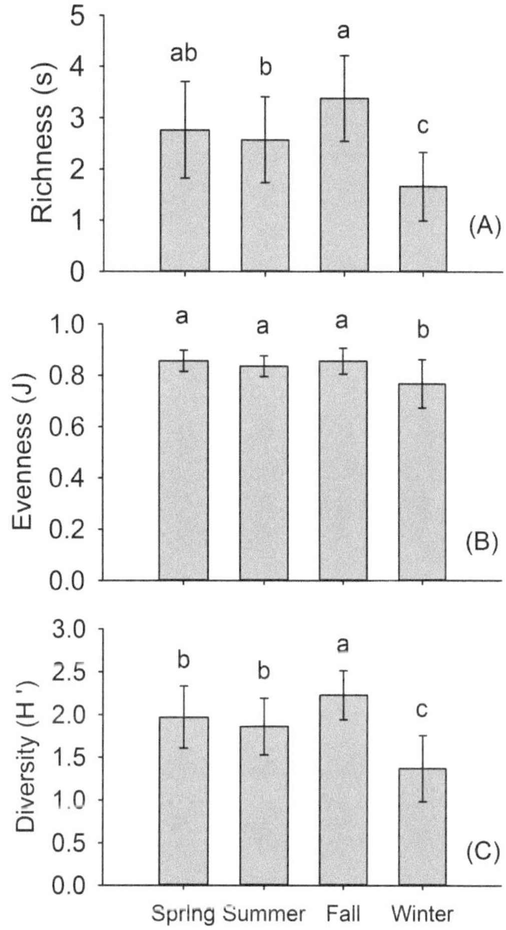

Figure 4. Comparison of diversity indices by season. **(A)** Species richness, **(B)** evenness, and **(C)** Shannon–Wiener diversity. Significant differences ($P < 0.05$) are indicated by different letters.

It was also expected that diversity would be greater during the fall. Again, this is a pattern that has been well documented in previous studies of artificial reef fish assemblages that showed seasonal changes on both large and small oil platforms (<4 m^3) reefs in the northern Gulf of Mexico (Stanley and Wilson 1991; Lingo and Szedlmayer 2006; Redman and Szedlmayer 2009). The increase in fish assemblage diversity during the fall is most likely related to increased water temperature, which facilitates an increase in tropical recruits (usually not observed at other times of the year). This temperature increase also aids the reef colonization of age-0 recruits of many reef fish species with more temperate and tropical affinities that spawn in the spring and summer and move to reef structure in the fall (Rooker et al. 1997; Szedlmayer and Conti 1999; Szedlmayer and Lee 2004; Hernandez et al. 2010; Szedlmayer 2011).

Table 8. Fish assemblage comparisons based on Bray–Curtis similarity by reef type, shelf location, and season from April 2012 to November 2015. Significant ($P \leq 0.05$) differences are indicated (*) for pairwise comparisons after significant effects were detected with three-way PERMANOVA. Only main effects are shown, as all two-way and three-way interactions were not significant.

Reef type	Pseudo-*t*	Exact *P*
Big, small	1.8	0.003*
Big, tank	0.6	0.936
Small, tank	2.0	0.001*
Shelf location		
Inner, mid	1.3	0.077
Mid, outer	1.8	0.004*
Inner, outer	2.0	0.001*
Season		
Fall, spring	1.8	0.002*
Fall, winter	1.6	0.014*
Fall, summer	1.1	0.256
Spring, winter	No test	
Spring, summer	1.1	0.272
Winter, summer	1.5	0.034*

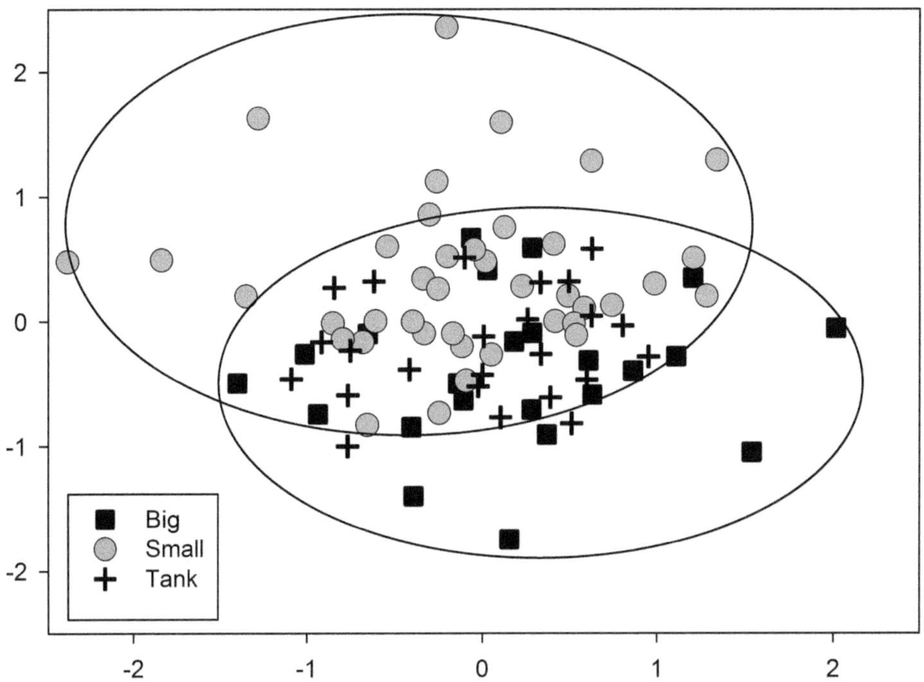

Figure 5. Ordination plot (multidimensional scaling) of fish assemblages based on Bray–Curtis similarity coefficients by reef type. Big reefs = squares, tank reefs = crosses, and small reefs = gray circles.

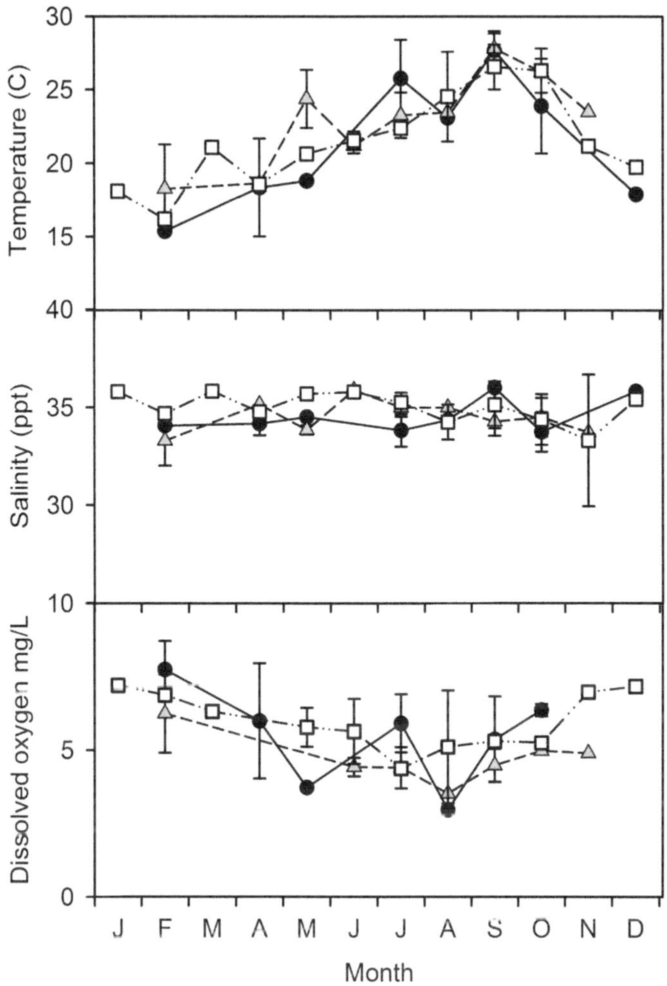

Figure 6. Mean ± SD temperature (°C), salinity (‰), and dissolved oxygen (mg/L) by shelf location and month from 2012 to 2015. Black dots with solid line = inner shelf, gray triangles with dashed lines = mid shelf, and open squares with dash-dot lines = outer shelf. Error bars = SD.

Fish Community Similarity Comparisons

Faunal similarity (i.e., Bray–Curtis similarity index) based on fish assemblage composition was similar to the assemblage diversity measures with regard to associations it had with reef attributes such as reef type, location, and season. Additionally, there were no interaction effects detected among the reef attributes relative to faunal similarity. Assemblage similarity identified groups associated with big and tank reefs as differing from small reefs, as with diversity measures. In the present study, tank reefs might be considered more similar to big reefs than small reefs in that tank reefs were large enough to attract and retain larger reef fishes. Also, as previously stated, longevity of big and tank reefs likely caused similarity differences from small reefs. For example, the northern Gulf of Mexico is subject to regular disruption from major tropical storms. These storms not only affect coastal areas, but large waves (>6 m) can reach the continental shelf sea floor (Wang et al. 2005) and may destroy smaller and generally more fragile reef structures or cause artificial reefs to subside and become buried in the substrate. Thus, small reefs (metal cages, sunken boats, concrete pyramids, etc.) might be expected to persist for shorter time periods while big and tank reefs are more permanent structures and fish communities can attain maximum carrying capacities due to longer reef existence. Further, confirmation of reef longevity is shown in the present study where tank reefs were still present and structurally intact 21 years after deployment. Similar conclusions were observed for fish communities on artificial reefs off Pensacola, Florida, where the decrease in larger piscivores (Serranidae) were associated with the reduction in smaller concrete artificial reefs, which were apparently became buried during tropical storms (Dance et al. 2011).

Another aspect of artificial reef fishes that became apparent in the present study was the differences among assemblages on outer-shelf locations from those on the mid- and inner-shelf locations. This pattern was consistent with the effect of reef type on fish communities and can be similarly explained with more consistent temperature, salinities, and structure longevity (deeper artificial reefs are less prone to wave action; i.e., greater stability). In addition, fish species that tend to occur in deeper water have a greater potential for contributing to fish diversity on the outer shelf due to the closer proximity of the extensive mesophotic deep reefs on the continental slope (Weaver et al. 2001; Beyea and Szedlmayer 2016). Similar patterns were observed off the Texas coast, where similarities in fish assemblages were related to depth (Ajemian et al. 2015). However, it is difficult to compare the Texas study to the present study because in that study, all reefs examined were deeper (34–84 m, similar to the outer-shelf locations in the present study) and all of reef types investigated were oil-gas platforms and classified as big reefs in the present study.

Total Fish Densities

Fish densities were greater on reefs at mid-shelf locations. This observation may be explained by the greater densities of unpublished small reefs inside the artificial reef permit zones on this part of the shelf (S. T. Szedlmayer and P. A. Mudrak, unpublished side-scan sonar surveys). Off coastal Alabama, artificial reef building started in 1953 with the deployment of 250 automobile bodies, and additional automobiles continued to be added at a steady rate for the next 30 years. Subsequently, reef building in the area substantially increased with the official designation of the Hugh Swingle General Permit Area in 1991 (Tatum 1993; Minton and Heath 1998). This mid-shelf reef building area has been in place for more than 60 years, and it is likely that its proximity to coastal ports has allowed fishers to deploy more artificial reefs in this area compared to other reef-building permit areas further offshore. Similar to assemblage diversity, total fish densities were greater in the fall, and as previously stated are most likely related to new age-0 recruits. However, there is a high mortality among these age-0 recruits during their first few months on the reef structure. Subsequently, their density will precipitously decline in the winter and spring (Szedlmayer 2011).

It is speculated that natural reefs in deeper waters on the continental slope should have higher diversity when compared to any shelf location examined in the present study. In fact, there is evidence that assemblages at these extensive offshore natural rock pinnacles have a higher species richness (113 fish species identified, Weaver et al. 2001). Oppositely, the population densities were lower on these deep reefs as fish were spread out over hundreds of meters of natural rocky habitat as opposed to being concentrated at a relatively small artificial reefs surveyed in the present shelf study.

Comparisons across the Northern Gulf of Mexico

Reef fish assemblages observed in the present study were similar to assemblages observed in a northwest Florida study in an adjacent area immediately east of the present study site (Dance et al. 2011). In both studies, Red Snapper and Tomtate were the dominant species. The top 10 species were also similar between the two studies, with the exception that Pinfish *Lagodon rhomboides* and Jackknifefish *Equetus lanceolatus* comprised 6–7% of the assemblage off Florida but represented less than 0.001% of the individuals observed in the present study. Although Red Snapper and Gray Triggerfish were dominant species off Alabama and northwest Florida, they did not display any preferences for artificial reef factors (reef type, location, and season) examined in the present study. Thus, community patterns were driven by other species (e.g., Tomtate, Vermilion Snapper, Atlantic Spadefish, and Greater Amberjack). Each of these species had different associations for main and interaction effects and none with the same associations for reef type, location, or season.

Comparisons to other artificial reef fish assemblages at greater distances from the present study (e.g., off Louisiana, Texas, and central Florida) had greater differences from the present study, with species differences ranging from 40% to 85% (Smith et al. 1979; Rooker et al. 1997; Wilson et al. 2003; Ajemian et al. 2015), and these differences were most likely attributed to geographical separation (Floeter et al. 2008). Consistent with greater differences in dominant fish species with increasing distance from the present study are changes in substrate type, with the Florida shelf showing more sand and rock rubble habitat, the Alabama shelf with greater sand-mud substrate, and the western shelf off Louisiana and Texas showing more mud-silt substrate (Dufrene 2005; Jenkins et al. 2008; Williams et al. 2012). This suggests that the changing fish compositions on artificial reefs across the northern Gulf of Mexico were related to these changing substrate types.

Fish Size Comparisons

Size (TL mm) of dominant species (>1.3 of the total density) was significantly different with regard to reef type, location, season, and two-way and three-way interactions, but the relations were inconsistent among species. Larger Red Snapper, Atlantic Spadefish, and Greater Amberjacks were associated with big and tank reefs, the outer-shelf location, and the spring season. These observations are similar to previous studies that suggested an ontogenetic habitat shift with age and size. For example, Red Snapper initially settle on the inner shelf on small reefs, and as they grow older and larger, they move to increasingly larger reefs (Gallaway et al. 2009; Jaxion-Harm and Szedlmayer 2012). These observations are further reinforced by the indication that there was a significant interactive effect between reef type and shelf location, with the smaller fish associated with small reefs at inner-shelf locations and larger fish associated with big reefs at outer-shelf locations. However, fishing mortality for Red Snapper and Greater Amberjacks could also explain the presence of larger-sized fish at outer-shelf locations. Artificial reefs that are closer inshore would be expected to experience higher fishing mortality rates, and removal of larger-sized fish on the inner shelf would result in similar distributional patterns (Gordon 1993; Kanamoto 1996). In contrast, we would not expect that fishing mortality would play a role for Atlantic Spadefish population dynamics as this species is not typically targeted by fishers or any part of the bycatch of any particular fishery.

Vermilion Snapper size (TL mm) associations with reef attributes most likely reflect an ontogenetic habitat shift as smaller fish occurred on small reefs at inner-shelf locations. These fish apparently moved to other reef types and locations as they matured. Atlantic Spadefish occurred at a larger size on big and tank reefs, but based on two-way interactions, these larger fish were also associated with mid-shelf locations. For these two species, the presence of larger fish on larger reefs likely resulted from predation pressure and habitat suitability, with little influence from fishing mortality.

The size of Tomtate, the second most dominant species, can be attributed to predation pressure alone. The smallest Tomtates were observed on big reefs at outer-shelf locations, and this coincided with the presence of larger Red Snapper and Greater Amberjacks on big reefs at outer-shelf locations. Also, the smallest Tomtates were observed during the summer. This seasonal occurrence likely resulted from a substantial increase in the number of age-0 fish on these reefs in late summer. This species was the smallest of any species associated with artificial reefs in the present study. It was likely that they were present on the reefs earlier as newly metamorphosed juveniles (<25 mm TL) in July, but unfortunately, these fish were too small for valid visual identification. Only after these new recruits grew larger and were repeatedly observed in later surveys were they identified as Tomtates. Tomtate is a confirmed prey for Red Snapper (Wells et al. 2008; Szedlmayer and Brewton, unpublished data) and most likely consumed by Greater Amberjack as well. We suggest that larger predators occurring on the big reefs at outer-shelf locations consumed Tomtate when it attained an optimum prey size, which resulted in smaller Tomtate on these reefs.

Conclusions

Fish assemblages associated with artificial reefs in the northern Gulf of Mexico displayed substantial differences in their habitat affinities or preferences. These differences were significantly associated with reef types, shelf locations, and seasons. In the present study, fish assemblages associated with big reefs (e.g., ships) were more similar to assemblages associated with tank reefs (i.e., U.S. Army tanks) when compared to assemblages associated with small reefs (e.g., metal cages and concreate pyramids). The larger size, longer life span, and relative stability of the big and tank reefs were most likely the drivers of these fish assemblage observations. Similarly, more stable conditions in greater depths (less affected by tropical storms) and proximity to deepwater reefs (pinnacles), and their distinctive associated fauna on the continental slope may have contributed to the increased diversity on the artificial reefs located on the outer shelf. As in previous studies, species diversity and number of individuals were highest during the fall as there was an increase in recruitment from more tropical species and new age-0 recruits that were spawned that same year. Geographically, the highest fish density occurred on reefs at mid-shelf locations. This higher density was most likely due to a combination of a higher number of artificial reefs and reduced fishing pressure in comparison to assemblages associated with inner-shelf locations. In general, artificial reefs have added significantly to both the diversity and abundance of reef-associated fish assemblages on the continental shelf off the coast of Alabama and most likely elsewhere in the northern Gulf of Mexico. However, not all artificial reefs are the same. Consequently, when artificial reefs are used by resource managers to help restore fish populations, it is important that they consider reef type, location, and survey season in their assemblage assessments. Also, individual target species may influence other assemblage members, and considerations should be made as to the overall objectives (faunal diversity, population abundance, species fitness) when proposing artificial reef deployments.

Acknowledgments

We thank M. Albins, R. Brewton, L. Grove, J. Herbig, D. Horn, C. Roberts, and J. Williams-Grove for assistance in the field and laboratory. This manuscript is a contribution of the Alabama Agricultural Experiments Station and the School of Fisheries, Aquaculture and Aquatic Sciences, Auburn University, and was supported by the Natural Resource Damages & Regulatory Affairs, Gulf Coast Restoration Organization, BP America Inc. (grant number GCRO-TX-9039).

References

Ajemian, M. J., J. J. Wetz, B. Shipley-Lozano, J. D. Shively, and G. W. Stunz. 2015. An analysis of artificial reef fish community structure along the northwestern Gulf of Mexico shelf: potential impacts of "rigs-to-reefs" programs. PLOS (Public Library of Science) ONE [online serial] 10(5):e0126354.

Anderson, M. J. 2001. Permutation tests for univariate or multivariate analysis of variance and regression. Canadian Journal of Fisheries and Aquatic Sciences 58:626–639.

Anderson, M. J., R. N. Gorley, and K. R. Clarke. 2008. PERMANOVA+ for PRIMER: guide to software and statistical methods. PRIMER-E, Plymouth, UK.

Baine, M. 2001. Artificial reefs: a review of their design, application, management and performance. Ocean and Coastal Management 44:241–259.

Beyea, R. T., and S. T. Szedlmayer. 2016. Remote video survey of the fish communities from deep water mesophotic reef habitats in the northern Gulf of Mexico. Proceedings of the Gulf and Caribbean Fisheries Institute 69:193–194.

Bohnsack, J. A. 1989. Are high densities of fishes at artificial reefs the result of habitat limitation or behavioral preference? Bulletin of Marine Science 44:631–645.

Bohnsack, J. A., D. E. Harper, D. B. McClellan, and M. Hulsbeck. 1994. Effects of reef size on colonization and assemblage structure of fishes at artificial reefs of southeastern Florida, USA. Bulletin of Marine Science 55:796–823.

Bohnsack, J. A., and D. L. Sutherland. 1985. Artificial reef research: a review with recommendations for future priorities. Bulletin of Marine Science 37:11–39.

Bolker, B. M., M. E. Brooks, C. J. Clark, S. W. Geange, J. R. Poulsen, M. H. H. Stevens, and J. S. S. White. 2009. Generalized linear mixed models: a practical guide for ecology and evolution. Trends in ecology and evolution 24:127–135.

Bortone, S. A., F. P. Brandini, G. Fabi, and S. Otake editors. 2011. Artificial reefs in fishery management. CRC Press, Boca Raton, Florida.

Bray, R. J., and J. T. Curtis. 1957. An ordination of the upland forest communities of southern Wisconsin. Ecological Monographs 27:325–349.

Chittaro, P. M. 2002. Species–area relationships for coral reef fish assemblages of St. Croix, U.S. Virgin Islands. Marine Ecology Progress Series 233:253–261.

Clarke, K. R., R. N. Gorley, P. J. Somerfield, and R. M. Warwick. 2014. Change in marine communities: an approach to statistical analysis and interpretation, 3nd edition. PRIMER-E, Plymouth, UK.

Clarke, K. R., and R. H. Green. 1988. Statistical design and analysis for a 'biological effects' study. Marine Ecology Progress Series 46:213–226.

Crowder, L. B., E. L. Hazen, N. Avissar, R. Bjorkland, C. Latanich, and M. B. Ogburn. 2008. The impacts of fisheries on marine ecosystems and the transition to ecosystem-based management. Annual Review of Ecology Evolution and Systematics 39:259–278.

Dance, M. A., W. F. Patterson, and D. T. Addis. 2011. Fish community and trophic structure at artificial reef sites in the northeastern Gulf of Mexico. Bulletin of Marine Science 87:301–324.

deReynier, Y. L., P. S. Levin, N. L. Shoji. 2010. Bringing stakeholders, scientists, and managers together through an integrated ecosystem assessment process. Marine Policy 34:534–540.

Dufrene, T. A. 2005. Geological variability and Holocene sedimentary record on the northern Gulf of Mexico inner to mid-continental shelf. Master's thesis. Louisiana State University, Baton Rouge.

Dupont, J. M. 2008. Artificial reefs as restoration tools: a case study on the west Florida shelf. Coastal Management 36:495–507.

Fabi, G., and L. Fiorentini. 1994. Comparison between an artificial reef and a control site in the Adriatic Sea: analysis of four years of monitoring. Bulletin of Marine Science 55:538–558.

Field, J. G., K. R. Clarke, and R. M. Warwick. 1982. A practical strategy for analysing multispecies distribution patterns. Marine Ecology Progress Series 8:37–52.

Floeter, S. R., L. A. Rocha, D. R. Roberton, J. C. Joyeus, W. F. Smith-Vaniz, P. Wirtz, A. J. Edwards, J. P. Barreiros, C. E. L. Ferreira, J. L. Gasparina, A. Brito, J. M. Falcon, B. W. Bowen, and G. Bernardi. 2008. Atlantic reef fish biogeography and evolution. Journal of Biogeography 35:22–47.

Gallaway, B. J., S. T. Szedlmayer, W. J. Gazey. 2009. A life history review for Red Snapper in the Gulf of Mexico with an evaluation of the importance of offshore petroleum platforms and other artificial reefs. Reviews in Fisheries Science 17:48–67.

Gordon, W. R. 1993. Travel characteristics of marine anglers using oil and gas platforms in the central Gulf of Mexico. Marine Fisheries Review 55:25–31.

Greene, L. E., and W. S. Alevizon. 1989. Comparative accuracies of visual assessment methods for coral reef fishes. Bulletin of Marine Science 44:899–912.

Gul, B., A. Lök, A. Özgül, A. Ula, F. Ozan Düzbastilar, and C. Metrin. 2011. Comparison of fish community structure on artificial reefs deployed at different depths on Turkish Aegean Sea coast. Brazilian Journal of Oceanography 59:27–32.

Hernandez, F. J., S. P. Powers, and W. M. Graham. 2010. Seasonal variability in ichthyoplankton abundance and assemblage composition in the northern Gulf of Mexico off Alabama. U.S. National Marine Fisheries Service Fishery Bulletin 108:193–207.

Hessler, R. R., and H. L. Sanders. 1967. Faunal diversity in the deep sea. Deep-Sea Research 14:65–78.

Hixon, M. A., and J. P. Beets. 1993. Predation, prey refuges, and the structure of coral reef fish assemblages. Ecological Monographs 63:77–101.

Hueckel, G. J., and R. M. Buckley. 1987. The influence of prey communities on fish species assemblages on artificial reefs in Puget Sound, Washington, USA. Environmental Biology of Fishes 19:195–214.

Huelsenbeck, J. P., and K. A. Crandall. 1997. Phylogeny estimation and hypothesis testing using maximum likelihood. Annual Reviews in Ecology and Systematics 28:437–466.

Jaxion-Harm, J., and S. T. Szedlmayer. 2012. Depth and artificial reef type effects on size and distribution of Red Snapper in the Northern Gulf of Mexico. North American Journal of Fisheries Management 35:86–96.

Jenkins, C., M. Bethel, S. Penland, and D. Mixon. 2008. A marine substrates geodatabase of the northern Gulf of Mexico, for benthic and fisheries habitats. Gulf States Marine Fisheries Commission, Ocean Springs, Mississippi.

Jordan, L. K. B., D. S. Gilliam, and R. E. Spieler. 2005. Reef fish assemblage structure affected by small-scale spacing and size variations of artificial patch reefs. Journal of Experimental Marine Biology and Ecology 326:170–18.

Kanamoto, Z. 1996. Daily changes in the catch of Parapristipoma trilineatum at the hook and line fishing ground of Toshima Fisherman's Union with the special reference to water conditions of Uwa Sea, Japan. Nippon Suisan Gakkaishi 62:565–571.

Lindberg, W. J., T. K. Frazer, K. M. Portier, F. Vose, J. Loftin, D. J. Murie, D. M. Mason, B. Nagy, and M. K. Hart. 2006. Density-dependent habitat selection and performance by a large mobile reef fish. Ecological Applications 16:731–746.

Lingo, M. E., and S. T. Szedlmayer. 2006. The influence of habitat complexity on reef fish communities in the northeastern Gulf of Mexico. Environmental Biology of Fishes 76:71–80.

Magurran, A. E. 1988. Ecological diversity and its measurement. Princeton University Press, Princeton, New Jersey.

McBride, R. S., F. J. Stengard, and B. Mahmoudi. 2002. Maturation and diel reproductive periodicity of Round Scad (Carangidae: *Decapterus punctatus*). Marine Biology 140:713–722.

McClain, C. R., and T. A. Schlacher. 2015. On some hypotheses of diversity of animal life at great depths on the sea floor. Marine Ecology 36:1–24.

McGurrin, J. M., S. B. Stone, and R. J. Sousa. 1989. Profiling United States artificial reef development. Bulletin of Marine Science 44:1004–1013.

Mills, K. A., P. A. Hamer, and G. P. Quinn. 2017. Artificial reefs create distinct fish assemblages. Marine Ecology Progress Series 585:155–173.

Milon, J. W. 1989. Artificial marine habitat characteristics and participation behavior by sport anglers and divers. Bulletin of Marine Science 44:853–862.

Minton, R. V., and S. R. Heath. 1998. Alabama's artificial reef program: building oases in the desert. Gulf of Mexico Science 16:105–106.

Molles, M. C. 1978. Fish species diversity on model and natural reef patches: experimental insular biogeography. Ecological Monographs 48:289–305.

Pickering, H., and D. Whitmarsh. 1997. Artificial reefs and fisheries exploitation: a review of the 'attraction versus production' debate, the influence of design and its significance for policy. Fisheries Research 31:39–59.

Pickering, H., D. Whitmarsh, and A. Jensen. 1998. Artificial reefs as a tool to aid rehabilitation of coastal ecosystems: investigating the potential. Marine Pollution Bulletin 37:505–514.

Redman, R. A., and S. T. Szedlmayer, S. T. 2009. The effects of epibenthic communities on reef fishes in the northern Gulf of

Mexico. Fisheries Management and Ecology 16:360–367.

Rooker, J. R., Q. R. Dokken, C. V. Pattengill, and G. J. Holt, G. J. 1997. Fish assemblages on artificial and natural reefs in the Flower Garden Banks National Marine Sanctuary, USA. Coral Reefs 16:83–92.

Sale, P. F., and W. A. Douglas. 1984. Temporal variability in the community structure of fish on coral patch reefs and the relation of community structure to reef structure. Ecology 65:409–422.

Sanders, H. L. 1968. Marine benthic diversity: a comparative study. American Naturalist 102:243–282.

Schroeder, R. E. 1987. Effects of patch size and isolation on coral reef fish recruitment. Bulletin of Marine Science 41:441–451.

Seavy, N. E., S. Quader, J. D. Alexander, and C. J. Ralph. 2005. Generalized linear models and point count data: statistical considerations for the design and analysis of monitoring studies. Pages 744–753 *in* C. J. Ralph and T. D. Rich, editors. Bird conservation implementation and integration in the Americas: proceedings of the third international partners in flight conference, 2nd volume. U.S. Forest Service, Pacific Southwest Research Station, General Technical Report PSW-GTR-191, Albany, California.

Sherman, R. L., D. S. Gilliam, R. E. Spieler. 1999. A preliminary examination of depth associated spatial variation in fish assemblages on small artificial reefs. Journal of Applied Ichthyology 15:116–121.

Shipp, R. L., and S. A. Bortone. 2009. A perspective of the importance of artificial habitat on the management of Red Snapper in the Gulf of Mexico. Reviews in Fishery Science 17:41–47.

Smith, G. B., D. A. Hensley, and H. H. Mathews. 1979. Comparative efficacy of artificial and natural Gulf of Mexico reefs as fish attractants. Florida Department of Natural Resources, Marine Research Laboratory, Marine Research Publication 35, St. Petersburg.

Smith, J. A., M. B. Lowry, and I. M. Suthers. 2015. Fish attraction to artificial reefs not always harmful: a simulation study. Ecology and Evolution 5:4590–4602.

Stanley, D. R., and C. A. Wilson. 1991. Factors affecting the abundance of selected fishes near oil and gas platforms in the northern Gulf of Mexico. U.S. National Marine Fisheries Service Fishery Bulletin 89:149–159.

Stanley, D. R., and C. A. Wilson. 1997. Seasonal and spatial variation in the abundance and size distribution of fishes associated with a petroleum platform in the northern Gulf of Mexico. Canadian Journal of Fisheries and Aquatic Sciences 54:1166–1176.

Szedlmayer, S. T. 2011. The artificial habitat as an accessory for improving estimates of juvenile reef fish abundance in fishery management. Pages 31–43 in S. A. Bortone, F. P. Brandini, G. Fabi, and S. Otake, editors. Artificial reefs in fishery management, CRC press, Boca Raton, Florida.

Szedlmayer, S. T., and K. W. Able. 1996. Patterns of seasonal availability and habitat use by fishes and decapods crustaceans in a southern New Jersey estuary. Estuaries 19:697–709.

Szedlmayer, S. T., and J. J. Conti. 1999. Nursery habitats, growth rates, and seasonality of age-0 Red Snapper *Lutjanus campechanus*, in the northeast Gulf of Mexico. U.S. National Marine Fisheries Service Fishery Bulletin 97:626–635.

Szedlmayer, S. T., and J. D. Lee. 2004. Diet shifts of juvenile Red Snapper (*Lutjanus campechanus*) with changes in habitat and fish size. U.S. National Marine Fisheries Service Fishery Bulletin 102:366–375.

Tatum, W. E. 1993. Artificial reef development and management in Alabama. Pages

9–16 *in* R. R. Lukens, editor. A profile of artificial reef development in the Gulf of Mexico. Gulf States Marine Fisheries Commission, No. 11-WB, Ocean Springs, Mississippi.

Wang, D. W., D. A. Mitchell, W. J. Teague, E. Jarosz, and M. S. Hulbert. 2005. Extreme waves under hurricane Ivan. Science 309:896.

Weaver, D. C., G. D. Dennis, and K. L. Sulak. 2001. Northeastern Gulf of Mexico Coastal and Marine Ecosystem Program: community structure and trophic ecology of demersal fishes on the pinnacles reef tract; final synthesis report. U.S. Geological Survey, USGS BSR-2001-0008, New Orleans, Louisiana.

Wells, R. J. D., J. H. Cowan, Jr., and B. Fry. 2008. Feeding ecology of Red Snapper Lutjanus campechanus in the northern Gulf of Mexico. Marine Ecology Progress Series 361:213–225.

Williams, S. J., J. Flocks, C. Jenkins, S. Khalil, and J. Moya. 2012. Offshore sediment character and sand resource assessment of the northern Gulf of Mexico, Florida to Texas. Pages 30–44 *in* S. M. Khalil, L. E. Parson, and J. P. Waters, editors. Technical framework for the Gulf regional sediment management master plan (GRSMMP). Journal of Coastal Research, Special Issue 60.

Wilson, C. A., A. Pierce, and M. W. Miller. 2003. Rigs and reefs: a comparison of the fish communities at two artificial reefs, a production platform, and a natural reef in the northern Gulf of Mexico. Coastal Fisheries Institute, School of the Coast and Environment. Louisiana State University. U.S. Department of the Interior, Minerals Management Service, Gulf of Mexico OCS Region, OCS Study MMS 2003-009, New Orleans, Louisiana.

Worm, B., E. B. Barbier, N. Beaumont, J. E. Duffy, C. Folke, B. S. Halpern, J. B. C. Jackson, H. K. Lotze, F. Micheli, S. R. Palumbi, E. Sala, K. A. Selkoe, J. J. Stachowicz, R. Watson. 2006. Impacts of biodiversity loss on ocean ecosystem services. Science 314:787–790.

Zar, J. H. 2010. Biostatistical analysis, 5th edition. Prentice-Hall, Englewood Cliffs, New Jersey.

American Fisheries Society Symposium 86:47–67, 2018

A Comparison of Fish and Epibenthic Assemblages on Artificial Reefs with and without Copper-Based, Anti-Fouling Paint

Stephen T. Szedlmayer*
School of Fisheries, Aquaculture, and Aquatic Sciences, Auburn University
8300 State Highway 104, Fairhope, Alabama 36532, USA

Dianna R. Miller
National Oceanic and Atmospheric Administration, Inouye Regional Center
1845 Wasp Boulevard, Building #176, Honolulu, Hawaii 96818, USA

Abstract.—Twenty artificial reefs were deployed early in October 2005 approximately 20 km south of Dauphin Island, Alabama (USA), in the Hugh Swingle General Permit Area. Each reef consisted of 12 concrete blocks (20 cm long × 20 cm wide × 41 cm high) arranged on a plywood base (1.5 m^2) and deployed on the bottom, 20 m deep. To quantify the epibenthic assemblage on the reefs, four removable bricks were attached to the reefs. Ten reefs were coated with copper-based, anti-fouling paint and 10 reefs were unpaint ed. Fish and epibenthic assemblages were compared between reef treatments (i.e., with and without copper-based paint). Reefs were surveyed 1 week after deployment in October 2005, then again in December 2005, May 2006, August 2006, and December 2006. During each survey, two scuba divers visually estimated the densities of all fish species and removed one of the removable bricks to identify and quantify the epibenthic organisms. The epibenthos (coverage area, biomass, diversity, species richness) and fish assemblages (total fish density, species diversity, species richness) were greater on unpainted reefs. Red Snapper *Lutjanus campechanus*, wrasses *Halichoeres* spp., Bank Sea Bass *Centropristis ocyurus*, and Atlantic Spadefish *Chaetodipterus faber* had higher densities on unpainted reefs. This study indicated that recruitment of fishes to artificial reefs was not just attraction to structure, but that growth of epibenthic assemblages had a significant influence on recruitment.

Introduction

Artificial structures have long been known to attract and aggregate fish species from surrounding waters. Artificial reefs provide shelter from predation, provide sites for orientation and spawning, and serve as substrate for epibenthic assemblages that may provide new and additional forage bases for fishes (Hueckel and Buckley 1987; DeMartini et al. 1994; Steele 1999; Szedlmayer and Lee 2004). Reef colonization typically is rapid, with many fish present before the in situ primary

* Corresponding author: szedlst@auburn.edu

and secondary trophic levels have had sufficient time to develop (Hueckel and Stayton 1982; Shulman 1984; Bohnsack 1989). This suggests that artificial reefs initially provide shelter or structure from which fish orient. Many studies have reported a positive correlation between refuge availability and juvenile fish survival or abundance (Shulman 1984, 1985; Hixon and Beets 1989, 1993; Caley and St John 1996; Steele 1999; Lingo and Szedlmayer 2006; Piko and Szedlmayer 2007; Mudrak and Szedlmayer 2012). For example, Hixon and Beets (1993) reported higher reef fish abundance on artificial reefs with holes in the reef surface in comparison to fish associated with reefs lacking holes. In addition to their function as a refuge from predators, artificial reefs may provide food resources that are important for fish recruited to the reefs (Buckley and Hueckel 1985; Relini et al. 2002; Szedlmayer and Lee 2004; Redman and Szedlmayer 2009). The trophic structure of the entire assemblage develops over time as algae, invertebrates, and fish colonize artificial reefs. However, the development of epibenthic assemblages is a highly variable process that depends on substrate type and reef complexity, time since deployment, environmental variables, competition, and grazing pressure (Osman 1977; Hixon and Brostoff 1985). Pioneer trophic groups such as algae, barnacles, and serpulid worms are typically among the first colonizers on artificial reefs (Fager 1971; Ardizzone et al. 1989; Relini et al. 1994; Boaventura et al. 2006). Although initial colonization is rapid, slower colonizing taxa continue to settle, gradually replacing early colonizers (Ardizzone et al. 1989; Wendt et al. 1989). Epibenthic assemblages on artificial reefs may converge towards that of natural reefs with time, but more often, they appear to differ. Consequently, factors influencing the differences between natural and artificial reef assemblages remain unclear (Hixon and Brostoff 1985; Wendt et al. 1989; Carr and Hixon 1997; Perkol-Finkel and Benayahu 2007).

Many studies have shown that reef fishes can obtain substantial food resources from biota directly associated with artificial reefs (Hueckel and Stayton 1982; Hueckel and Buckley 1987; Vose 1990; Vose and Nelson 1994; Relini et al. 2002; Ouzts and Szedlmayer 2003; Szedlmayer and Lee 2004). For example, Szedlmayer and Lee (2004) reported that Red Snapper *Lutjanus campechanus* increased their dependency on reef-associated prey as they moved from open sand to artificial reef habitats. Gray Triggerfish *Balistes capriscus* also forage on artificial reefs, consuming a variety of epibenthic invertebrates including barnacles and bivalves (Vose 1990; Vose and Nelson 1994; Blitch 2000). Similarly, in the Mediterranean Sea, three of the four fish species examined fed on reef-associated prey, and 91% of the prey items consumed by Annular Seabream *Diplodus annularis* were exclusive to artificial reefs (Relini et al. 2002). Further, the contribution of reef-associated prey to the diets of reef fishes may depend on both fish species and their size. For example, medium and large Striped Seaperch *Embiotoca lateralis* and Quillback Rockfish *Sebastes maliger* fed chiefly on artificial reefs, whereas small fish foraged in the nearby sand (Hueckel and Stayton 1982).

In contrast, other studies have suggested fish depend little on food resources from artificial reefs (Randall 1963; Shulman 1984; Lindquist et al. 1994; Ibrahim et al. 1996; Nelson and Bortone 1996). Nelson and Bortone (1996) showed that many commercially important reef-associated fishes in the northern Gulf of Mexico fed on fishes, crabs, squids, polychaetes, and shrimps that were obtained from surrounding sand-bottom habitat.

Similarly, examination of stomach content showed that fishes associated with artificial habitat in St. John, Virgin Islands foraged in adjacent sea grass beds rather than on reef prey (Randall 1963).

It is clear that artificial reefs attract fish, but whether or not such reefs lead to increased fish production (e.g., g $\cdot$ m^{-2} $\cdot$ d^{-1}) remains unclear (Bohnsack and Sutherland 1985; Alevizon and Gorham 1989; Bohnsack 1989; Polovina 1989; Polovina and Sakai 1989; Bohnsack et al. 1994; DeMartini et al. 1994). Polovina (1989) suggested that artificial reefs may attract fish to locations where they are more easily harvested, potentially decreasing overall fish biomass through increased fishing mortality. Polovina and Sakai (1989) examined time series, catch, and effort data for flatfish catches near Shimamaki, Japan and reported that they were related to attraction to artificial reefs, not enhanced fish production. Bohnsack (1989) suggested that artificial habitats function as fish attractors when natural food and shelter resources are plentiful, and fish abundance is most likely limited by recruitment or exploitation. However, in areas lacking natural reef habitat, artificial reefs may increase fish carrying capacity by providing limiting factors such as food and shelter and thereby enhancing fish production.

Most studies on artificial reefs have focused on the fishery resources with limited attention to the development of epibenthic assemblages (Relini et al. 1994; Svane and Petersen 2001). Also, the few studies that have examined both fish and epibenthic assemblages involved variable reef complexity or low sample sizes (Buckley and Hueckel 1985; Hueckel and Buckley 1987; Relini et al. 2002). Usually, newly deployed artificial reefs provide new substrate for the settlement and recruitment of epibenthic species (Svane and Petersen 2001), especially those organisms with an affinity for hard substrate. Since epibenthos may be important in the diets of many reef fishes, the development of epibenthic assemblages on artificial reefs may affect fish recruitment and the retention of fish on the reef once recruited.

The present study examined the influence of epibenthic assemblages on the recruitment of reef fishes to artificial habitats by comparing fish assemblages on reefs with and without epibenthos. To make this comparison, copper-based paint was applied to inhibit the development of epibenthos on artificial reefs. Several studies have used copper paint to manipulate the development of epibenthic assemblages (Bosman and Hockey 1988; Farrell 1988), but few have used this approach to examine the associated effect of copper-based paint on reef fish recruitment (Redman and Szedlmayer 2009).

Methods

Laboratory Experiment

The potential for fish avoidance of painted reefs due to the presence of copper was examined in the laboratory. Two common species of reef fish, Red Snapper and Gray Triggerfish, were placed in 1,200-L circular tanks, which were part of a 14,080-L recirculating seawater system. Each experimental tank contained three copper-painted blocks (34% copper, 214-7070, Ameron International; 20 cm long × 20 cm wide × 41 cm high concrete construction blocks) on one side of the tank and three unpainted blocks on the opposite side. For each trial, six Red Snapper, six Gray Triggerfish, or a combination (three Red Snapper and three Gray Triggerfish) were placed in the circular tank (1,200 L) and their behaviors were video recorded for 1 h. Eleven trials with six fish per trial were completed (n = 66 total fish examined). Video recordings were reviewed with Image Pro-Plus 4.5 software and fish positions recorded in relation to the painted and unpainted blocks.

Field Study

Artificial reefs (n = 20) were deployed on October 10 and 12, 2005, 20 km south of Dauphin Island, Alabama (USA) in the Hugh Swingle General Permit Area (Figure 1). Each reef consisted of 12 concrete blocks (20 × 20 × 41 cm) arranged on a plywood base (1.5 m^2) and placed on the bottom at 20-m depths (Figure 2). Concrete blocks were secured to the plywood base using 1.2-m plastic cable ties. Additionally, four small, removable concrete bricks (9 × 6 × 20 cm) were secured to the larger blocks with 30-cm plastic cable ties to allow assessment of the associated epibenthic assemblage (Figure 2). Reefs were anchored in place with a 1.5-m nylon rope attached to a 1.2-m metal ground anchor embedded in the substrate. All reefs were labeled with numbered metal or plastic tags to facilitate reef identification. Concrete blocks in 10 reefs were painted with copper-based, anti-fouling paint (painted treatment, same copper paint as above), and concrete blocks in 10 additional (control) reefs were not painted (unpainted treatment). Reefs were

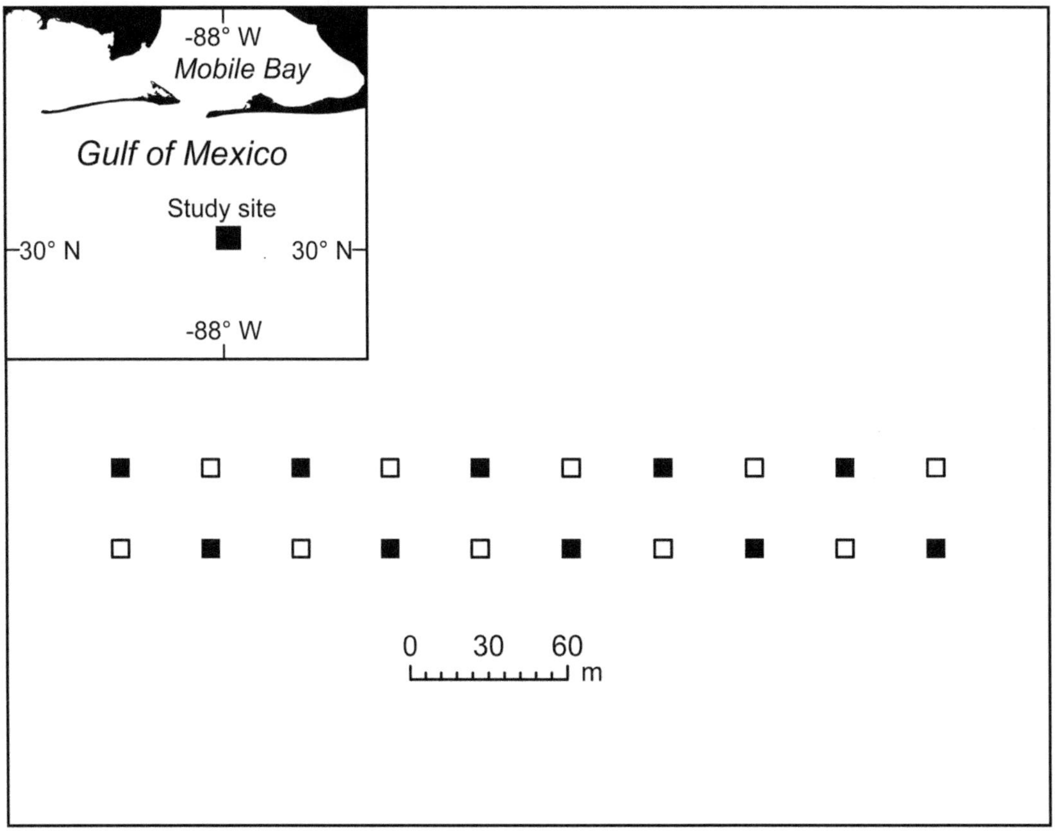

Figure 1. Location of study sites in the northern Gulf of Mexico and reef array design alternating between unpainted (open squares) and painted (black squares) reefs at 30-m intervals.

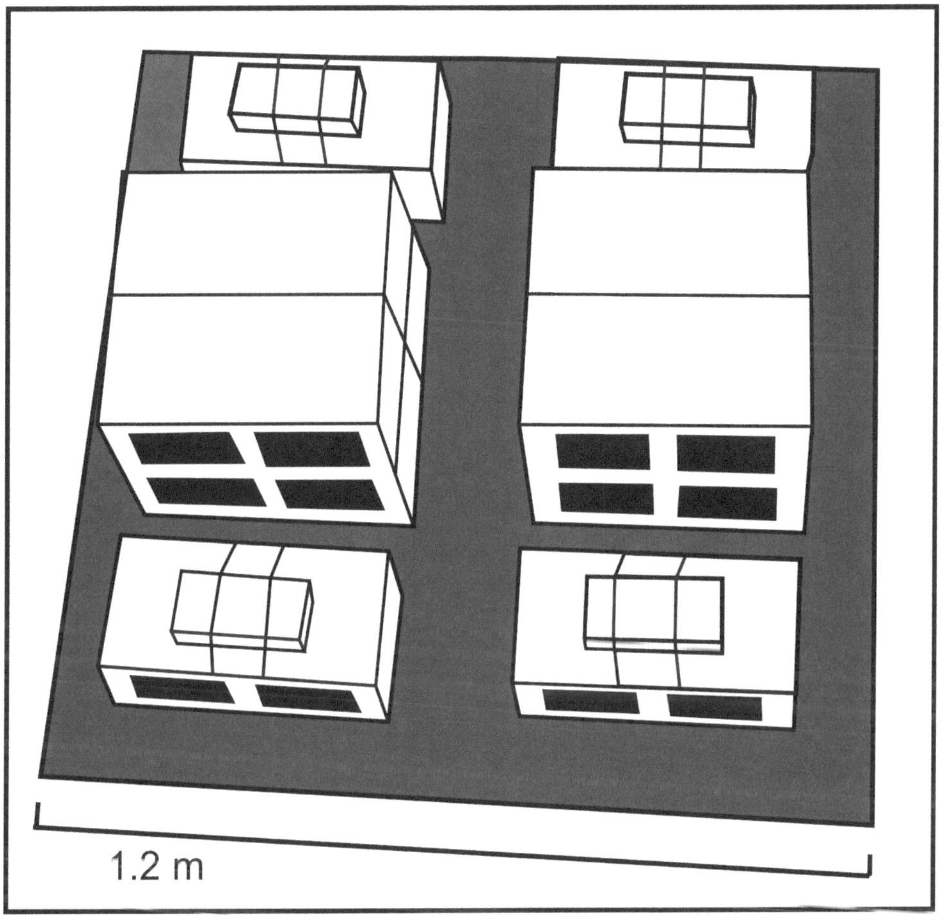

Figure 2. Artificial reef design indicating the placement of 12 concrete blocks on the plywood base and four small "break-away" bricks.

deployed along two transects, with 10 reefs per transect, alternating between painted and unpainted reefs at 30-m intervals (Figure 1).

Reef Surveys

All reefs were surveyed on October 20, 2005, December 1, 2005, May 29, 2006, August 2, 2006, and December 14, 2006. During each survey, two scuba divers used the discrete group census method (Greene and Alevizon 1989) to identify fish species densities and size-class (total length in 25-mm intervals). Since reefs were small (1.5 m^2), all fish were counted. Reefs and the surrounding area were also video recorded to later verify species identifications and record rare or unknown species. A small removable concrete brick was collected per reef during each survey. Bricks were placed in cloth sample bags (25 × 43 cm, #5250, Hubco Protexo) and preserved in alcohol-based preservative (NOTOXhisto fixative, Scientific Device Laboratory) for later analysis of epibenthic assemblages. Water samples were collected within 1 m of each reef to determine copper levels in the surrounding environment. Salinity (‰), temperature (°C) and dissolved oxygen (mg/L) were recorded during each survey with a YSI 6920 meter.

Epibenthos Analysis

All epibenthic organisms on the upper surface of removed bricks were counted, measured, and identified to the lowest taxa possible. The upper surface of each brick was photographed with a digital camera (Canon PowerShot A530). Images in the photographs were size-calibrated and analyzed with Image Pro-Plus 4.5 software to determine surface area (mm^2) of each individual organism or taxonomic group. The area covered by each taxon was recorded relative to the brick area measured as $cm^2/100\ cm^2$. Areas of both mobile and sessile epibenthic organisms were determined. In some cases, individuals overlapped spatially so that the total epibenthic area per brick may be greater than the total brick area (i.e., percent coverage per brick may be >100%). Areas of colonial or encrusting species (e.g., Sabellidae worms, sea mat bryozoan *Membranipora tenuis*, bugula bryozoan *Bugula neritina*, and Demospongiae sponges) were measured in clusters since it was difficult to distinguish individual organisms. The sample bags used to contain the bricks contained many mobile epibenthic organisms and additional sessile organisms that had become dislodged from the bricks. To count and measure these organisms, bag contents were rinsed through a 500-μm sieve to remove sand and mud and retrieve any organisms that had become unattached from the bricks. These organisms were counted, photographed, measured with Image Pro-Plus software, and identified to the lowest taxa possible.

Following measurement of areas, all epibenthos were removed from the top surface of each brick. These samples were then combined with the organisms from the corresponding sample bags and then dried in a drying oven at 60°C for 24 h. Samples were weighed to the nearest 0.01 g, and weight was reported as dry weight $g/100\ cm^2$.

Statistical Analyses

Reefs were surveyed multiple times; therefore, repeated measures generalized linear models (rmGLM) were used with negative binomial distributions to compare mean fish density per reef (number of fish/m^2), dominant fish species densities (i.e., >1% of the total density), epibenthos area ($cm^2/100\ cm^2$), epibenthos biomass (dry weight $g/100\ cm^2$), and dominant epibenthos taxa area (>1% of the total area, $cm^2/100\ cm^2$) among reef treatments and survey date or time after deployment (Seavy et al. 2005; Bolker et al. 2009).

Fish species densities and epibenthic taxa areas were square-root-transformed prior to diversity measure calculations. Fish and epibenthos assemblages were compared using the Shannon–Weiner diversity index (H'), species richness or the total number of species (S) present in the sample, and evenness ($J = H'/H'\text{max}$; Magurran 1988). These community variables (H', S, and J) for fish and epibenthos assemblages were compared among reef treatments and survey date with repeated measures analysis of variance (rmANOVA). If significant differences ($P \leq 0.05$) were detected with rmGLM or rmANOVA, specific differences were assessed with Tukey's post-hoc test (Zar 2010).

Fish and epibenthic assemblages were compared among reef treatments and survey dates using nonparametric multidimensional scaling (MDS; Szedlmayer and Able 1996; Lingo and Szedlmayer 2006). Bray–Curtis similarity coefficients were calculated as the similarity between all samples based on square-root-transformed fish densities and epibenthos areas for all species (Field et al. 1982). These similarities were calculated among individual reef samples and plotted as MDS ordinations labeled by reef treatment and survey date. We used two-way permutational multivariate analysis of variance

(PERMANOVA) to statistically compare fish assemblage similarities and separately epibenthic assemblage similarities among reef treatments, survey date, and interactions (Anderson et al. 2008).

Results

Laboratory Experiment

In the avoidance behavior experiment, fish had a significant preference for painted compared to unpainted block treatments. Thus, it was concluded that fish would not avoid the reefs owing to the presence of copper paint in field studies.

Field Unpainted versus Painted Artificial Reefs

During the reef surveys, water samples were collected from the vicinity of each reef and tested for the presence of copper. Copper was detected in seawater 0.5 m from painted reefs (mean + SE = 0.27 + 0.1 parts per million) during the first survey (October 2005), after which copper was not detected in any subsequent water samples. Water temperature (°C) was typically low in the winter while dissolved oxygen (mg/L) and salinity (‰) changed little during the surveys (Figure 3).

Fish Assemblages

There were 33 fish taxa identified (lowest taxa: 26 species, five genera, two families) among 1,119 individuals on unpainted reefs and 766 individuals on painted reefs during the first four survey periods (80 reef surveys). Survey data recorded during the last survey, December 2006—14 months after deployment, were not included in the analyses because the reefs were broken apart and

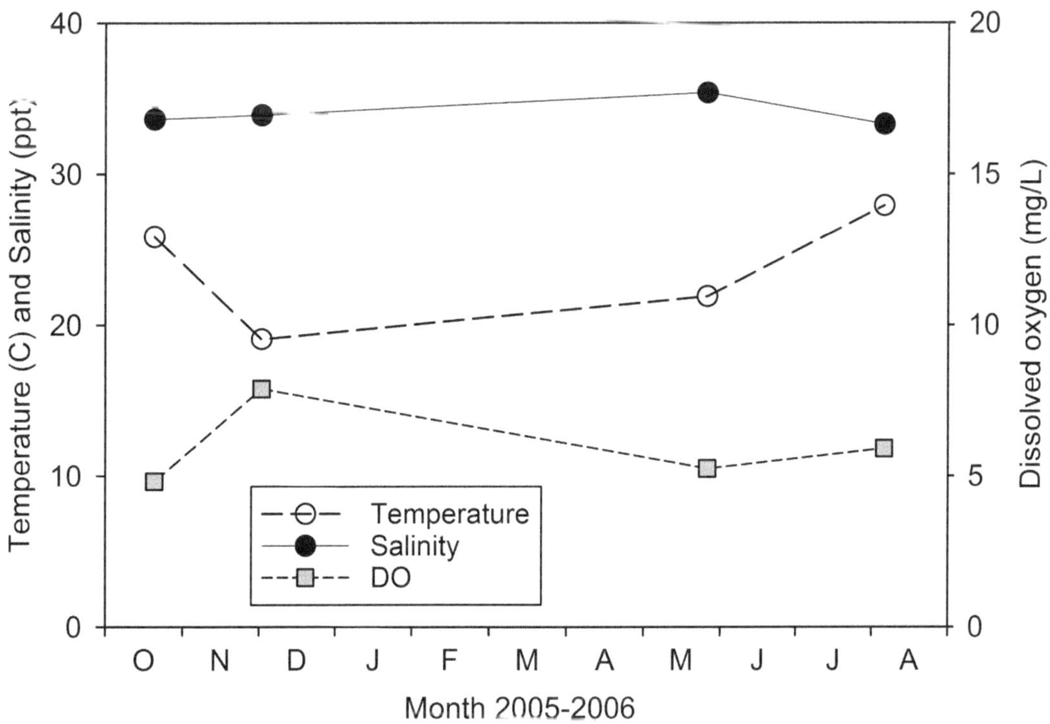

Figure 3. Temperature (°C), salinity (‰) and dissolved oxygen (mg/L) by survey date in 2005–2006.

more than 50% of the reef materials were buried in the substrate. Mean fish densities per reef were higher on unpainted compared to painted reefs ($F_{1, 18}$ = 5.9, P = 0.026) and increased with survey date ($F_{3, 54}$ = 93.7, P < 0.001; Table 1). Fish densities were not associated with reef treatment and survey date interactions ($F_{3, 54}$ = 1.2, P < 0.307; Table 1). Dominant species on the reefs included Red Snapper, wrasses *Halichoeres* spp., Tomtate *Haemulon aurolineatum*, Gray Triggerfish, Lane Snapper *Lutjanus synagris*, sand perch *Diplectrum* spp., Cocoa Damselfish *Stegastes variabilis*, Pigfish *Orthopristis chrysoptera*, Bank Sea Bass *Centropristis ocyurus*, Blenniidae, Rock Sea Bass *Centropristis philadelphica*, Greater Amberjack *Seriola dumerili*, and Atlantic Spadefish *Chaetodipterus faber*. Each of these 13 taxa were more than 1% of the total density and together accounted for 95.3% of all the fish observed during the study (Table 1).

Densities of Red Snapper, wrasses, Bank Sea Bass, and Atlantic Spadefish were higher on unpainted reefs compared to painted reefs (Table 1). Densities of Red Snapper, Tomtate, Gray Triggerfish, Gray Snapper, Pigfish, Bank Sea Bass, Blenniidae, and Atlantic Spadefish increased with time after deployment (i.e., survey date; Table 1). Also, fish densities for wrasses, Pigfish, Bank Sea Bass, and Atlantic Spadefish were significantly associated with an interaction between reef treatment and survey date (Figure 4). No significant differences in fish densities were detected for the other common fish species relative to reef treatment or survey date (Table 1).

All fish observed were either juveniles or small-sized adults (<300 mm total length [TL]; Table 2). Few species had different sizes (TL mm) between reef treatments, but wrasses and Greater Amberjack were larger on unpainted reefs and Atlantic Spadefish were larger on painted reefs. Red Snapper were associated with an interaction of larger fish (TL mm) on unpainted reefs during the December 2005 survey ($F_{3, 612}$ = 4.2, P < 0.006). There was a general trend of increasing size (TL mm) with survey date for several species that included Red Snapper, Lane Snapper, sand perch, and Gray Triggerfish (Table 2).

Higher fish diversity (H') and species richness (S) were observed among fish associated with unpainted reefs. Species diversity and richness increased with survey date, but these assemblage characteristics were not associated with interactions between reef treatment and survey date (Table 3). Examination of the MDS plot and Bray–Curtis similarities for fish assemblages allowed visual recognition of assemblages and PERMANOVA analyses detected significance differences by reef treatment for survey dates in December 2005 (pseudo-$F_{1, 18}$ = 2.9, P = 0.034) and August 2006 (pseudo-$F_{1, 18}$ = 1.8, P = 0.041; Figure 5).

Epibenthos

Epibenthos samples (n = 57) were collected in December 2005, May and August 2006, and 33 epibenthic taxa were identified (lowest taxa: 12 species, six genera, six families, and nine higher taxonomic groups). Mean coverage area was greater on unpainted reefs, increased with survey date (Table 3), and was associated with an interaction between reef treatment and survey date (Figure 6). Epibenthos dry weight was greater on unpainted reefs, and this increased with survey date (Figure 7).

Barnacles (*Balanus* spp.) covered the greatest area, followed by Sabellidae worms, purse-oysters *Isognomon* spp., mud crabs *Panopeus* spp., Demospongiae sponges, brown bryozoan, and Florida dovesnail *Costoana-*

Table 1. Mean and percent densities (number/m^2) for all fish and dominant fish taxa (≥1% of the total density) for all surveys and mean ± SE densities for all fish and dominant fish taxa by reef treatment and survey date. Different letters indicate significant differences ($P \le 0.05$) between reef treatments (UP = unpainted, P = painted) and among survey dates.

			Reef treatment		Survey date			
Fish taxa	Mean	%	UP	P	Oct 2005	Dec 2005	May 2006	Aug 2006
All fish			28.0±4.1 z	19.2±2.6 y	4.0±1.0 w	12.7±1.3 x	23.1±1.5 y	54.5±4.6 z
Red Snapper	7.8	32.9	8.9±1.2 z	6.6±0.9 y	0.4±0.1 x	7.1±0.9 y	10.7±1.1 z	12.9±1.8 z
Halichoeres spp.	3.6	15.1	5.3±1.9 z	1.8±0.7 y				14.3±3.1
Tomtate	3.5	14.9	3.7±0.9	3.3±1.1	0.1±0.1 x	0.1±0.1 x	3.7±0.7 y	10.2±2.0 z
Lane Snapper	1.5	6.2	1.5±0.2	1.4±0.2	0.8±0.2 yx	0.7±0.1 x	1.6±0.3 zy	2.8±0.5 z
Diplectrum spp.	1.1	4.5	1.1±0.2	1.0±0.2	1.3±0.3	1.0±0.2	1.0±0.2	1.1±0.5
Gray Triggerfish	1.0	4.4	1.0±0.2	1.0±0.2	0.1±0.1 w	0.7±0.2 x	1.6±0.3 yx	1.8±0.3 y
Cocoa Damselfish	0.9	3.6	1.0±0.4	0.8±0.2				3.4±0.6 z
Pigfish	0.7	2.8	0.8±0.2	0.5±0.1	0.1±0.1 y	1.1±0.3 z	1.2±0.2 z	0.3±0.1 y
Blenniidae	0.6	2.4	0.9±0.3	0.3±0.1			0.7±0.5	1.7±0.5
Bank Sea Bass	0.6	2.4	0.7±0.1 z	0.5±0.1 y		0.2±0.1 y	1.0±0.2 z	1.2±0.2
Greater Amberjack	0.5	1.9	0.6±0.6	0.3±0.2				1.8±1.3
Mycteroperca sp.	0.4	1.6	0.3±0.3	0.5±0.5	1.0±1.0		0.5±0.5	
Rock Sea Bass	0.4	1.5	0.3±0.1	0.4±0.1			0.8±0.3	0.7±0.2
Atlantic Spadefish	0.3	1.1	0.4±0.2 z	0.1±0.1 y		0.9±0.3z	0.2±0.1 y	

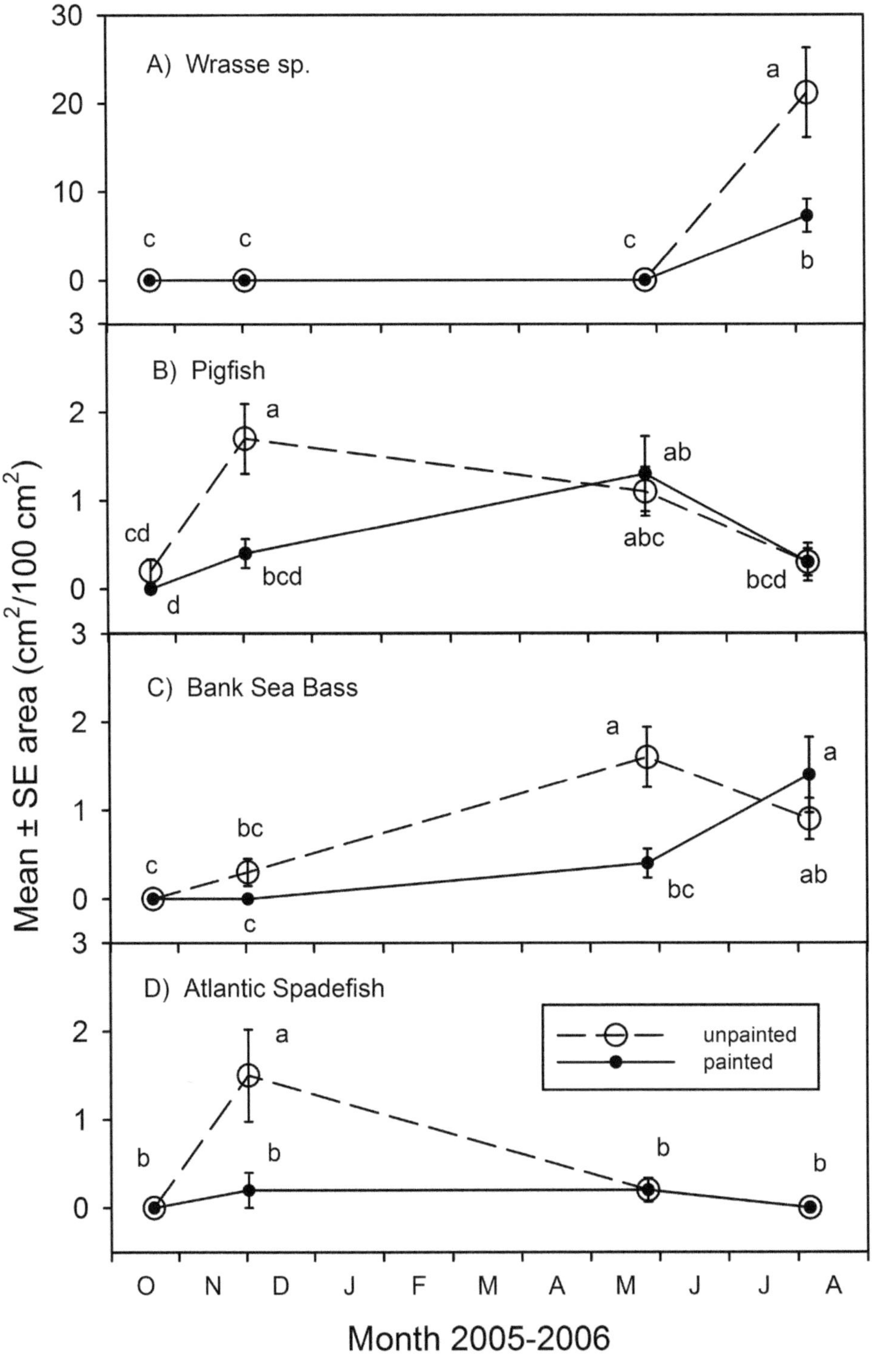

Figure 4. Mean ± SE densities (number/m^2) of **(A)** wrasses, **(B)** Pigfish, **(C)** Bank Sea Bass, and **(D)** Atlantic Spadefish by reef treatment and survey date. Different lowercase letters indicate significant differences ($P \leq 0.05$) between reef treatments and among survey dates.

Table 2. Mean size (TL mm) by reef treatment and survey date, for fish taxa ≥1% of the total density. Different letters indicate significant differences ($P \leq 0.05$) between reef treatments (UP = unpainted, P = painted) and among survey dates.

	Reef treatment		Survey date			
Fish taxa	UP	P	Oct 2005	Dec 2005	May 2006	Aug 2006
Red Snapper	129±2.0	123±2.1	100±13.5 y	102±3.2 y	118±2.2 y	148±1.5 z
Halichoeres spp.	39±0.4 z	37±0.7 y				39±6.2
Tomtate	142±2.4	144±1.8			138±3.1	145±1.8
Lane Snapper	132±6.2	136±6.2	102±9.4 y	77±5.2 y	138±9.2 z	155±4.6 z
Diplectrum spp.	92±8.5	87±7.8	65±4.6 y	86±4.3 y	156±13.9 z	61±8.6 y
Gray Triggerfish	253±8.9	227±11.7	166±25.5 zy	211±17.9 y	230±11.4 zy	265±10.8 z
Cocoa Damselfish	32±1.1	32±1.2				32±0.8
Pigfish	168±6.3	155±9.4	153±12.5	165±8.2	170±8.4	136±12.1
Blenniidae	53±3.3	42±6.1			25±0.0	61±2.4
Bank Sea Bass	122±12.7	125±14.2		106±8.7	102±15.7	143±11.8
Greater Amberjack	292±0.0 z	212±8.9 y				267±6.8
Mycteroperca sp.	216±0.0	89±0.0	89±0.0		216±0.0	
Rock Sea Bass	152±15.7	143±13.1			135±9.6	162±18.0
Atlantic Spadefish	112±5.3 y	153±24.1 z		112±4.8 y	153±26.4 z	

chis sertulariarum (Table 4). Areas for each of these taxa were more than 1% of the total area, and these seven taxa together accounted for 93.1% of the total area on the removable bricks. The remaining 26 taxa comprised 6.9% of the total area.

Most (88%) of the dominant epibenthos taxa covered more area on unpainted reef surfaces, and all taxa increased in area with survey date (Table 4). Changes in epibenthos surface area coverage were associated with interactions of reef treatment and survey date for several taxa, with higher areas for barnacles in December 2005, May and August 2006 ($F_{2,33}$ = 36.8, P < 0.001), Sabellidae worms in December 2005 and August

Table 3. Mean fish and epibenthos assemblage diversity (H'), species richness (S), and evenness (J), by reef treatment and survey date. Different letters indicate significant differences ($P \leq 0.05$) between reef treatments (UP = unpainted, P = painted) and among survey dates.

	Type		Survey			
Fish	UP	P	Oct 2005	Dec 2005	May 2006	Aug 2006
H'	1.6±0.1 z	1.5±1 y	0.8±0.1 z	1.3±0.1 y	1.7±0.1 x	2.0±0.1 w
S	6.0±0.4 z	5.0±0.5 y	2.1±0.2 z	4.3±0.3 y	6.5±0.3 x	9.0±0.5 w
J	0.93±0.01	0.93±0.02	0.89±0.04	0.94±0.01	0.94±0.01	0.94±0.01
Epibenthos						
H'	2.0±0.1 z	1.5±0.1 y		1.3±0.1 x	1.8±0.1 y	2.0±0.1 z
S	12.5±0.9 z	5.8±0.7 y		3.3±0.7 x	10.4±1.0 y	13.0±0.9 z
J	0.82±0.01	0.76±0.01		0.81±0.04	0.81±0.01	0.77±0.02

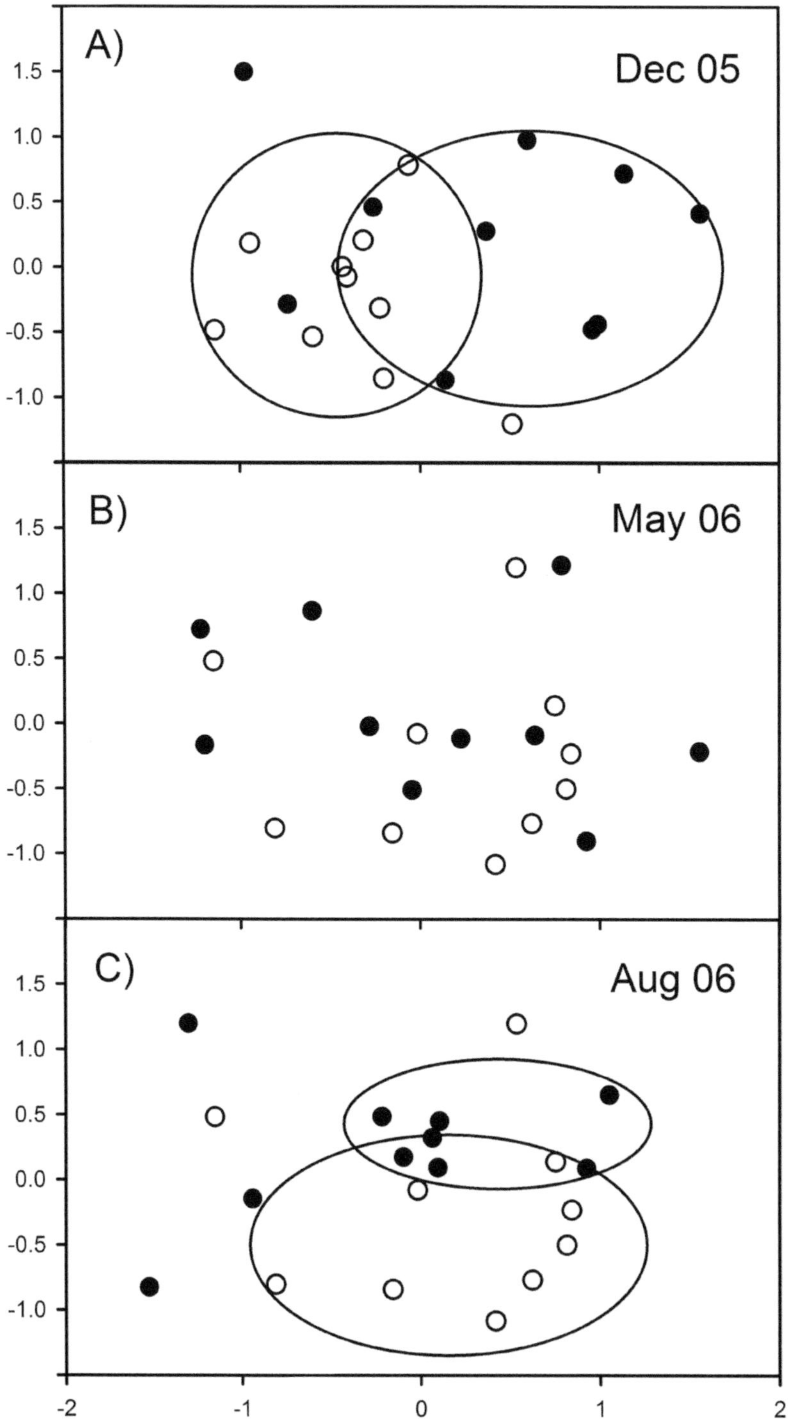

Figure 5. Nonmetric multidimensional scaling plots of fish assemblages based on Bray–Curtis similarities by reef treatment and survey date. Unpainted reefs = open circles and filled circles = painted reefs.

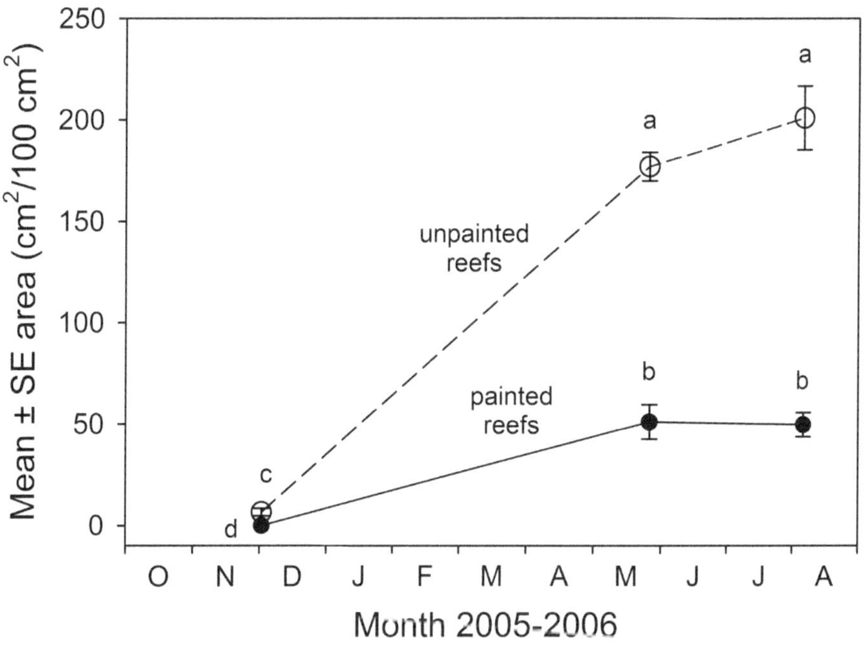

Figure 6. Mean ± SE epibenthos area (cm²/100 cm²) by reef treatment and survey date. Different lowercase letters indicate significant differences ($P \leq 0.05$) between reef treatments and among survey dates.

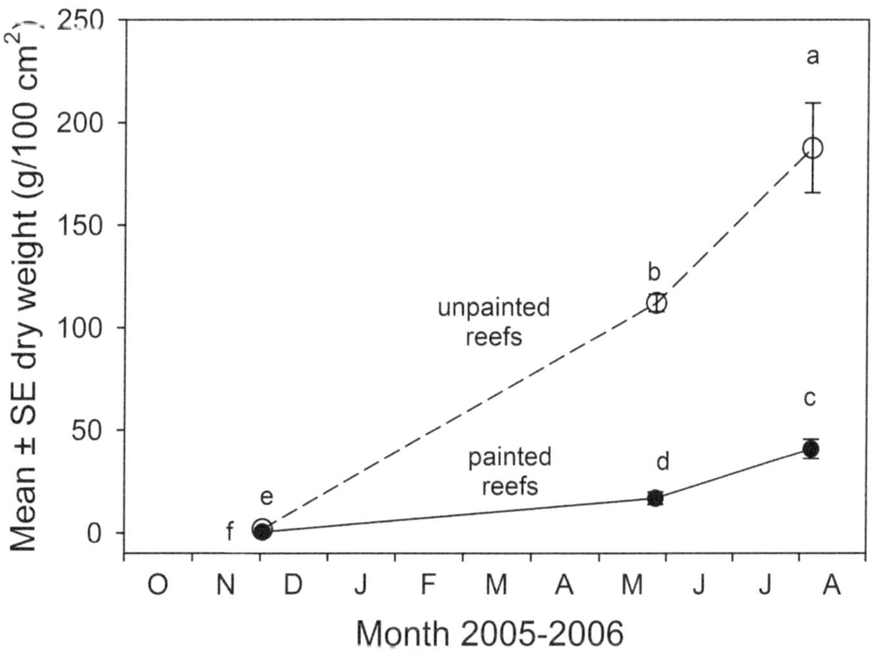

Figure 7. Mean ± SE epibenthos dry weights (g/100 cm²) by reef treatment and survey date. Different loiwercase letters indicate significant differences ($P \leq 0.05$) between reef treatments and among survey dates.

Table 4. Mean and percent area for all epibenthos and dominant epibenthos taxa (≥1% of the total density) for all surveys and mean ± SE areas for all epibenthos and dominant taxa by reef treatment and survey date. Different letters indicate significant differences ($P \leq 0.05$) between reef treatments (UP = unpainted, P = painted) and among survey dates.

			Reef type		Survey		
Epibenthos taxa	Mean	%	UP	P	Dec 2005	May 2006	Aug 2006
All epibenthos			132.4±17.5 z	35.9±5.7 y	3.5±1.3 y	114.0±15.4 z	125.3±19.2 z
Balanus spp.	56.5	66.5	87.7±11.1 z	24.2±4.0 y	2.8±1.2 y	72.9±11.7 z	85.8±10.9 z
Sabellidae	8.8	10.4	8.8±2.0	8.9±2.8	0.1±0.05 x	22.7±3.0 z	2.4±0.5 y
Isognomon spp.	4.9	5.7	9.4±1.6 z	0.1±0.05 y	0.1±0.05 x	5.1±1.6 y	8.7±2.1 z
Mud crabs	3.7	4.3	6.4±0.9 z	0.9±0.2 y	0.01±0.003 x	4.5±0.9 y	6.0±1.1 z
Demospongiae	2.7	3.2	5.3±2.3 z	0.03±0.04 y			7.8±3.2
Bugula bryozoan	1.6	1.9	3.0±1.2 z	0.1±0.1 y			4.5±1.7
Florida dovesnail	0.9	1.1	1.1±0.2 z	0.7±0.1 y	0.01±003 y	1.3±0.1z	1.4±0.2 z

2006 ($F_{2,33}$ = 11.5, P < 0.001), purse oysters in December 2005, May and August 2006 ($F_{2,33}$ = 58.1, P < 0.001), mud crabs in May and August 2006 ($F_{2,33}$ = 11.4, P < 0.001), Demospongiae in August 2006 ($F_{2,51}$ = 18.9, P < 0.001), and brown bryozoans in August 2006 ($F_{2,51}$ = 12.7, P < 0.001).

Species diversity (H') was higher for the epibenthos on unpainted reefs and H' increased with survey date. Similarly, there were more species (S) on unpainted reefs and S increased with survey date. There were no interaction effects detected between reef treatment and survey date for epibenthos H' or S (Table 3).

Visual inspection of the epibenthos MDS plot indicates almost complete separation by reef treatment and survey date. These differences were statistically significant based on the PERMANOVA analyses of Bray–Curtis similarities for epibenthos areas by reef treatment (pseudo-$F_{1,51}$ = 31.1, P = 0.001), survey date (pseudo-$F_{2,51}$ = 46.0, P = 0.001), and reef treatment and survey date interaction (pseudo-$F_{2,51}$ = 26.2, P = 0.001; Figure 8).

Discussion

Epibenthos and Fish Assemblages

The function of artificial reefs has long been debated (Bohnsack 1989; Shipp and Bortone 2009; Gallaway et al. 2009). It is obvious that artificial reefs attract many fish species, but whether or not any new production can be attributed to artificial reefs is the center of controversy. If production (biomass pro-

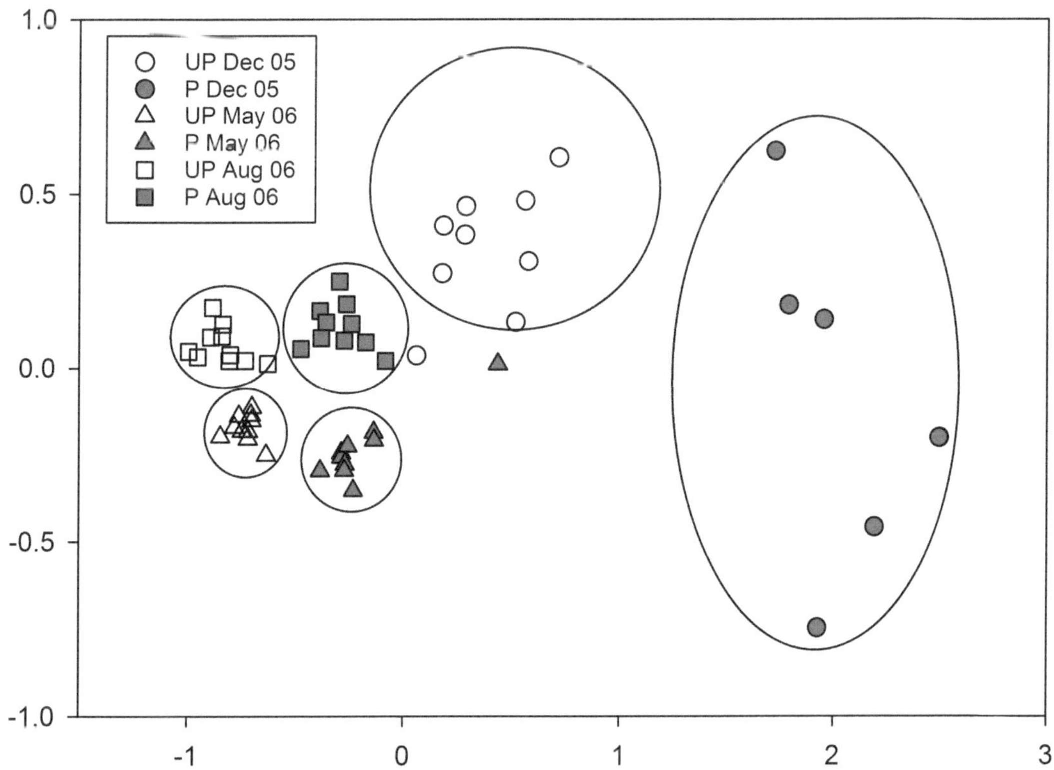

Figure 8. Nonmetric multidimensional scaling plots of epibenthos areas based on Bray–Curtis similarities by reef treatment and survey date. Unpainted reefs = UP open symbols and painted reefs = P closed symbols.

duced per unit time) is occurring, it must result from two features: increased shelter from predation (decrease in natural mortality) or increased food resources from the presence of benthic organisms associated with the addition of new hard structure that would not exist in shifting sand and mud substrates. The connection between protection from predation and artificial reefs has been reported (Topolski and Szedlmayer 2004; Mudrak and Szedlmayer 2012), but demonstration of increased fish abundance due to increased prey base has been more difficult to document (Hueckel and Buckley 1987; Redman and Szedlmayer 2009). The present study attempted to evaluate the food resource aspect of artificial reefs by limiting the growth of epibenthos on artificial reefs. This attempt was successful as the unpainted reefs clearly had higher epibenthos areas, biomass, diversity, and species compared to copper-treated or painted artificial reefs. Epibenthos areas and biomass increased with survey date for both reef treatments, but there were persistent substantial differences between these treatments over the 10-month survey period. These results were consistent with two previous studies that reported lower epibenthos coverage on copper-painted surfaces (Lee and Trott 1973; Redman and Szedlmayer 2009).

In the present study, clear differences in fish assemblages were detected between unpainted and painted reefs. The densities of all fish and of Red Snapper, wrasses, Bank Sea Bass, and Atlantic Spadefish were higher on unpainted reefs. Possible criticisms of the present results are that some other factor (or factors) could have caused the observed differences in fish assemblages between treatments. Possible factors may be differences in temperature, salinity, dissolved oxygen, substrate, or proximity to other reefs. However, we can discount such possibilities due to the study design, as reefs were alternated between unpainted and painted reefs over relatively short distances (30 m); thus, any changes in such variables would be equally distributed among both reef treatments.

An important consideration is the connection between the increased amounts of epibenthos to the diets of fishes. Prey for Red Snapper and Gray Triggerfish were part of the epibenthos in the present study and were consistent with the present study findings. These prey items included Ascidiacea and Polychaeta for Red Snapper (Ouzts and Szedlmayer 2003; Szedlmayer and Lee 2004) and barnacles for Gray Triggerfish (Vose 1990; Vose and Nelson 1994; Blitch 2000). However, many smaller epibenthos that can serve as potential prey items may have been missed in the present study, particularly the more motile Decapoda that would tend to escape as the removable bricks were retrieved from the reefs. For example, conspicuously absent from epibenthos samples were caprellid amphipods (Caine 1991; Woods 2009). These amphipods typically attach to scuba divers by the hundreds and remain attached after divers surface yet were absent in the breakaway samples (S. T. Szedlmayer, personal observation). In future epibenthos studies on artificial reefs, it is suggested that some type of net capture method should be employed to assure a more complete assessment.

Other studies have also reported connections between the epibenthos and fish assemblages on artificial reefs. Diet analyses indicate that increases in Copper Rockfish *Sebastes caurinus* and Quillback Rockfish *S. maliger* were correlated to successional epibenthos development on artificial reefs (Buckley and Hueckel 1985). Further, many fish species became more abundant on artificial reefs that have a more developed epibenthos

(Hueckel and Buckley 1987). Perhaps the most appropriate previous study for comparison to the present study was a survey of fish assemblages also on unpainted versus copper-painted concrete block reefs in the same study area of the northern Gulf of Mexico (Redman and Szedlmayer 2009). In this previous study, similar relations were reported for Red Snapper and Gray Triggerfish and their association with epibenthos. An important difference was that the Redman and Szedlmayer (2009) study did not collect epibenthos or attempt to identify the epibenthic taxa, but simply relied on underwater photographs to document the epibenthos on the concrete block surfaces. Thus, an important aspect in the present study was the identification and quantification of epibenthos on artificial reefs.

Increased shelter provided by the reef can also be linked to increases in the epibenthos. For example, the density of four species of Blenniidae on artificial structures in the north-central Gulf of Mexico were positively related to densities of barnacles as these fishes used empty barnacle shells as shelter (Topolski and Szedlmayer 2004). In the present study, blenniid densities may also have been related to barnacle areas on artificial reefs. Although blenniids did not differ significantly between reef types, their densities increased with the increase in barnacle area over survey date.

Copper Paint Treatment

A potential issue in the present study was that copper toxicity may inhibit fish recruitment. However, following submersion painted surfaces soon become covered with a thin biofilm that slows the release of copper to low levels (8–22 $\mu g \cdot cm^{-2} \cdot d^{-1}$) while still preventing epibenthos development (Dempsey 1981; Valkirs et al. 2003). Also, the constant water flow in the open Gulf of Mexico prevents accumulation of copper in the water immediately surrounding the painted reef surfaces. Water samples collected from the seawater surrounding the reefs indicated that copper was present at low concentrations (<1 parts per million) near the painted reefs 1 week after deployment, but copper was not detected on any reefs on subsequent surveys. Also, fish in the laboratory experiment showed no avoidance of copper-painted blocks. Therefore, any differences in fish density between reef treatments were not related to the presence of copper paint on the reef. More likely, the differences in epibenthos assemblages were responsible for the differences in associated fish assemblage parameters.

Conclusions

Copper-paint treatment of artificial reefs was associated with greater epibenthos areas, biomass, species diversity, and species richness on unpainted versus painted reefs. The increased epibenthos likely provided an increased forage base for fish that resulted in increased total fish density and increased densities of four fish species on the unpainted reefs compared to painted reefs. In addition, increased epibenthos may have increased the shelter aspect of artificial reefs, particularly for small cryptic species such as Blenniidae, thus leading to lower natural mortality. To further confirm such patterns, that an increase in the artificial reef associated fish assemblage was the result of an increase in the epibenthic assemblage, future studies should attempt to quantify fish diets simultaneously with unpainted and painted reef treatments. In addition, obtaining fish biomass estimates for both painted and unpainted reefs may provide greater insights on the value of artificial reefs.

Acknowledgments

We thank S. Beyer, C. Simmons, and D. Topping, for help with field and data collection. This project was funded by the Marine Resources Division, Alabama Department of Conservation and Natural Resources and the Sportfish Restoration Fund. This study is a contribution of the Alabama Agricultural Experiment Station and School of Fisheries, Aquaculture and Aquatic Sciences, Auburn University.

References

Alevizon, W. S., and J. C. Gorham. 1989. Effects of artificial reef deployment on nearby resident fishes. Bulletin of Marine Science 44:646–661.

Anderson, M. J., R. N. Gorley, and K. R. Clarke. 2008. PERMANOVA+ for PRIMER: guide to software and statistical methods. PRIMER-E, Plymouth, UK.

Ardizzone, G. D., M. F. Gravina, and A. Belluscio. 1989. Temporal development of epibenthic communities on artificial reefs in the central Mediterranean Sea. Bulletin of Marine Science 44:592–608.

Blitch, K. A. 2000. The feeding habits of Gray Triggerfish, *Balistes capriscus* (Gemlin), from the northeast Gulf of Mexico. Master's thesis. Auburn University. Auburn, Alabama.

Boaventura, D., A. Moura, F. Leitão, S. Carvalho, J. Cúrdia, P. Pereira, L. Cancela da Fonseca, M. Neves dos Santos, and C. Costa Monteiro. 2006. Macrobenthic colonisation of artificial reefs on the southern coast of Portugal (Ancão, Algarve). Hydrobiologia 555:335–343.

Bohnsack, J. A. 1989. Are high densities of fishes at artificial reefs the result of habitat limitation or behavioral preference? Bulletin of Marine Science 44:631–645.

Bohnsack, J. A., D. E. Harper, D. B. McClellan, and M. Hulsbeck. 1994. Effects of reef size on colonization and assemblage structure of fishes at artificial reefs off southeastern Florida, USA. Bulletin of Marine Science 55:796–823.

Bohnsack, J. A., and D. L. Sutherland. 1985. Artificial reef research: a review with recommendations for future priorities. Bulletin of Marine Science 37:11–39.

Bolker, B. M., M. E. Brooks, C. J. Clark, S. W. Geange, J. R. Poulsen, M. H. H. Stevens, and J. S. S. White. 2009. Generalized linear mixed models: a practical guide for ecology and evolution. Trends in Ecology and Evolution 24:127–135.

Bosman, A. L., and P. A. R. Hockey. 1988. Life-history patterns of populations of the limpet *Patella granularis*: the dominant roles of food supply and mortality rate. Oecologia 75:412–419.

Buckley, R. M., and G. L. Hueckel. 1985. Biological processes and ecological development on an artificial reef in Puget Sound, Washington. Bulletin of Marine Science 37:50–69.

Caley, M. J., and J. St John. 1996. Refuge availability structures assemblages of tropical reef fishes. The Journal of Animal Ecology 65:414–428.

Carr, M. H., and M. A. Hixon. 1997. Artificial reefs: the importance of comparisons with natural reefs. Fisheries 22:28–33.

DeMartini, E. E., A. M. Barnett, T. D. Johnson, and R. F. Ambrose. 1994. Growth and production estimates for biomass-dominant fishes on a Southern California artificial reef. Bulletin of Marine Science 55:484–500.

Dempsey, M. J. 1981. Colonisation of antifouling paints by marine bacteria. Botanica Marina 24:185–191.

Caine, E. A. 1991. Caprellid amphipods: fast food for the reproductively active. Journal of Experimental Marine Biology and Ecology 148:27–33.

Fager, E. W. 1971. Pattern in the development

of a marine community. Limnology and Oceanography 16:241–253.

Farrell, T. M. 1988. Community stability: effects of limpet removal and reintroduction in a rocky intertidal community. Oecologia 75:190–197.

Field, J. G., K. R. Clarke, and R. M. Warwick. 1982. A practical strategy for analysing multispecies distribution patterns. Marine Ecology Progress Series 8:37–52.

Gallaway, B. J., S. T. Szedlmayer, and W. J. Gazey. 2009. A life history review for Red Snapper in the Gulf of Mexico with an evaluation of the importance of offshore petroleum platforms and other artificial reefs. Reviews in Fisheries Science 17:48–67.

Greene, L. E., and W. S. Alevizon. 1989. Comparative accuracies of visual assessment methods for coral reef fishes. Bulletin of Marine Science 44:899–912.

Hixon, M. A., and J. P. Beets. 1989. Shelter characteristics and Carribean fish assemblages: experiments with artificial reefs. Bulletin of Marine Science 44:666–680.

Hixon, M. A., and J. P. Beets. 1993. Predation, prey refuges, and the structure of coral-reef fish assemblages. Ecological Monographs 63:77–101.

Hixon, M. A., and W. N. Brostoff. 1985. Substrate characteristics, fish grazing, and epibenthic reef assemblages off Hawaii. Bulletin of Marine Science 37:200–213.

Hueckel, G. J., and R. M. Buckley. 1987. The influence of prey communities on fish species assemblages on artificial reefs in Puget Sound, Washington, USA. Environmental Biology of Fishes 19:195–214.

Hueckel, G. J., and R. L. Stayton. 1982. Fish foraging on an artificial reef in Puget Sound, Washington Marine Fisheries Review 44:38–44.

Ibrahim, S., M. A. Ambak, L. Shamsudin, and M. Z. Samsudin. 1996. Importance of fish aggregating devices (FADs) as substrates for food organisms of fish. Fisheries Research 27:265–273.

Lee, S. W., and L. B. Trott. 1973. Marine succession of fouling organisms in Hong Kong, with a comparison of woody substrates and common, locally-available, antifouling paints. Marine Biology 20:101–108.

Lindquist, D. G., L. B. Cahoon, I. E. Clavijo, M. H. Posey, S. K. Bolden, L. A. Pike, S. W. Burk, and P. A. Cardullo. 1994. Reef fish stomach contents and prey abundance on reef and sand substrata associated with adjacent artificial and natural reefs in Onslow Bay, North Carolina. Bulletin of Marine Science 55:308–318.

Lingo, M. E., and S. T. Szedlmayer. 2006. The influence of habitat complexity on reef fish communities in the northeastern Gulf of Mexico. Environmental Biology of Fishes 76:71–80.

Magurran, A. E. 1988. Ecological diversity and its measurement. Princeton University Press, Princeton, New Jersey.

Mudrak, P. A., and S. T. Szedlmayer. 2012. Proximity effects of larger resident fishes on recruitment of age-0 Red Snapper in the northern Gulf of Mexico. Transactions of the American Fisheries Society 141:487–494.

Nelson, B. D., and S. A. Bortone. 1996. Feeding guilds among artificial-reef fishes in the northern Gulf of Mexico. Gulf of Mexico Science 2:66–80.

Osman, R. W. 1977. The establishment and development of a marine epifaunal community. Ecological Monographs 47:37–63.

Ouzts, A. C., and S. T. Szedlmayer. 2003. Diel feeding patterns of Red Snapper on artificial reefs in the north-central Gulf of Mexico. Transactions of the American Fisheries Society 132:1186–1193.

Perkol-Finkel, S., and Y. Benayahu. 2007. Differential recruitment of benthic communities on neighboring artificial and natural reefs. Journal of Experimental Marine Biology and Ecology 340:25–39.

Piko, A. A., and S. T. Szedlmayer. 2007. Effects of habitat complexity and predator exclusion on the abundance of juvenile Red Snapper. Journal of Fish Biology 70:758–769.

Polovina, J. J. 1989. Artificial reefs: nothing more than benthic fish aggregators. Reports of California Cooperative Oceanic Fisheries Investigations 30:37–39.

Polovina, J. J., and I. Sakai. 1989. Impacts of artificial reefs on fishery production in Shimamaki, Japan. Bulletin of Marine Science 44:997–1003.

Randall, J. E. 1963. An analysis of the fish populations of artificial and natural reefs in the Virgin Islands. Caribbean Journal of Science 3:31–47.

Redman, R. A., and S. T. Szedlmayer. 2009. The effects of epibenthic communities on reef fishes in the northern Gulf of Mexico. Fisheries Management and Ecology 16:360–367.

Relini, G., M. Relini, G. Torchia, and G. De Angelis. 2002. Trophic relationships between fishes and an artificial reef. ICES Journal of Marine Science 59:S36–S42.

Relini, G., N. Zamboni, F. Tixi, and G. Torchia. 1994. Patterns of sessile macrobenthos community development on an artificial reef in the Gulf of Genoa (northwestern Mediterranean). Bulletin of Marine Science 55:745–771.

Seavy, N. E., S. Quader, J. D. Alexander, and C. J. Ralph. 2005. Generalized linear models and point count data: statistical considerations for the design and analysis of monitoring studies. Pages 744–753 *in* C. J. Ralph and T. D. Rich, editors. Bird conservation implementation and integration in the Americas: proceedings of the third international partners in flight conference, 2nd volume. U.S. Forest Service, Pacific Southwest Research Station, General Technical Report PSW-GTR-191, Albany, California.

Shipp, R. L., and S. A. Bortone. 2009. A prospective of the importance of artificial habitat on the management of Red Snapper in the Gulf of Mexico. Reviews in Fisheries Science 17:41–47.

Shulman, M. J. 1984. Resource limitation and recruitment patterns in a coral reef fish assemblage. Journal of Experimental Marine Biology and Ecology 74:85–109.

Shulman, M. J. 1985. Recruitment of coral reef fishes: effects of distribution of predators and shelter. Ecology 66:1056–1066.

Steele, M. A. 1999. Effects of shelter and predators on reef fishes. Journal of Experimental Biology and Ecology 233:65–79.

Svane, I., and J. K. Petersen. 2001. On the problems of epibioses, fouling and artificial reefs, a review. Marine Ecology 22:169–188.

Szedlmayer, S. T., and K. W. Able. 1996. Patterns of seasonal availability and habitat use by fishes and decapod crustaceans in a southern New Jersey estuary. Estuaries 19:697–709.

Szedlmayer, S. T., and J. D. Lee. 2004. Diet shifts of juvenile Red Snapper (Lutjanus campechanus) with changes in habitat and fish size. U.S. National Marine Fisheries Service Fishery Bulletin 102:366–375.

Topolski, M. F., and S. T. Szedlmayer. 2004. Vertical distribution, size structure, and habitat associations of four Blenniidae species on gas platforms in the northcentral Gulf of Mexico. Environmental Biology of Fishes 70:193–201.

Valkirs, A. O., P. F. Seligman, E. Haslbeck, and J. S. Caso. 2003. Measurement of copper release rates from antifouling paint under laboratory and in situ conditions: implications for loading estimation to marine water bodies. Marine Pollution Bulletin 46:763–779.

Vose, F. E. 1990. Ecology of fishes on artificial and rock outcrop reefs off the central east coast of Florida. Doctoral dissertation. Florida Institute of Technology, Melbourne.

Vose, F. E., and W. G. Nelson. 1994. Gray Triggerfish (*Balistes capriscus* Gmelin) feeding from artificial and natural substrate in shallow Atlantic waters of Florida. Bulletin of Marine Science 55:1316–1323.

Wendt, P. H., D. M. Knott, and R. F. Van Dolah. 1989. Community structure of the sessile biota on five artificial reefs of different ages. Bulletin of Marine Science 44:1106–1122.

Woods, C. M. C. 2009. Caprellid amphipods: an overlooked marine finfish aquaculture resource? Aquaculture 289:199–211.

Zar, J. H. 2010. Biostatistical analysis, 5th edition. Prentice-Hall, Englewood Cliffs, New Jersey.

American Fisheries Society Symposium 86:69–80, 2018

Environmental Factor Analysis of Snow Crab Associated with Artificial Reefs

Nobuhisa Kohno
Fukui Prefectural Office, Fisheries Division
3-17-1 Ote, Fukui, Fukui 910-8580, Japan

Sohei Okubo and Shinya Otake*
Fukui Prefectural University, Faculty of Marine Bioscience
1-1 Gakuen, Obama, Fukui 917-0003, Japan

Wataru Fujiie
Kokusai Kogyo Co., Ltd.
3-6-3 Hakataekihigashi, Hakata-ku, Fukuoka 812-0013, Japan

Toshihiko Kaneko
Kokusai Kogyo Co., Ltd.
1-1-15 Nishinagasu, Amagasaki, Hyogo 660-0805, Japan

Tomoki Miyamukai
Marine Biological Research Institute of Japan Co., Ltd.
3-9-2 Koubai, Syowa-hu, Nagoya, Aichi 466-0031, Japan

Kazumasa Imao
Marine Biological Research Institute of Japan Co., Ltd.
18-28 Toyotsu, Suita, Osaka 564-0051, Japan

Abstract.—Enhancing the environment through deployment of artificial reefs begins with determining those environmental factors that are limiting to target organisms. Properly designed artificial reefs can increase, enhance, improve, or, in some instances, interfere with those environmental factors that benefit some fishery-associated species. Consequently, it is essential that fishery managers identify the species' specific limiting factors that are influenced by an artificial reef deployment. Subsequently, a cost–benefit analysis can be conducted to determine the feasibility of designing, constructing, and deploying artificial reefs to benefit a fishery. Here, we examine the effect of these artificial reefs deployments on the snow crab *Chionoecetes opilio* off Japan and the environmental factors with which it associates. This study was conducted to allow an analysis of the environmental response to the deployment of artificial reefs through an examination of the life history and population parameters of the snow crab.

* Corresponding author: otakes@fpu.ac.jp

Introduction

The process for conducting projects to establish fishing grounds in Japan follows a four-step protocol: (1) selection of target fish species, (2) investigation of the environment, (3) environmental deployment, and (4) assessment of the project. Fisheries engineering is conducted at each level to provide environmental protection measures. However, the fundamental objective when creating fishing grounds is to maintain a sound ecosystem for fisheries resources and to stabilize the management and status of the fishery. In order to protect marine resources and ensure stable fishery management relative to artificial reef deployments, the above protocol is insufficient and it necessary to add an additional step to the process. In light of the economic perspective regarding past fishing ground construction, the economic value of the project becomes difficult to evaluate. For example, when the maintenance cost of the facility is not included in the project costs and only the production value of the resource increase is included, the calculated true cost of the project is inaccurate. Also, despite the fact that there are often many field surveys conducted in a region, because they are not specific to the organized project, additional surveys are often required that are similar to ongoing and previous surveys. Thus, there is considerable waste in the project cost. We decided to alter the protocol and add additional steps to form what is now termed "the fisheries resources management system." Figure 1 shows the 10 steps in the modified protocol to establish fishing grounds in Japan. The basic outline of this new protocol is described below.

1 Selection of site
2 Selection of target organisms
3 Classification of environmental factors
4 Removal/extinction of limiting factors or creation of optimal environment
5 Development/choice of technical method
6 Prediction of fish catch
7 LCC calculation of technical method
8 Development of business model
9 Implementation of the project
10 Handling of environmental changes or building of adaptive management

Figure 1. Flow of the fisheries resources management system. LLC = life cycle cost.

Selection of site (i.e., Step 1) and target organisms (i.e., Step 2) is greatly affected by the economies and/or demands of fishermen. Thus, they are each determined politically or administratively. The potential utilization of a site can be broad. Specifically, nearshore sites in urban areas are often affected by economic activities. This broad diversification of uses increases with the complexity of the problems associated with the fishery. Consequently, the environmental concepts associated with fisheries engineering become very important.

Once the target organisms are determined, we assess the environment of the site in relation to the ecological behavior of the organisms (i.e., Step 3: classification of environmental factors). Here, we need biologists as well as ecologists to work interactively to classify the environmental factors. Furthermore, we conduct data collection or field observations based on previous knowledge. After the assessment of current and past data, we determine the optimal environmental factors and conditions that are favorable for the target organisms. Nakamura (1980) postulated three patterns of the relationship between environmental factors (abiotic) and biomass (biotic): (1) a favorable factor, (2) a factor with no influence, and (3) a factor that limits or constrains biomass. Here, a limiting factor implies an environmental factor that reduces biomass when the factor is suboptimal, even if the overall environment is favorable to the fishery. Environmental assessment projects are conducted with the objective of removing or reducing those limiting factors. Of course, environmental stabilization or creation of a favorable environment is key to environment assessment projects. It is necessary, however, to establish an analytical method that can determine what constitutes a favorable environment or limiting factors. As mentioned previously, Nakamura (1980) proposed using environmental factor analysis so that a limiting factor can be determined.

To conduct the environmental analysis, it is necessary to collect data on environmental factors, including already existing data as well as numerically computational (i.e., derived) study results. This collection of data is normally conducted for each project. This is a labor-intensive effort that must be conducted for each situation. However, even for studies conducted in Tokyo Bay or the Ariake Sea, it was reported that the environmental data considered had been updated using concurrent data collection or observational buoy systems. Field studies on the associated fisheries would also be required in the near future. These intensive data collections are necessary to help establish an environmental platform or baseline. Here, platform refers to the concept of comprehensively evaluating targeted sea areas by including experts and users such as investigators, business personnel, fishermen, and so forth, in addition to past environmental databases. In Japan, the Ministry of Land, Infrastructure and Transport focuses on the marine environment. In this research, the objective is to create a database for constructing this baseline.

Through the previously mentioned processes to determine environmental factors that are favorable or limiting, we propose a technical method to create/preserve or remove/eliminate these factors. For instance, the creation of artificial reefs, seawater exchange facilities, artificial upwelling structures, additional sand coverage, waterway dredging, and artificial tidal flats are some of the actions that could be taken to reduce, alter, or enhance specific environmental factors. Kohno et al. (2016) investigated the distribution range of the snow crab *Chionoecetes opilio* by analyzing the operation records of a fishing boat. It was found that snow crab is distributed around artificial reefs, but a lot of distribution was found in places where it seems that there is nothing topographically. As it turned out, there was an old sunken ship there. This suggests that topographical changes like artificial reef affect the distribution of crabs. There are other structures to be built or projects that could be undertaken that may prove promising in the future as well. Regardless, any outcome of action taken to increase fisheries production in a specific fishery should be predictable. For instance, quantification of production using an ecological productivity model or a stock biomass estimation model could be used to arrive at the predictive response of the fishery after implementation of any structural modification to the environment. At present, the concept of stock management is used to direct the efforts of various projects toward the maintenance and management efforts regarding constructed structures. Ideas concerning stock management have been proposed at previous Conferences on Artificial Reefs and Related Artificial Habitats (CARAHs). However, thus far, papers presented at the CARAHs have not dealt with specific technologies of fish stock maintenance and management. In the future, we expect that a determination of life cycle cost (www.businessdictionary.com/

definition/life-cycle-cost.html) would be required that incorporates issues concerning technologies related to stock maintenance and management. It is suspected that the life cycle cost will preempt the business cost approach in the future. These concepts have not been considered for most artificial reef projects until recently. We conducted an environmental assessment project that integrates the benefits that are important to environmental planning, including the increases in stock biomass or economic value. However, it is difficult to incorporate the depreciation of some structures or the degradation of functions as the structures deteriorate, even if life cycle cost is included as a fundamental concept. In addition, we were unable to consider business costs to fishermen who directly earn their benefits from the fishery. Therefore, we have only evaluated the investment effect of public works when the benefits from the deployed structures relative to economic costs are clear. That is, an economic test calculation should calculate cost–benefit by adding the investment cost of the artificial reefs as well. This type of analysis will help determine if it is reasonable to conduct the project or not.

Nevertheless, we included this basic concept into previous construction projects on fishing grounds. As a result, we are now required to use an adaptive management approach to design a deployment project that includes a consideration of environmental changes on a global and local scale. Thus, when requesting approval, the overall environmental organization of the project is taken into consideration by the Fisheries Agency of Japan. In Japan, coastal fisheries ground enhancement and development projects based on the Coastal Fisheries Grounds Enhancement and Development Program Act have been implemented for 40 years. This is sufficient time to evaluate the overall effectiveness of the projects by focusing on the progress of the projects and their management based on the few projects that have been assessed to date. With the improvement of deployment technology, it is also important to consider installation of the structures in the management assessment.

The above background information is important as we attempt to assess the effect of protection artificial reefs for the capture of snow crab in the Fukui Prefecture. This background information also allowed us to establish a fisheries resources management system directed toward clarifying how these structures affected the fisheries management of the snow crab. Here, we introduce the establishment of environmental platform concept and also report observational results about the effect of utilizing protection artificial reefs.

Methods

Building an Environment Assessment Protocol

Target area.—The environmental assessment protocol begins with a collection of existing data to establish a database. Figure 2 shows a study area for environmental assessment. The area includes 35.4501° N–36.86067° N, 135.1116° E–136.1139° E (Figure 2). Each grid for data input of water depth is a rectangular shape (grid size 500 × 500 m, simulation model named Japan Sea Data Assimilation Experiment (JADE2: http://jade2.dc.affrc.go.jp/jade2/). Data were obtained from the Fukui Prefectural Fisheries Experimental Station. Previously, we used a grid size of about 10,000 m; however, using a smaller grid size allowed us to more accurately determine the seabed gradient and distance from the artificial reefs.

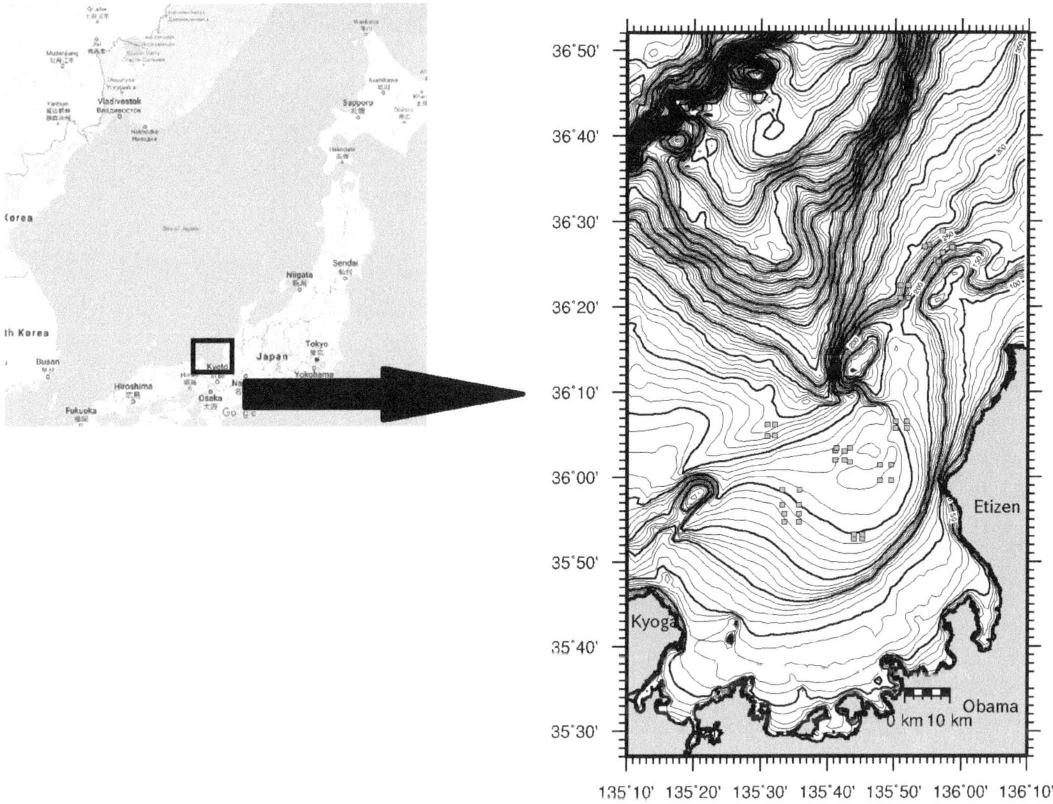

Figure 2. Topography of Wakasa Bay and off Wasaka Bay. The gray cubic symbols show the position of protection reefs.

Target organism and protection reef.—The target organism in this study is the snow crab as snow crab is an important organism for Fukui Prefecture fisheries. Figure 3 indicates that snow crab production relative to the total catch in Fukui Prefecture. However, in recent years, the catch has greatly declined. Consequently, Fukui Prefecture has implemented catch regulations that include not taking small crabs and a strict observance of catch season. In addition, protection artificial reefs were installed to prevent fishing. Recently, the snow crab catch has gradually increased as a result of these catch regulations. Figure 4 indicates a protection artificial reef, 3.25 m^3, installed at 13 sites, as shown in Figure 2. The reef in the figure is similar to reefs deployed at all sites. The reefs were installed at a density of 55–320 reefs per site, an average of 185 reefs in a 9,200,000 m^2 area (average reef site dimension was 3,000 × 3,000 m).

Collection of fish catch data.—Data collected during this study were target fish catch, fish size, water depth, location of the protection reef, and type of reef. The contour of water depth is indicated in Figure 5. Much of the catch data were acquired from the Fukui Prefectural Fisheries Experimental Station and Fukui Prefecture during their investigations of snow crab stock biomass.

The information on the protection reefs deployed in Wakasa Bay (i.e., location, mass of artificial reefs, and type of artificial reef)

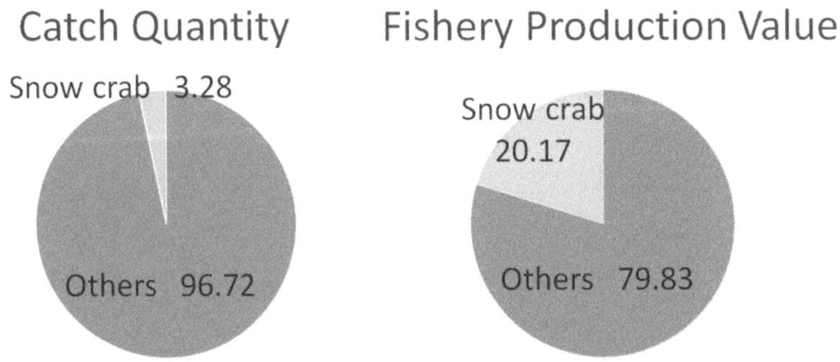

Figure 3. The ratio of snow crab to overall fishery production in Fukui Prefecture, Japan. The left figure shows the ratio of fishery production of snow crab and the right figure shows the fishery production value of snow crab in Fukui Prefecture, Japan. The ratio of fishery production value is very high, although the ratio of fishery production is low in Fukui Prefecture.

was obtained from the surveys Fukui Prefecture conducted from 1985 until 2013. The distance from the protection reefs to every grid was also determined.

Fish catch survey data from Fukui Prefecture surveys were used in association with the environmental data in the analysis. These data were obtained with the assistance of commercial fishing boats that included information on fishing locations, catch methods, fishing date, duration of fishing, and species. Because the snow crab and its various life stages are classified by local names, determining the catch amount was complicated. For example, the local name of snow crab at various life stages or forms is *akako*—female larva of snow crab colored red or orange, 1–8 months after spawning; *kuroko*—female larva of snow crab that has changed from red or orange to brown-black and will soon hatch; *zenmaru*—female juvenile with an ovary with eggs are premature; *seiko*—a female snow crab with eggs; *yakeseiko*—a brown-colored snow crab that has not molted but bears cumulated abrasions; *ogani, chugani*—an adult snow crab that, a year after a final molting, is divided from each other by each ship without an exact measurement; and *yamagani*—a male snow crab just prior to final molting. In addition, the unit of fish catch is expressed as follows: akako—1 box = 50 crabs; *zenmaru*—1 box = 70 crabs; *mizugani*—1 box = 9 cages, 1 cage = 18 ≈ 27crabs; seiko—1 box = 30 crabs; and *tigani*—1 box = undefined by number but includes juvenile snow crabs too small to be counted.

Figure 4. Artificial reefs used as protection reefs in Fukui Prefecture. The height of a reef is 3.25 m.

The detail of local name classification is very important because the price is greatly affected by maturity, size, and number of crabs molting. These data represent valuable records acquired in collaboration with local

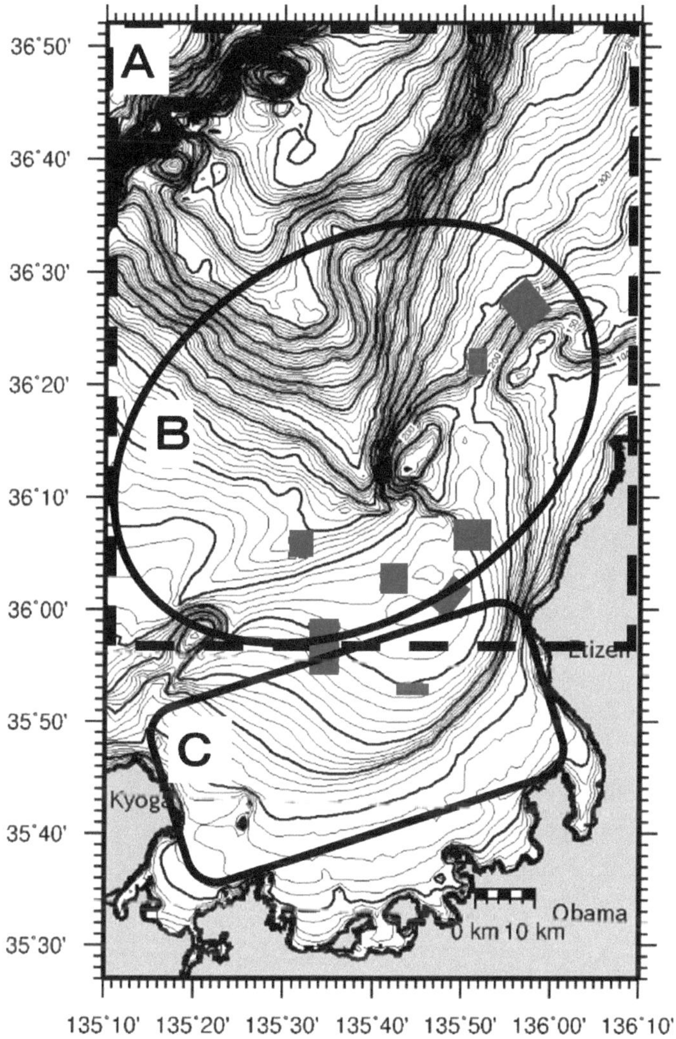

Figure 5. Environment platform in Wakasa Bay. **(A)** shows the environmental platform range. **(B)** shows the fishing ground of the snow crab. **(C)** shows the measurement range of temperature and salinity by depth investigated by Fuiui Prefectural Experiment Station.

fishermen. Also, this survey was continuous from 2002 to 2013. Other data gathered included information on address, name of ship, ID (identification) of ship, ID of fish catch, and ID of species so that we could recognize who used what fishing boat and what species they caught and how much. These obtained data allowed us to summarize the numerical value of the fishery data used in our study. Names of the species here are names used by local fishermen. The names chiefly refer to sizes, sex, and life stage of snow crab. Local names and their sizes are indicated below.

Subsequently, we entered these data into the database. Table 1 indicates the data included in our analysis. These fish catch data were converted to catch per unit effort (CPUE), and this value was calculated from the ratio of fish catch to the number of times the net was set.

Table 1. Data of fish catch input to environment platform.

1. Date of operation (trawl fishing)
2. Operation location (longitude and latitude)
3. Depth
4. Catch of snow crab (distinction by sex and local name), including immature snow crab, larval snow crab
5. Catch of discarded snow crab (amount of snow crabs not covered by catch)
6. Catch by fish species other than snow crab

Multiple Regression Analysis

Extracting the data from the environmental platform, we analyzed the data based on their life stage name after dividing a relationship between environmental data and fish catch into three patterns and organizing them in terms of the environmental factor assessment method. In this study, we also performed a multiple regression analysis using both the topographic (i.e., site) data and the CPUE data. The topographical data are the distance from protection reefs, seabed gradient, and water depth. The multiple regression analysis used the analytical algorithm in Microsoft Excel.

Results and Discussion

Building the Environmental Platform

The data used to construct the environmental platform were considered as data taken at the center of the computational grid. Thus, we judged that the fish catch by CPUE was conducted at the center of the grid. Figure 6 shows distributions of fish catch recorded using the data obtained from a platform. The data used here were for large male snow crab. As a result, the effect of protection reefs is promising as there is a fishing ground near reefs.

Analytical Results

We investigated a relationship between fish catch and each environmental factor Here, the environmental factors are water depth, sea bottom gradient, and distance from protection reefs. Also, the fish catch was converted to CPUE. The results are indicated in Figure 7. Here, the seabed gradient indicates the maximum depth difference of square grid against the length of the grid (500 m). The height of the protection reefs was not included in the analyses. In addition, the distance of protection reefs indicates the distance between the position of a protection reef and a fishing ground. As a result, we note that catch depth is concentrated between 200 and 300 m. We also note that there were three peaks in the relationship between protection reefs and the catch (CPUE). The first peak occurred a distance from the protection reef close to the reef. The occurrence of the second peak in abundance is difficult to explain topographically but is in a region with a steep seabed gradient. The third peak occurred further from the platform but was closer to other structures (i.e., artificial reefs, undiscovered sunken ships, etc.). The relationship with the seabed gradient shows that CPUE was higher where there was a greater seabed gradient.

For female snow crab, the relationships between CPUE and the environment were slightly different. Figure 8 shows the relationship between topography and catch for females. The relationship with catch depth and seabed gradient was almost the same as for males. However, only two peaks in

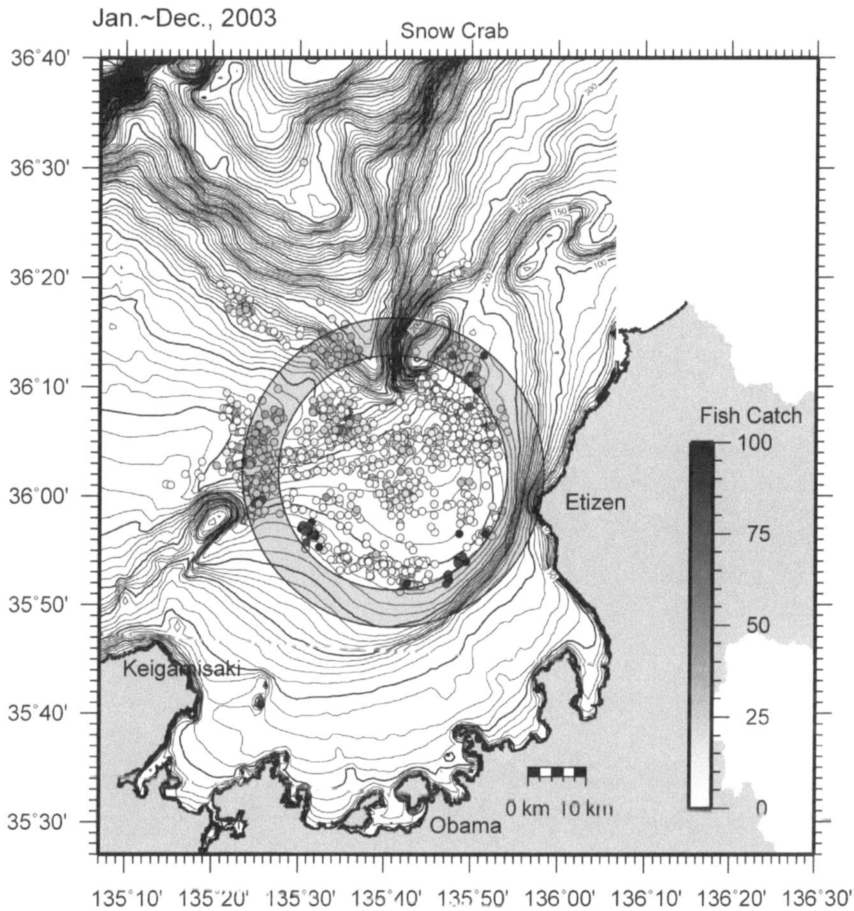

Figure 6. Fish catch distribution. of snow crab. The gray dots show fish catch and fish catch amount by investigation ship. The thick band shows a distance of 20,000 to 25,000 m from the central reef.

CPUE were noted for the relationship with the protection reef. The first peak was near the protection reef, and the second peak in CPUE occurs about 20 km away from the protection reefs. More-distant regions of catch were associated with steep seabed gradients, similar to the situation observed for males.

Consequent to the observations above, we decided to analyze males and females separately. In the case of males, we decided to examine the conditions when the distance from the reef is close (i.e., 0 to ~2,500 m) and where CPUE is 3,000 or more, to analyze for the relationship between the topographical conditions and CPUE. In the case of females, multiple regression analysis was conducted for catches made close to fishing reefs.

Table 2 shows results of multiple regression analysis for male snow crab using Microsoft Excel. For male snow crab, the CPUE was statistically different according to distance from protection reefs. When the distance is greater than 2,500 m, the CPUE has a negative correlation with the distance from the protection reefs and the CPUE is a positive correlation with the seabed gradient. When the distance is 10 to ~20 km, the

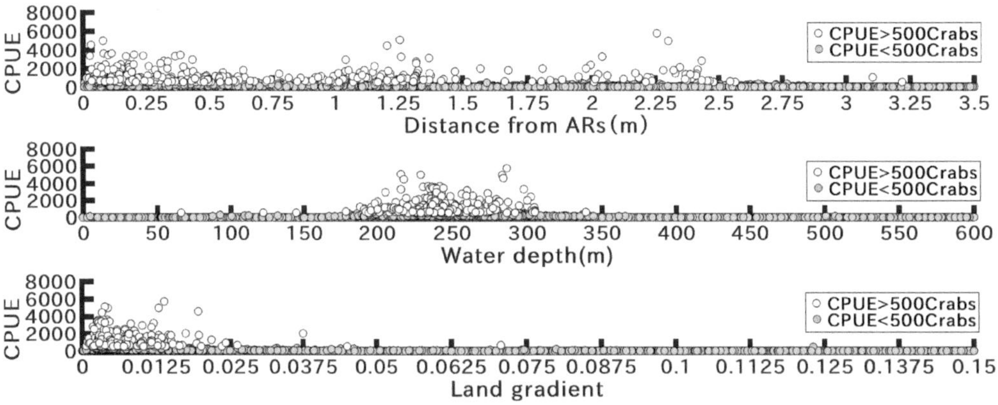

Figure 7. Relationships between topographical conditions and catch per unit effort (CPUE) of male snow crab. The *x*-axis of the upper figure shows the distance where crab were captured from the protection reef, the *x*-axis of the middle figure shows water depth where crab were caught by trawl gear, and the *x*-axis of the lower figure shows seabed gradient where snow crab were caught. All *y*-axes show CPUE of snow crab. As Figure 6 shows, the closer the distance is, the more effective it is; the CPUE is larger at a distance of 20,000–25,000 m from protection reefs.

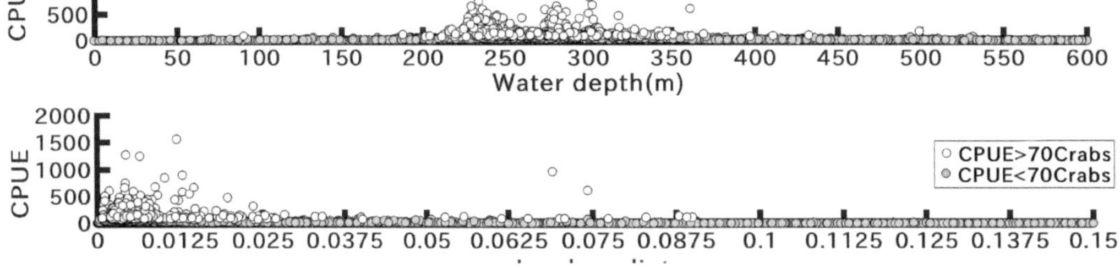

Figure 8. Relationships between topographical conditions and catch per unit effort (CPUE) of female snow crabs. The *x*-axis of the upper figure shows distance from protection reef, the *x*-axis of the middle figure shows water depth where females were caught by trawl gear, and the *x*-axis of the lower figure indicates the seabed gradient where females were caught. All *y*-axes show CPUE of snow crab. As Figure 6 shows, the closer the distance is, the more effective it is; the CPUE is larger at a distance of 20,000–25,000 m from protection reefs.

Table 2. Results of multiple regression analysis for male snow crab.

Distance from artificial reefs	0~2,500 m	10,000~20,000 m
Coefficient of determination	45%	63%
Analysis of variance	$p < 0.05$	$p < 0.05$
Environmental factor		
Distance from artificial reefs	–0.2235	0.541
Land gradient	0.1287	–0.3307

correlation with the distance from the protection reefs was positive and was negatively correlated with the seabed gradient. When the distance to the protection reef is close, our analysis indicates that there are more snow crabs closer to the protection reef. If the distance from the protection reef is far, our analysis indicates that CPUE for males is highly associated with the seabed gradient. These results help explain the distribution (Figure 6).

Table 3 shows the results of multiple regression analysis for female snow crabs using Microsoft Excel. In the case of female snow crabs, the relation to the distance from the protection reefs and the relationship with the seabed gradient were not clear.

Conclusions

With the availability of modern technology and analytical methods, a fishery resource management system is required to effectively manage a fishery. The reason is that it is necessary to consider the economic value contributed by the artificially created ecosystem. Artificially created ecosystems respond variably but naturally in response to changes of global environment. Consequently, it is necessary for fishery and environmental managers to gather large-scale and detailed ecosystem data like that indicated in the environmental platform discussed here.

The currently used environmental platform's database is inadequate; however, to improve its utility, we adapted or modified the environmental platform to accommodate the topographical conditions associated with the fishing grounds of snow crabs in Wakasa Bay. As a result of the multiple regression analysis, there is an indication that male snow crabs aggregate near the protection reefs and on seabeds with steep gradients. In the future, we will add other environmental factors, such as water temperature and salinity, to the database.

Male snow crabs may take 7 years to grow to maturity. If we are planning to increase this resource, it may be necessary to reduce the current consumption of eggs and floating larvae. Moreover, where the larva settle on the ocean floor and where the larvae are distributed have not been elucidated as yet, as there is great difficulty in capturing these life stages. Therefore, if we character-

Table 3. Results of multiple regression analysis for female snow crab.

Coefficient of determination	20%
Analysis of variance	$p < 0.01$
Coefficient of environmental factor	
Distance from protection reefs	–1.504
Seabed gradient	532,000

ized the environment 7 years ago, it is understood that the present environment may not be as favorable to snow crabs. So, future efforts will determine the environmental conditions that should allow the development of a ground in the future. This will include evaluating water mass behavior, identified by water temperature and salinity, in the past when the current year-class was developing.

Previous efforts to elucidate the relationships of the snow crab fishery have not been successful, largely because of an inadequate and incomplete database. If we are able to include additional detailed information in the environmental database, it is likely that we can conduct additional analyses in greater detail.

Acknowledgment

We thank Professor Moon-Ock Lee, Chonnam National University, Korea for his assistance in translating the content of this manuscript into English.

References

Kohno, N., Y. Kodaira, and S. Otake. 2016. GIS database making from operation record by small trawl gear fishing boat. Pages 107–108 *in* Proceedings of the annual meeting of the Japanese Society of Fisheries Engineering. The Japanese Society of Fisheries Engineering, Fukui.

Nakamura, M. 1980. Fisheries civil engineering. Kogyo-jiji-tsushin-sya Co., Ltd., Tokyo.

American Fisheries Society Symposium 86:81–95, 2018

A Comparison of Population Dynamics from Red Snapper Associated with Inshore and Offshore Artificial Reefs in the Northwestern Gulf of Mexico

C. Y. M. Froehlich, A. M. Lee, R. Oquita, and C. E. Cintra-Buenrostro*
School of Earth, Environmental and Marine Sciences, University of Texas Rio Grande Valley
One West University Boulevard, Brownsville, Texas 78520, USA

J. D. Shively and J. B. Shipley
Texas Parks and Wildlife Department, Artificial Reef Program
4200 Smith School Road, Austin, Texas 78744, USA

Abstract.—Identifying morphometrics, age composition, and growth of Red Snapper *Lutjanus campechanus* (Poey, 1860) in the Gulf of Mexico is crucial for management of this valuable species. Red Snapper are highly associated with artificial reefs situated off the coast of Texas. Artificial reefs, two in federal waters (offshore) and two in state waters (inshore), were sampled quarterly from December 2014 through December 2016. Characteristic differences in age and growth patterns at inshore and offshore sites and across sampled sites were compared. Mean ± standard deviation was calculated for total length (TL), weight, and biological age. Red Snapper collected averaged 464.5 ± 92.6 mm in TL, 1.62 ± 1.03 kg, and 3.8 ± 1.5 years. Significant differences in TL, weight, and age were observed across sites but not between inshore and offshore locations. In comparing 12 previous Red Snapper length-at-age (0–25 years) studies across the Gulf of Mexico, results were similar except for two studies significantly different from one another, but none with respect to our study. Our findings suggest that Red Snapper populations are young and restricted to one or two spawning years before capture on artificial reefs. Continued management of fisheries associated with these reefs are necessary to increase the stock and eventually lead to robust and sustainable fisheries.

Introduction

The northern Red Snapper *Lutjanus campechanus* (Poey, 1860) is a demersal predatory teleost fish found throughout the Gulf of Mexico, Caribbean Sea, and Atlantic coast from northern Ceara in Brazil to southern Massachusetts in the United States (Camber 1955). Commonly observed at 15–30 m, Red Snapper are associated with limestone outcroppings, natural embankments, live bottom, and artificial reef habitats (Gallaway et al. 2009; Streich et al. 2017). Considered

* Corresponding author: carlos.cintra@utrgv.edu

a moderate- to long-lived species, individuals occasionally surpass 50 years, but most (90%) individuals collected from fisheries range from 2 to 6 years of age (Render 1995; Wilson and Nieland 2001). Sexual maturity is attained for 50% of individuals at 4 years of age and 100% of individuals at 5–8 years (Schirripa and Legault 1999; White and Palmer 2004; Jackson et al. 2007).

Throughout their life history, Red Snapper are observed at artificial reefs, from low-relief reefs during postlarval stages and subsequently to larger and deeper artificial reefs as they age (Gazey et al. 2008; Gallaway et al. 2009). The use of artificial reefs during all life stages suggests that these reefs serve as adequate habitat for Red Snapper (Gallaway et al. 2009). The northwestern Gulf of Mexico is dominated by soft sediments, and natural hard-bottom habitat is rare, making up less than 0.01% (Mueller 2012); thus, artificial reefs act as preferred Red Snapper habitat in a habitat-limited region. Artificial reefs are deemed a useful management tool to mitigate reduction in fish stock and add essential fishing habitat (Stephan et al. 1990).

Over the past decades, several age-and-growth studies of Red Snapper have been conducted across the Gulf of Mexico at natural and artificial reefs to understand the growth patterns of the species, as well as to determine the existing population structures (Futch and Bruger 1976; Szedlmayer and Shipp 1994; Patterson et al. 2001a; Wilson and Nieland 2001; Woods et al. 2003; Fischer et al. 2004; Gazey et al. 2008; Syc and Szedlmayer 2012; Saari et al. 2014; Streich et al. 2017). A critically understudied area (south Texas) has been investigated for Red Snapper movement, size, and recruitment (Garcia 2013; Froehlich and Kline 2015; Arney et al. 2017; Streich et al. 2017), but only Alexander (2015) has provided morphometric comparisons among four artificial reefs in the area, which encompasses inshore and offshore waters. Other studies have examined morphometric data of Red Snapper in the northern Gulf of Mexico (Wilson and Nieland 2001; Fischer et al. 2004; Syc and Szedlmayer 2012), but there has been little focus on the differences in the morphometric parameters determined between inshore and offshore waters.

Thus, the present study was conducted quarterly to determine the age and growth of Red Snapper at inshore and offshore artificial reefs in the northwestern Gulf of Mexico from December 2014 through December 2016. The current study provided a detailed investigation of the Red Snapper populations on the border of Mexico, where Texas (inshore) water fisheries are open year-round, federal (offshore) fisheries are open for a few weeks a year, and local and Mexican fishermen illegally fish on federal and/or cross maritime national waters. We hypothesized that (1) larger and older fish would be found at offshore sites because a reduced fishing season may allow for more growth and because of previously reported ontogenetic shifts (Gazey et al. 2008; Gallaway et al. 2009), and (2) there would be no difference in age and size of individuals among sampling quarters, based upon findings from previous studies in the Gulf of Mexico (Gallaway et al. 2009; Syc and Szedlmayer 2012; Alexander 2015). This study was designed to contribute to a more complete understanding of Red Snapper artificial reef usage in the northwestern Gulf of Mexico, specifically southwest Texas. To achieve a holistic approach to species management, a comprehensive picture of the morphometric characteristics, age dynamics, and growth is needed as it incorporates differences in localized fishing efforts.

Methods

Red Snapper were collected at four artificial reefs sites along the northwestern Gulf of Mexico off the south Texas coast (Figure 1). Two reefs were located in Texas State (inshore) waters within 8–10 km from shore: PS-1047 South Padre Island Reef (inshore north [INNO]) and PS-1169 L Port Isabel Reef (inshore south [INSO]); and two were located in federal (offshore) waters 20–32 km from shore: PS-1122 Texas Clipper Artificial Reef (offshore south [OFSO]) and PS-1070 Port Mansfield Liberty Ships (offshore north [OFNO]). The INNO reef is at a depth of 19.7 m and is 10.37 km from the coast. The INNO reef has a 30-m-long tugboat deployed in 2009 and includes more than 4,800 concrete culverts (1 m diameter × 3 m length) deployed in 2011. In July 2011, the INSO reef was deployed it comprised 2 three-pile jackets, 53 reef balls, and 3 vessels ranging from 18 to 45 m in length. The INSO reef is 21.8 m deep and 7.96 km from shore. The OFSO reef is a retired maritime training vessel that measures ~145 m in length and was deployed in 2007. The OFSO ship is located at a depth of 40 m and is 31.70 km from the coast. The OFNO reef is a compilation of 3 World War II Liberty ships (~134 × 17 m) and 15 assorted four-pile jackets deployed at 30.9 m depth and is 20.19 km from shore. For detailed images and description of reef sites, see Bollinger and Kline (2017).

National Oceanic and Atmospheric Administration Southeast Monitoring and Assessment Program (NOAA SEAMAP) vertical longline protocols (SEAMAP 2012) were used to collect Red Snapper. Commercial bandit rigs were used to collect individuals on a quarterly basis from December 2014 to December 2016. Modifications to the protocol were implemented to accommodate the rigging on the research vessel. Gear deployment intervals remained unchanged; however, bandit rigs were deployed from either port or starboard one at a time to avoid tangling, with no gear deployments from the stern. Vertical longline rigs were employed midship starboard and portside interchangeably, with side designated by random number table generated in SPSS Statistics v.23 (IBM Corporation). Fish were collected using three Mustad 39960D circle hooks sizes (8/0, 11/0, and 15/0) on a 10-hook backbone attached to each bandit rig. A backbone is a single monofilament line with one terminus attached to the bandit rig and the other to a weight. On the backbone, there were 10 equally spaced (60 cm) crane swivels that provided the attachment point for 38-cm

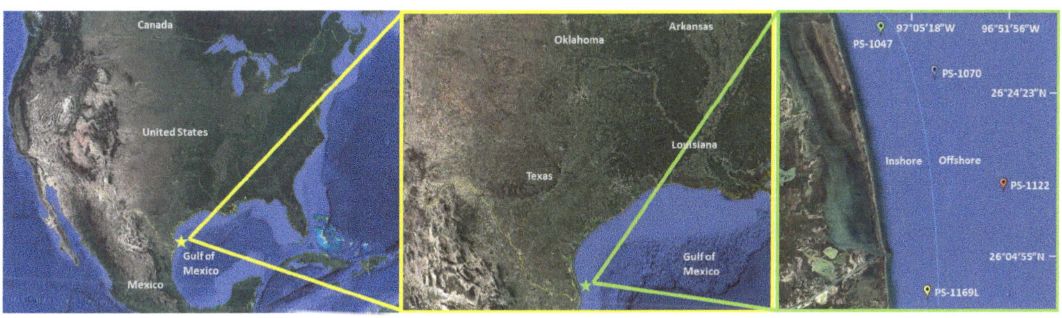

Figure 1. Geographic location of artificial reef sites for collection of Red Snapper. Modified from Google Earth (www.google.com/Earth).

leader line. Hooks were completely covered with Atlantic Mackerel *Scomber scombrus* (Linnaeus, 1758) used as bait. Backbones with the same hook size were immersed and set 1 m above the bottom for a soaking time of 5 min. All three hook sizes were randomly deployed based on a random number table generated from SPSS Statistics, with no repeated deployments. This sampling protocol was followed at three different areas within each site (INNO, INSO, OFNO, and OFSO) on the sampling day. Red Snappers were stored on ice and processed within 48 h or less of collection.

Individual morphometric measurements collected from dissected specimens comprised total length (±1.0 mm) and total weight (±0.0001 kg). Otoliths were removed, dried, bifurcated, and mounted onto a glass slide with clear, two-part epoxy for age characterization. The mounted otoliths were cross-sectioned (0.5 mm thickness) with an MTI Corp 150 low-speed diamond saw. Three readers used a multiphase microscope to independently identify the number of annuli and margin code for each otolith following established protocols (VanderKooy and Guindon-Tisdel 2003). There was more than a 99% agreement among readers on ring counts (average cumulative variance [ACV] = 0.49%) and margin code (ACV = 0.14%). Consequently, no otoliths were excluded from analyses. The number of annuli and the edge margin, month of capture, and a common shared date of birth for Gulf of Mexico Red Snapper were utilized to assign biological age (±0.1 year), following methods in VanderKooy and Guidon-Tisdel (2003).

Biological age and total length (TL in mm) of individuals were utilized to produce a three-parameter von Bertalanffy growth curve:

$$L_t = L_\infty 1 - e^{-k(t - t_0)},$$

where L_t = calculated TL at age, L_∞ = theoretical average maximum TL, k = growth coefficient, t = age (biological age), and t_0 = time at which mean length is 0.

Unless otherwise noted, all statistical analyses were performed in SPSS Statistics v.23. Means were presented with standard deviations (SD). Although no sexual dimorphism exists in Red Snapper (Goodyear 1995), a one-way analysis of variance (ANOVA) was performed to verify that morphometric data did not vary between sexes (one-way ANOVAs of sex versus TL, weight, and age were included). Another one-way ANOVA was performed to test for differences in TL among hook sizes. Location (inshore versus offshore) was used as the nesting factor and nested ANOVAs were performed on TL, weight, and age separately among sites and quarters. This approach was deemed necessary because Red Snapper were expected to be larger, heavier, and older at offshore locations (Gallaway et al. 2009; Syc and Szedlmayer 2012), and differences were anticipated among sites (i.e., variable vertical relief and materials). Normality was assessed with Q–Q plots and by Kolmogorov–Smirnov prior to running analyses on TL, weight, and biological age while homoscedasticity was verified with Levene's test (Sokal and Rohlf 2012). All data, except weights, were homoscedastic; thus, a $\log_{10} + 1$ transformation was applied to weight to achieve homoscedasticity. A Tukey honestly significant difference post-hoc test was performed manually because the SPSS program requires more than three variables to perform post-hoc tests. The test was used here to determine which factors were responsible for any significant differences.

Two independent general linear model regression analyses were performed on TL and transformed weight data to determine if they could be accurate predictors of biological age. Assumptions for the analyses (i.e., residual, independence, normality, and homoscedasticity, as well as linearity relationships) were previously assessed by normal probability and residual plots (Sokal and Rohlf 2012). Total lengths at age were calculated for ages 0–25 (years) across 12 length-at-age von Bertalanffy models from the Gulf of Mexico (Nelson and Manooch 1982; Szedlmayer and Shipp 1994; Patterson et al. 2001a; Wilson and Nieland 2001; Mueller 2012; Syc and Szedlmayer 2012; Saari et al. 2014). Calculated lengths at age were compared among the 12 models utilizing a one-way ANOVA.

Results

A total of 522 Red Snapper were collected from December 2014 to December 2016. Of those individuals, 503 (~96%) were successfully measured and analyzed for TL and total weight, and 420 (~81%) were retained and analyzed for age readings. No significant difference in TL ($F_{0.05,\ 1,419}$ = 0.854, P = 0.356), transformed weight ($F_{0.05,\ 1,419}$ = 0.576, P = 0.448), or biological age ($F_{0.05,\ 1,419}$ = 1.754, P = 0.186) was found between sexes in the sampled population; thus, males and females were combined for all analyses. From all samples (n = 503), Red Snapper, on average, were 464.5 ± 92.6 mm TL, 1.62 ± 1.03 kg, and 3.8 ± 1.5 years old.

Total lengths ranged from 266 to 879 mm; ~47% of the fish were caught with the medium hook size (11/0), followed by ~35% with the smallest (8/0) and the rest (~18%) with the largest hook (15/0) (Table 1). There was a significant difference in TL among hook sizes ($F_{0.05,\ 2,502}$ = 40.155, P < 0.0001). The nested ANOVA yielded no significant differences between offshore and inshore reefs (Figure 2A, $F_{0.05,\ 1,487}$ = 9.301, P = 0.088) or quarters sampled (Figure 2B, $F_{0.05,\ 12,487}$ = 1.096, P = 0.361). However, TLs significantly differed among sampling sites (Figure 2C, $F_{0.05,\ 2,487}$ = 6.429, P = 0.005). All sites had significantly different TL from one another (P < 0.001), except the OFNO and OFSO reefs (P = 0.073).

Total weights ranged from 0.31 to 7.80 kg. There was no significant difference in transformed total weights of individuals between inshore and offshore reefs (Figure 3A, nested ANOVA: $F_{0.05,\ 1,487}$ = 8.530, P = 0.095) or quarters sampled (Figure 3B, $F_{0.05,\ 12,487}$ = 1.692, P = 0.065). Sampling sites had significantly different transformed total weights of Red Snapper (Figure 3C, $F_{0.05,\ 2,487}$ = 4.707, P = 0.019). The offshore sites (OFNO and OFSO) were not significantly different (P = 0.498) while all other sites had significantly different transformed weights from one another. As expected—because of their interdependence—a positive relationship was

Table 1. Sample size (n, number of fish), mean, standard deviation (SD), standard error (SE), and minimum and maximum total lengths (mm) by hook size of Red Snapper captured at four artificial reefs off the coast of Brownsville, Texas.

Hook size	n	Mean	SD	SE	Minimum	Maximum
8/0	176	430.02	79.96	6.03	266.00	768.00
11/0	235	465.01	92.22	6.02	291.00	825.00
15/0	92	529.36	81.48	8.49	373.00	879.00

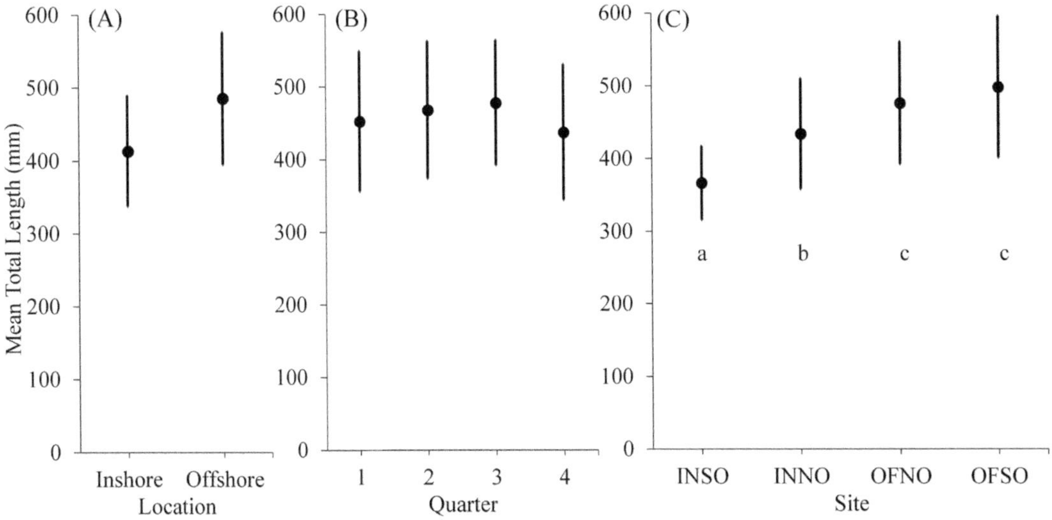

Figure 2. Mean total length ± SD (mm) of Red Snapper collected from December 2014–2016 at four artificial reefs off the coast of Brownsville, Texas comparing **(A)** inshore (within 15 km from Texas coast) and offshore (federal waters), with no significant difference observed; **(B)** quarters collected, with no significant difference observed; and (**C**) sampling sites, with nonmatching lowercase letters indicating sites that were significantly different from one another. Sample size (n) = 503, inshore n = 148, offshore n = 355. Quarter (Q) Q1 = September–November, Q2 = December–February, Q3 = March–May, Q4 = June–August; Q1 n = 71, Q2 n = 196, Q3 n = 160, Q4 n = 76; PS-1169 L (inshore south [INSO]) n = 45, PS-1070 (offshore north [OFNO]) n = 200, PS-1122 (offshore south [OFSO]) n = 155, PS-1047 (inshore north [INNO]) n = 103.

found between TL and transformed weight of all individuals measured (linear regression: $F_{0.05,\ 1,501} = 12013092.638$, $r^2 = 0.969$, $P < 0.0001$, $y = 357.05x + 60.311$).

Red Snapper ages ranged from 0 to 10.8 years and were not significantly different for fish between offshore and inshore reefs (Figure 4A, nested ANOVA: $F_{0.05,\ 1,408} = 3.835$, $P = 0.186$) or quarters sampled (Figure 4B, $F_{0.05,\ 8,408} = 1.817$, $P = 0.056$). However, age significantly differed among sampling sites (Figure 4C, $F_{0.05,\ 2,408} = 12.138$, $P = 0.001$). All sites differed significantly in age from one another ($P < 0.05$), with OFNO and INNO reefs being marginally significant ($P = 0.046$). As TLs increased, biological ages significantly increased (Figure 5A, linear regression: $F_{0.05,\ 1,418} = 357.796$, $r^2 = 0.460$, $P < 0.001$). Total transformed weight and biological ages also had a significantly positive relationship (Figure 5B, $F_{0.05,\ 1,418} = 349.069$, $r^2 = 0.454$, $P < 0.001$). The age and TL of individuals were utilized to generate a three-parameter von Bertalanffy length-at-age growth curve with the following equation:

$$L_t = 800 * 1 - e^{\left[-0.19 * (t - 1.00)\right]}.$$

The von Bertalanffy growth curve suggested that continued fish growth would be reduced after 10 years. The expected length at age from the growth curve was not significantly different from the observed length-at-age measures (t-test: $t_{0.05,\ 1,419} = 0.410$, $P = 0.344$), which suggests the calculated von Bertalanffy growth curve suited the data col-

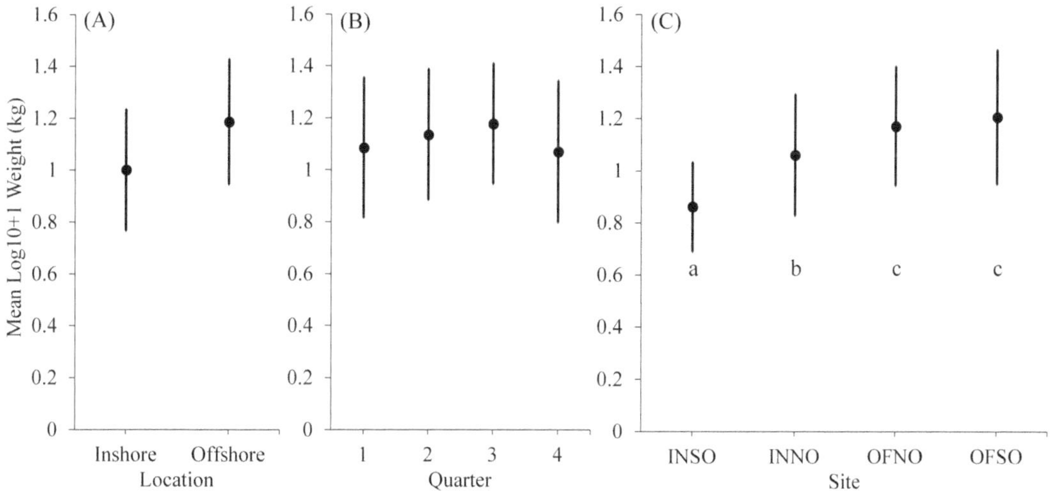

Figure 3. Mean-transformed ($\log_{10} + 1$) total weight ± SD (kg) of Red Snapper collected from December 2014–2016 at four artificial reefs off the coast of Brownsville, Texas comparing **(A)** inshore (within 15 km from Texas coast) and offshore (federal waters), with no significant difference observed; **(B)** quarters collected, with no significant difference observed; and **(C)** sampling sites, with nonmatching lowercase letters indicating sites that were significantly different from one another. Sample sizes and abbreviations as in Figure 2.

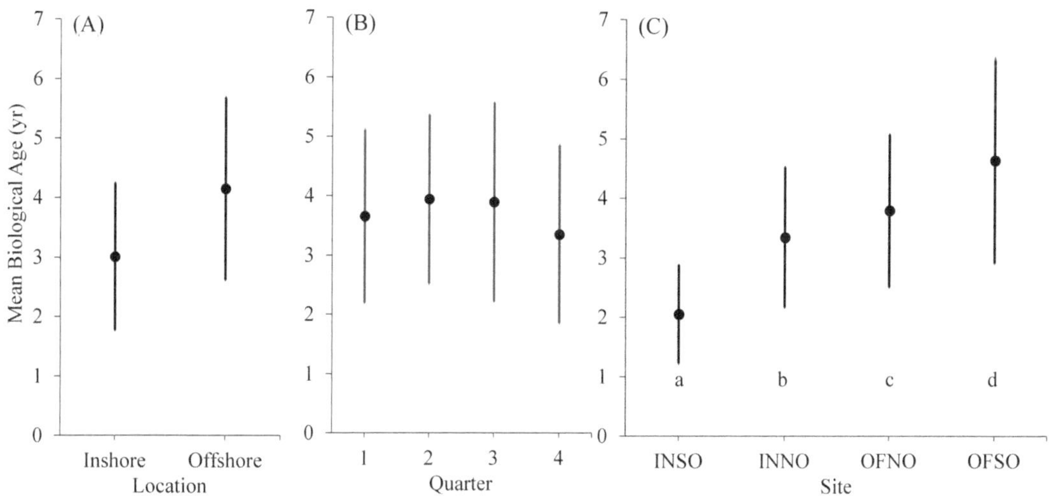

Figure 4. Mean biological age ± SD (years) of Red Snapper collected from December 2014–2016 at four artificial reefs off the coast of Brownsville, Texas comparing **(A)** inshore (within 15 km from Texas coast) and offshore (federal waters), with no significant difference observed; **(B)** quarters collected, with no significant difference observed; and **(C)** sampling sites, with nonmatching lowercase letters indicating sites that were significantly different from one another. Sample sizes and abbreviations as in Figure 2.

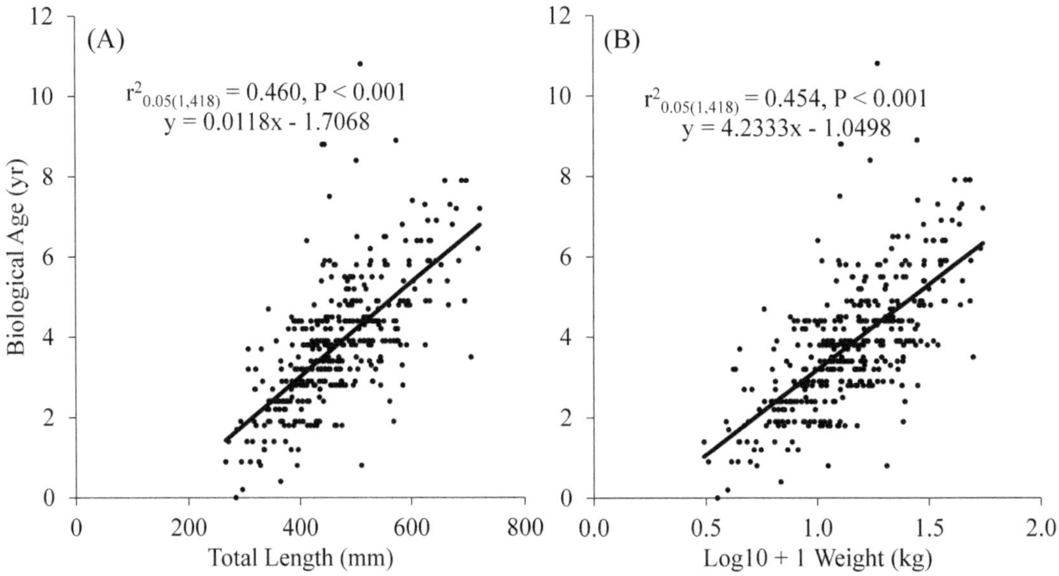

Figure 5. Relationship between biological age (years) and **(A)** total length (mm) and **(B)** transformed ($\log_{10}$ + 1) total weight (kg) of Red Snapper collected from December 2014 to 2016 at four artificial reefs off the coast of Brownsville, Texas. Sample size = 420.

lected. Note, $t_0 = 0$ was not forced to estimate for the growth of individuals smaller than 250 mm TL as it produced values inconsistent with observed TLs; a forced $t_0 = 0$ is known to contribute to differences when models are compared. The maximum age calculated for individuals collected in this study was 15 years; however, individuals of this age were not collected and did not reflect known age ranges. At about half the observed maximum age, 25 years was determined as an appropriate cutoff for the length-at-age projections.

There were significant differences observed among the 12 von Bertalanffy models from the Gulf of Mexico (Table 2; Figure 6, $F_{0.05,\ 1,12} = 2.075$, $P = 0.018$). However, upon closer look, the only difference was between the Saari et al. (2014) southern Texas model and the Szedlmayer and Shipp (1994) Alabama model, which are the smallest and largest growth curves, respectively ($P = 0.05$, which is marginally significant).

Discussion

Red Snapper captured in the current study were relatively small, young fish. Some studies have reported similar size and age findings (Futch and Bruger 1976; Wilson and Nieland 2001; Fischer et al. 2004; Syc and Szedlmayer 2012) while others reported slightly larger and older individuals in the Gulf of Mexico (Saari et al. 2014; Streich et al. 2017). From the present study, 83% of Red Snapper were over the minimum TL size limit (381 mm) invoked in Texas and 30% below those invoked in federal waters (406 mm TL). Early studies suggested that Red Snapper attain maturity at TLs between 300 and 320 mm (Camber 1955) and only after 2 years of age (Futch and Bruger 1976). Woods et al. (2003) collected Red Snapper with minimum spawning capabilities of individuals at 2 years and measuring 267 mm TL. White and Palmer (2004) found the smallest female and male Red Snapper to reach maturity were at 287 and 200 mm TL, respectively, whereas there were

Table 2. A selection of three-parameter von Bertalanffy growth curves for estimated length at age of Red Snapper in the Gulf of Mexico. Theoretical maximum length = L_∞, growth coefficient = k, and age at which length is zero = t_0. All lengths reported as total length (mm); italics = study from which all fish were collected from artificial reefs.

Study	Study area	L_∞	k	t_0
Current study	Southern Texas	800	0.19	–1
Saari et al. 2014	Southern Texas	645	0.419	–0.37
Saari et al. 2014	Northern Texas	908	0.191	–1.14
Saari et al. 2014	Louisiana	771	0.299	–0.45
Saari et al. 2014	Alabama	840	0.275	0.42
Saari et al. 2014	Northern Florida	690	0.322	–1.13
Saari et al. 2014	Central Florida	761	0.31	–0.73
Syc and Szeldmayer 2012	Alabama	936	0.205	–0.14
Wilson and Nieland 2001	Louisiana, Mississippi, and northern Texas	938	0.175	–0.53
Patterson et al. 2001a	Alabama	969	0.192	0.02
Szeldmayer and Shipp 1994	Alabama	1,025	0.15	–0.79
Nelson and Manooch 1982	Louisiana and northern Florida	941	0.17	–0.1
Mueller 2012	Gulf of Mexico	911	0.24	–0.33

immature females and males reported at 435 and 378 mm TL, respectively. Such findings suggest that most fish collected may have only spawned for 1 or 2 years before they were removed via the fishery. Because Red Snapper spawning capacity increases with age (Collins et al. 1996), the fishery removes the potentially most fecund members of the population before they have an opportunity to spawn.

When comparing Red Snapper growth across different areas of the Gulf of Mexico, it is important to acknowledge that sampling methodology, habitat variety, and geographic region vary across studies and have strong selective effects for Red Snapper (Campbell et al. 2014; Garner et al. 2014; Saari et al. 2014). For example, hook sizes used to catch fish in the present study showed that medium hooks (11/0) caught the most individuals and the largest range of fish lengths, followed by the smallest (8/0) and then the largest hooks (15/0) (Table 1); such findings are not surprising based on the size either by length or weight and age of the studied individuals and their known feeding habits (voracious and opportunistic at all times of day) (Gallaway et al. 2009). Even with the aforementioned selective effects, some comparative conclusions can be deduced from different studies. The von Bertalanffy growth curve for Red Snapper in the current study shows individuals growing at a rate of ~100 mm/year until age 4, after which their growth slows to ~50 mm annually. Approximately 90% of their theoretical maximum size is achieved by age 12 (Figure 6A). When including other artificial reef studies in Texas, Red Snapper in the northwestern Gulf of Mexico achieve legal size (406 mm) in federal waters at ~age 4 years, but as early as age 1 year in observed data. Other studies from south Texas waters predicted a smaller average maximum TL (Fischer et al. 2004; Saari et al. 2014); however, it is unknown if sampling in these studies was from areas with complex, high-relief structures, established as the preferred habitat of larger and older Red Snapper (Galla-

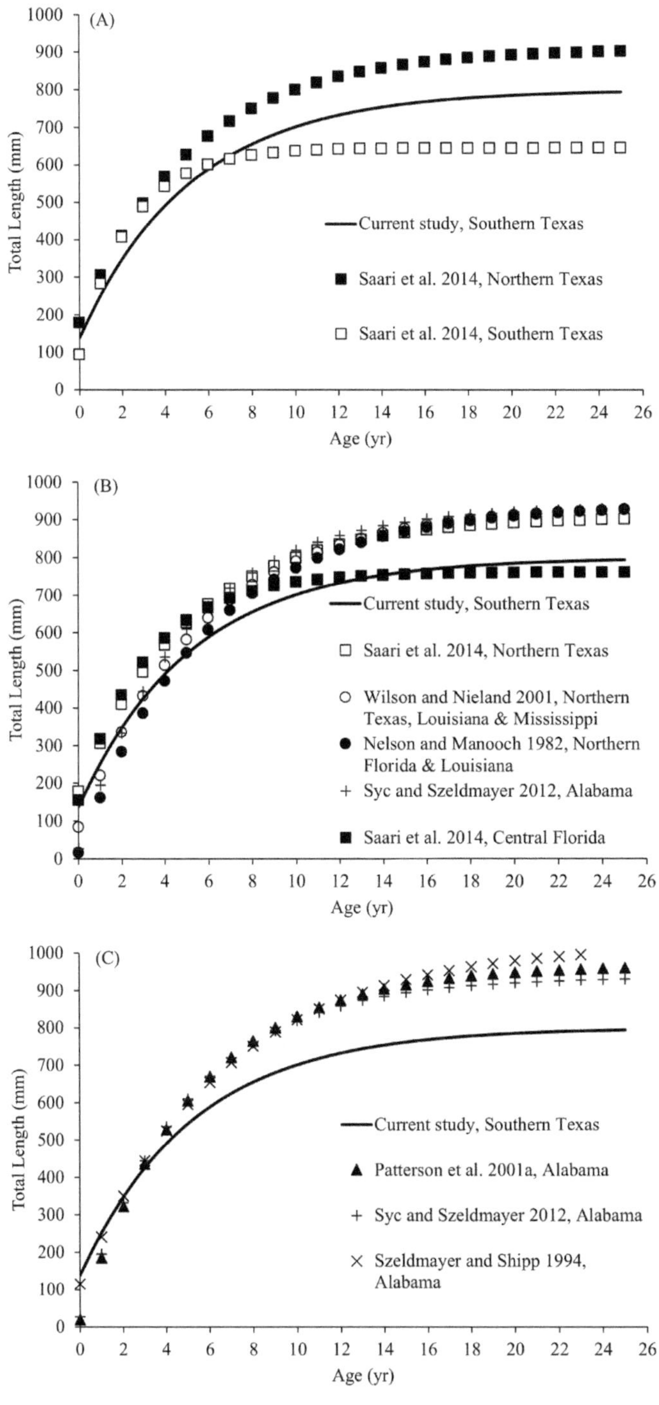

Figure 6. Three-parameter length-at-age von Bertalanffy growth curves for Red Snapper collected from December 2014 to 2016 on artificial reefs in Texas waters in the northwestern Gulf of Mexico plotted alongside observed age (years) at length (mm) data compared to other studies that collected Red Snapper **(A)** in Texas waters, **(B)** across different Gulf of Mexico states, and **(C)** on artificial reefs only.

way et al. 2009). Theoretical average maximum TL estimations are also similar to those for populations at similar latitude in central Florida and very similar to populations from Alabama waters (Fischer et al. 2004; Saari et al. 2014).

Other studies from the Gulf of Mexico have calculated similar growth coefficients (±0.01) for Red Snapper as the present study (Patterson et al. 2001a; Wilson and Nieland 2001; Saari et al. 2014) while another study calculated almost double the growth rates (Saari et al. 2014). Growth rates are impacted by a host of factors, yet an absence of preferred habitat and food availability can result in a selection for slower-growing individuals. The mean TLs of age-0 fish collected for this study are larger than described in other studies for either Texas waters or the Gulf of Mexico (Nelson and Manooch 1982; Szedlmayer and Shipp 1994; Patterson et al. 2001a; Wilson and Nieland 2001; Mueller 2012; Syc and Szedlmayer 2012; Saari et al. 2014); however, no individuals below 250 mm TL were collected, which provides an incomplete picture of early life history for this area as smaller individuals usually represent younger individuals.

Depending on environmental conditions, cohorts can vary in age and growth by year, but the present study found no differences between the two studied years. There were also no differences in morphometric data among quarters, which may be due to high site fidelity year-round. Garcia (2013) found strong site fidelity in individuals throughout the year at two reefs—one offshore and one inshore—that were part of the present study (OFSO and INNO, respectively); this is similar to findings from other studies (Szedlmayer and Schroepfer 2005; Piraino and Szedlmayer 2014). Yet, several studies suggest low fidelity of Red Snapper among sites (Patterson et al. 2001b; Peabody 2004; Topping and Szedlmayer 2011), including one study at the INNO and neighboring natural reefs (Getz 2017).

Gazey et al. (2008) and Gallaway et al. (2009) suggested that Red Snapper move further offshore as they age, from small nearshore reefs to larger deeper structures, and then to areas lacking significant reef structure. In contrast, there were no differences in age and growth in Red Snapper between inshore and offshore sites in the present study. Instead, specific sites were different from one another, with the smallest and youngest fish closest to the Mexican border. It is important to note that offshore sites in the current study were still relatively close to shore (~20–30 km); thus, larger fish may be moving even further from shore. With the fishing regulations in Texas open year-round and the added pressure of illegal fishing from across and within the border, greater fishing effort may be occurring both in nearby and offshore waters. Enforcing fisheries regulations is difficult in this area; thus, greater fishing effort may be occurring in both inshore and offshore waters, selecting for populations with young and less-fecund fish (Manooch 1976).

Some authors have suggested that artificial reefs are not properly designed to sustain healthy populations of Red Snapper and are often associated with heavy fishing efforts (Bohnsack and Sutherland 1985; Cowan et al. 2010). Differences in habitat type can be a factor explaining differences in growth and size of Red Snapper; however, there are no cohesive findings observed in the current study. There have been reports that a higher percentage of larger fish associate with artificial reefs compared to natural structure (Saari 2011), while Streich et al. (2017) found a larger proportion of big fish at natural as opposed to artificial structures. Sampling individuals at nearby natural sites may add information that helps explain the size structure of Red Snapper in the area, yet Froehlich and Kline (2015) found no significant difference in length estimates of Red Snapper at natural sites compared to nearby culvert reef patches also sampled in the INNO reef of the present study.

The estimated life span of Red Snapper is 50–55 years (Render 1995; Wilson and Nieland 2001). However, our estimates place most (84%) fish in Texas waters at less than 5 years, as also seen by Wilson and Nieland (2001), Syc and Szedlmayer (2012), and Saari et al. (2014). Other studies indicate age for most Red Snapper in the northwestern Gulf of Mexico at less than 7 years (Szedlmayer and Shipp 1994; Woods et al. 2003; Streich et al. 2017). This information indicates that these Red Snapper populations should have relatively low fecundity. The present study adds important information in an area targeted from several fishing industries and two nations that suggest that better management practices are required to augment Red Snapper populations and recover from depleted fisheries.

Acknowledgments

The authors thank the University of Texas Rio Grande Valley dive team, composed of volunteers, undergraduates, and graduate students, for help collecting samples and enduring rough weather conditions, in particular R. Alexander during the earlier sampling period. The Texas Parks and Wildlife Department (TPWD) Artificial Reef Program provided funding for research, and TPWD also granted Scientific Permit SPR-0314–050 to collect Red Snapper in Texas waters. The National Oceanic and Atmospheric Administration–National Marine Fisheries Service for a letter of acknowledgment recognizing biological monitoring and research activities as being outside the boundaries of fishing regulations.

References

Alexander, R. 2015. Comparing reproductive capacity of nearshore and offshore Red Snapper, *Lutjanus campechanus*, reproduction on artificial reefs in the northwestern Gulf of Mexico. Master's thesis. University of Texas Rio Grande Valley, Brownsville.

Arney, R. N., C. Y. M. Froehlich, and R. J. Kline. 2017. Recruitment patterns of juvenile fish at an artificial reef area in the Gulf of Mexico. Marine and Coastal Fisheries: Dynamics, Management, and Ecosystem Science [online serial] 9:79–92.

Bohnsack, J. A., and D. L. Sutherland. 1985. Artificial reef research: a review with recommendations for future priorities. Bulletin of Marine Science 37:11–39.

Bollinger, M. A., and R. J. Kline. 2017. Validating sidescan sonar as a fish survey tool over artificial reefs. Journal of Coastal Research 33:1397–1407.

Camber, C. I. 1955. A survey of the Red Snapper fishery of the Gulf of Mexico, with special reference to the Campeche Banks. Florida State Board of Conservation, Marine Laboratory, Technical Series 12, St. Petersburg.

Campbell, M. D., A. G. Pollack, W. D. Driggers, and E. R. Hoffmayer. 2014. Estimations of hook selectivity of Red Snapper and Vermilion Snapper from fishery-independent surveys of natural reefs in the northern Gulf of Mexico. Marine and Coastal Fisheries: Dynamics, Management, and Ecosystem Science [online serial] 6:260–273.

Collins, L. A., A. G. Johnson, and C. P. Keim. 1996. Spawning and annual fecundity of the Red Snapper (*Lutjanus campechanus*) from the northeastern Gulf of Mexico. Pages 174–188 *in* F. Arreguin-Sanchez,

J. L. Munro, M. C. Balgos, and D. Pauly, editors. Biology, fisheries, and culture of tropical groupers and snappers. ICLARM (International Center for Living Aquatic Resources Management) Contribution 1323.

Cowan, J. H., Jr., C. B. Grimes, W. F. Patterson, C. J. Walters, A. C. Jones, W. J. Lindberg, D. J. Sheehy, W. E. Pine, J. E. Powers, M. D. Campbell, K. C. Lindeman, S. L. Diamond, R. Hilborn, H. T. Gibson, and K. A. Rose. 2010. Red Snapper management in the Gulf of Mexico: science- or faith-based? Reviews in Fish Biology and Fisheries 21:187–204.

Fischer, A. J., M. S. Baker, Jr., and C. A. Wilson. 2004. Red Snapper (*Lutjanus campechanus*) demographic structure in the northern Gulf of Mexico based on spatial patterns in growth rates and morphometrics. U.S. National Marine Fisheries Service Fishery Bulletin 102:593–603.

Froehlich, C. Y. M., and R. J. Kline. 2015. Using fish population metrics to compare the effects of artificial reef density. PLOS (Public Library of Science) ONE [online serial] 10(9):e0139444.

Futch, R. B., and J. M. Bruger. 1976. Age, growth and reproduction of Red Snapper in Florida waters. Pages 165–184 *in* H. R. Bullis and A. C. Jones, editors. Colloquium on snapper-grouper fishery resources of the western central Atlantic Ocean 1976. Florida Sea Grant College Program, University of Florida, Report 17, Gainesville.

Gallaway, B. J., S. T. Szedlmayer, and W. J. Gazey. 2009. A life history review for Red Snapper in the Gulf of Mexico with an evaluation of the importance of offshore petroleum platforms and other artificial reefs. Reviews in Fisheries Science 17:48–67.

Garcia, A. 2013. A comparison of site fidelity and habitat use of Red Snapper (*Lutjanus campechanus*) to evaluate the performance of two artificial reefs in South Texas utilizing acoustic telemetry. Master's thesis. University of Texas at Brownsville, Brownsville.

Garner, S. B., W. F. Patterson, C. E. Porch, and J. H. Tarnecki. 2014. Experimental assessment of circle hook performance and selectivity in the northern Gulf of Mexico recreational reef fish fishery. Marine and Coastal Fisheries: Dynamics, Management, and Ecosystem Science [online serial] 6:235–246.

Gazey, W. J., B. J. Gallaway, J. G. Cole, and D. A. Fournier. 2008. Age composition, growth, and density-dependent mortality in juvenile Red Snapper estimated from observer data from the Gulf of Mexico penaeid shrimp fishery. North American Journal of Fisheries Management 28:1828–1842.

Getz, E. 2017. Utilizing accelerometers to compare Red Snapper behavior on artificial and natural reefs. Master's thesis. University of Texas Rio Grande Valley, Brownsville.

Goodyear, C. P. 1995. Red Snapper stocks in U.S. waters of the Gulf of Mexico. National Marine Fisheries Service, Southeast Fisheries Science Center, Miami Laboratory.

Jackson, M. C., J. H. Cowan, Jr., and D. L. Nieland. 2007. Demographic differences in northern Gulf of Mexico Red Snapper reproductive maturation: implications for the unit stock hypothesis. Pages 217–228 *in* W. F. Patterson, III, J. H. Cowan, Jr., G. R. Fitzhugh, and D. L. Nieland, editors. Red Snapper ecology and fisheries in the U.S. Gulf of Mexico. American Fisheries Society, Symposium 60, Bethesda, Maryland.

Manooch, C. S. I. 1976. Reproductive cycle, fecundity, and sex ratios of the Red Porgy, *Pagrus pagrus*, (Pisces: Sparidae) in North Carolina. U.S. National Marine Fisheries Service Fishery Bulletin 74:775–781.

Mueller, M. 2012. Artificial structure and hard-bottom spatial coverage in the Gulf of Mexico. SEDAR, SEDAR 31-DW-29, North Charleston, South Carolina.

Nelson, R. S., and C. S. I. Manooch. 1982. Growth and mortality of Red Snappers in the west-central Atlantic Ocean and northern Gulf of Mexico. Transactions of the American Fisheries Society 111:465–475.

Patterson, W. F., J. H. Cowan, Jr., C. A. Wilson, and R. L. Shipp. 2001a. Age and growth of Red Snapper, *Lutjanus campechanus*, from an artificial reef area off Alabama in the northern Gulf of Mexico. U.S. National Marine Fisheries Service Fishery Bulletin 99:617–627.

Patterson, W. F., J. C. Watterson, R. L. Shipp, and J. H. Cowan, Jr. 2001b. Movement of tagged Red Snapper in the northern Gulf of Mexico. Transactions of the American Fisheries Society 130:533–545.

Peabody, M. B. 2004. The fidelity of Red Snapper (*Lutjanus campechanus*) to petroleum platforms and artificial reefs in the northern Gulf of Mexico. Master's thesis. Louisiana State University and Agricultural and Mechanical College, Baton Rouge.

Piraino, M. N., and S. T. Szedlmayer. 2014. Fine-scale movements and home ranges of Red Snapper around artificial reefs in the northern Gulf of Mexico. Transactions of the American Fisheries Society 143:988–998.

Render, J. H. 1995. The life history (age, growth and reproduction) of Red Snapper (*Lutjanus campechanus*) and its affinity for oil and gas platforms. Master's thesis. Louisiana State University and Agricultural and Mechanical College, Baton Rouge.

Saari, C. R. 2011. Comparison of the age and growth of Red Snapper (*Lutjanus campechanus*) amongst habitats and regions in the Gulf of Mexico. Doctoral dissertation. Louisiana State University and Agricultural and Mechanical College, Baton Rouge.

Saari, C., J. H. Cowan, Jr., and K. Boswell. 2014. Regional differences in the age and growth of Red Snapper (*Lutjanus campechanus*) in the U.S. Gulf of Mexico. U.S. National Marine Fisheries Service Fishery Bulletin 112:261–273.

Schirripa, M. J., and C. M. Legault. 1999. Status of the Red Snapper in US waters of the Gulf of Mexico: updated through 1998. Southeast Fisheries Science Center, Sustainable Fisheries Division, Contribution SFD-99/00-75, Miami.

SEAMAP (Southeast Monitoring and Assessment Program). 2012. Southeast Monitoring and Assessment Program vertical line survey protocol. Gulf States Marine Fisheries Commission, Ocean Springs, Mississippi.

Sokal, R. R., and F. J. Rohlf. 2012. Biometry: the principles and practice of statistics in biological research, 4th edition. Peter Marshall, New York.

Stephan, C. D., B. G. Dansby, H. R. Osburn, G. C. Matlock, R. K. Riechers, and R. Rayburn. 1990. Texas artificial reef fishery management plan. Texas Parks and Wildlife Department, Fishery Management Plan Series 3, Austin.

Streich, M. K., M. J. Ajemian, J. J. Wetz, J. A. Williams, J. B. Shipley, and G. W. Stunz. 2017. A comparison of size structure, age, and growth of Red Snapper from artificial and natural habitats in the Western Gulf of Mexico. Transactions of the American Fisheries Society 146:762–777.

Syc, T. S., and S. T. Szedlmayer. 2012. A comparison of size and age of Red Snapper (*Lutjanus campechanus*) with the age of artificial reefs in the northern Gulf of Mexico. U.S. National Marine Fisheries Service Fishery Bulletin 110:458–469.

Szedlmayer, S. T., and R. L. Schroepfer. 2005. Long-term residence of Red Snapper on artificial reefs in the northeastern Gulf of Mexico. Transactions of the American Fisheries Society 134:315–325.

Szedlmayer, S. T., and R. L. Shipp. 1994. Movement and growth of Red Snapper, *Lutjanus campechanus*, from an artificial reef area in the northeastern Gulf of Mexico. Bulletin of Marine Science 55:887–896.

Topping, D., and S. Szedlmayer. 2011. Site fidelity, residence time and movements of Red Snapper *Lutjanus campechanus* estimated with long-term acoustic monitoring. Marine Ecology Progress Series 437:183–200.

VanderKooy, S., and K. Guindon-Tisdel. 2003. A practical handbook for determining the ages of Gulf of Mexico fishes. Gulf States Marine Fisheries Commission, Publication No. 111, Ocean Springs, Mississippi.

White, D. B., and S. M. Palmer. 2004. Age, growth, and reproduction of the Red Snapper, *Lutjanus campechanus*, from the Atlantic waters of the Southeastern U.S. Bulletin of Marine Science 75:335–360.

Wilson, C. A., and D. L. Nieland. 2001. Age and growth of Red Snapper, *Lutjanus campechanus*, from the northern Gulf of Mexico off Louisiana. U.S. National Marine Fisheries Service Fishery Bulletin 99:653–664.

Woods, M. K., A. J. Fischer, J. H. Cowan Jr., and D. L. Nieland. 2003. Size and age at maturity of female Red Snapper *Lutjanus campechanus* in the northern Gulf of Mexico. Proceedings of the 54th Gulf and Caribbean Fisheries Institute 54:526–537.

American Fisheries Society Symposium 86:97–116, 2018

Investigating Reproductive Characteristics of Gray Triggerfish on Three Artificial Reefs in the Northwest Gulf of Mexico

A. M. Lee and C. E. Cintra-Buenrostro*

School of Earth, Environmental and Marine Sciences, University of Texas Rio Grande Valley One West University Boulevard, Brownsville, Texas 78520, USA

J. D. Shively

Texas Parks and Wildlife Department, Artificial Reef Program 4200 Smith School Road, Austin, Texas 78744, USA

Abstract.—Artificial reefs in the northwest Gulf of Mexico act as preferential habitat for Gray Triggerfish *Balistes capriscus* (Gmelin, 1789) and are heavily targeted by local and international fishing efforts. Three artificial reefs off the south Texas coast were studied to investigate the reproductive characteristics of this ubiquitous reef member whose importance to both commercial and recreational fishing has resulted in a dramatic reduction in Gulf of Mexico-wide estimated biomass for the past 20 years. Characteristics examined included sex ratio, peak spawning season, batch fecundity of females (1×10^6 oocytes), fork length (FL, mm) at maturity, and batch fecundity as a function of FL or weight (i.e., $\Delta 1 \times 10^6$ oocytes = ΔFL). These characteristics were investigated to build a foundational understanding of an understudied region of the Gulf and a relatively new artificial reef habitat. Reef sites managed by the Texas Parks and Wildlife Department Artificial Reef Program were sampled for 1 year, yielding 112 individuals. The sex ratio was 1:1 males to females but varied among sample sites, suggesting that an unidentified factor may have influenced aggregations of males and females. Spawning season was similar in other regions of the Gulf of Mexico (i.e., between June and August with a peak in July). Female batch fecundity was $7.36 \times 10^5 \pm 2.69 \times 10^5$ mature oocytes, and the batch fecundity to FL relationship was $14.353 \times FL^{1.88}$, similar to other populations of Gray Triggerfish in the Gulf of Mexico.

Introduction

Gray Triggerfish *Balistes capriscus* (Gmelin, 1789) are found on artificial reefs throughout their ontogeny and have been identified spawning at artificial reefs (Moore 1997; Ingram 2001; MacKichan and Szedlmayer 2007; Simmons and Szedlmayer 2011). Artificial reefs are estimated to have a footprint of <1% of 1% (0.01%) of all available habitat in the Gulf of Mexico, making their

* Corresponding author: carlos.cintra@utrgv.edu

presence in south Texas rare in an otherwise sand and mud-dominated benthic expanse (Mueller 2012). Site fidelity (where individuals have a strong affinity for a single area selected after postlarval recruitment) and site specificity (individuals found associated with a limited habitat with unique features) are hallmarks of Gray Triggerfish. Consequently, individuals maintain a limited home range after recruitment (Addis et al. 2007; Herbig and Szedlmayer 2016). These behaviors make them vulnerable to fishing efforts (both intentional and accidental; Simmons 2001; Smith et al. 2015). With Texas State waters open year-round to fishing and a bag limit of 20 Gray Triggerfish per day (TPWD 2017), understanding the reproductive characteristics of Gray Triggerfish can provide a basis for a robust holistic management approach. The Gray Triggerfish's status as an incidental or secondary target for local and commercial interests in the northwest Gulf further complicates management efforts. Local population can appear robust and continue to be treated as such irrespective of regional populations declines. Because of an aggressive collection allowance and a vulnerable life history regime, Gray Triggerfish are likely to experience stricter fishery management regulation in the future, similar to those imposed on Red Snapper *Lutjanus campechanus* (Poey, 1860) and Red Grouper *Epinephelus morio* (Valenciennes, 1828) (Gulf of Mexico Fishery Management Council 2014; SEDAR 2015).

Other studies on Gray Triggerfish have focused on artificial and natural habitat in the northern and eastern Gulf of Mexico (Johnson and Saloman 1984; Escorriola 1991; Moore 1997; Ingram 2001; Burton et al. 2014) where many states in the USA have either prohibited or severely limited their Gray Triggerfish harvest. Genetic studies have determined that the Gulf of Mexico population is a single stock (Antoni et al. 2011; Antoni and Saillant 2017). However, separate Gray Triggerfish populations residential to the Gulf of Mexico may differ in ontogeny due to environmental conditions by latitudinal distribution. Hence, obtaining essential life history information from an understudied region within its range and from a relatively new habitat is important information for future stock assessments. Key reproductive and life history characteristics of sex ratio, spawning period, size and age at maturity, and reproductive output were investigated in this novel population.

Gray Triggerfish form reproductive harems of one male and up to five females, and while this suggests a population dominated by females, assessments of sex ratio Gulf-wide identify sex ratios dominated by both males (Wilson et al. 1995; Hood and Johnson 1997) and females (Moore 1997; Fitzhugh et al. 2015); however, for management purposes, the sex ratio is set to 1:1 (SEDAR 2015), which suggests that we should find a similar ratio Gulf-wide. The identification of a single genetic population and utilization of a specific habitat type across the Gulf of Mexico suggests that Gray Triggerfish off south Texas should exhibit reproductive characteristics and traits, such as spawning period and size and age at maturity, similar to populations elsewhere in the Gulf of Mexico. Furthermore, we suspect that Gray Triggerfish from new and established reefs would not differ with regard to reproductive characteristics because the maturation (1–2 years) and recruitment (0–1 year) ages shared by the Gulf of Mexico population suggest that even young reefs would support individuals that are reproductively capable (Ingram 2001; Bernardes 2002; Simmons and Szedlmayer 2011). Additionally, it is well established that fishes increase their reproductive capability with an increase in size and age (Hixon et al. 2014). Other studies of Gray Triggerfish in the Gulf of Mexico have also shown increases in size are associated with higher reproductive output (Ingram 2001; Fitzhugh et al. 2015).

Previous evidence withstanding, we hypothesize that Gray Triggerfish on artificial reefs in the northwest Gulf of Mexico would (1) exhibit a 1:1 sex ratio, (2) experience a peak spawning season similar to other populations found on other artificial reefs in the Gulf of Mexico, (3) share a size at maturation with other populations found in the Gulf of Mexico, and (4) display an increase in batch fecundity (and subsequent increase of annual fecundity) with age, total weight, and fork length.

Methods

From January 1, 2015 until January 1, 2016, Gray Triggerfish were collected quarterly from two inshore (<15 km from shore) and one offshore (>15 km from shore) artificial reef sites managed and monitored by the Texas Parks and Wildlife Department (TPWD) Artificial Reef Program from latitude 26.5255831° N, longitude –97.153587° W to latitude 25.968407° N, longitude –97.066917° W. The southernmost artificial reef site PS-1169 L (hereafter, "Inshore A") is 7.96 km from shore, composed of 2 three-pile petroleum jackets, 53 reef balls, and 3 vessels (ranging from 18 to 45 m in length), and an average depth of 21.8 m. All material for this reef was deployed in July 2011. The northernmost reef site PS-1047 (hereafter, "Inshore B") is 10.37 km from shore, contains >4,800 concrete culverts distributed in dense and loose aggregates, has a 30-m reefed tugboat, and has an average depth of 19.7 m. Materials here were deployed between 2009 and 2011. The only reef located in federal waters (20.19 km from shore) is PS-1070 (hereafter, "Offshore C"), composed of 3 ships and 15 assorted four-pile petroleum jackets at a depth of 30.9 m. For detailed site descriptions, see Bollinger and Kline (2017).

Our initial fishing regime consisted of a standardized vertical longline protocol on a quarterly basis adapted and modified from SEAMAP (2012), detailed in Froehlich et al. (2018, this volume), but to warrant proper characterization of the studied populations reproduction, additional sampling effort occurred during the Gulf of Mexico known spawning season (June to August). However, collecting Gray Triggerfish with a hook-and-line system designed to target prey that tend to swallow reef fishes like Red Snapper and Red Grouper presented challenges. Large fishing hooks and a voracious reef assemblage prevented efficient collection of Gray Triggerfish, which are generally weak swimmers with a small mouth gape (Nelson and Bortone 1996; Burns and Froeschke 2012). Consequently, in this study, the collection of individuals was restricted to spearfishing. While this method presents inherited biases, such as an inability to collect small individuals and conditioning fish for avoidance behavior (Dickens et al. 2011), it is not an uncommon collection method for this species (Ingram 2001; Simmons 2001; MacKichan and Szedlmayer 2007) and has been suggested as less biased than other collection methods by Randall (1967). All collected individuals were stored on ice and processed within 48 h of capture.

Morphometric characteristics recorded from specimens included fork length (FL) in mm (±1), weight in g (*W*, ±0.1), gonad weight (GW) in g (±0.01) and sex. Mean and standard deviation (SD) morphometric statistics were calculated. Sex ratio was evaluated using a chi-square test (χ^2; Sokal and Rohlf 2012).

Otoliths are inconspicuous and unreliable for aging Gray Triggerfish while first dorsal spines have been validated as an alternative body part for aging Gray Triggerfish (Allman et al. 2015). First dorsal spines were excised, excess tissue removed and immediately frozen at –40°C to restrict vascularized tissue damage. Before processing, first dorsal spines were dried for a minimum of 6 h at 60°C before the peduncle was removed and the spine mounted on a glass slide with a two-part clear epoxy resin. The spine was cross-sectioned at 0.5 mm utilizing an MTI Corp 150 low-speed diamond-blade saw. Age (years ± 0.1) was determined from annuli and marginal increments on the first dorsal spine following Allman et al. (2015). Age was determined by two independent readers with an average cumulative variance in age result of 2.79%. Consequently, no spines were excluded from age analyses.

The gonadosomatic index (GSI) of individuals was calculated to determine peak spawning season and batch fecundity after Branco et al. (2013), where

$$\text{GSI} = \text{GW}/W \times 100.$$

A GSI of 1 or greater indicates that individuals are capable of spawning (Kjesbu 2009) while the highest average GSI in females during the month of capture was used to determine both peak spawning season and a specific month of peak spawn.

To determine peak spawning season, values for batch fecundity, annual fecundity, age at maturity, and length at maturity, a histological characterization of gonadal tissue is required. Gonad tissue was dissected, trimmed of excess fat, and immediately immersed in Prefer (glyoxal fixative) for at least 14 but no more than 30 d. After fixation, a random lobe and quartile of the gonad was selected for histological sampling. Each gonad lobe (right and left) and each lobe region (top, middle, lower, and bottom) was assigned a value. A random number table generated in SPSS v. 23 (IBM 2013) within the range of assigned values was used to select a region. After selection, a 4–6-mm-thick section was excised. A KD-TS3D tissue cassette processing system impregnated tissue with paraffin wax; then, the paraffin blanks were sectioned (5–10 μm) using a Richter-Jung 2030 rotary microtome. Tissue sections were mounted on a glass slide and stained with Gill's 2 hematoxylin and EOSIN-Y using a Tissue-Tek II slide staining set.

Reproductive status of gonadal tissue was determined following methods adapted from Kacem et al. (2015) by two independent readers, with no gonad excluded from the analyses. Gametogenic and accessory tissue reproductive stage characterization was adapted from Brown-Peterson et al. (2011; Table 1). Maturity status (mature or immature) was assigned to males or females who exhibited a spawning-capable, actively spawning, and regressing reproductive status.

Batch fecundity (1×10^6 oocytes) was calculated by selecting females histologically identified as either spawning capable or actively spawning, with a GSI >1. Ovarian tissues from these individuals ($n = 12$) were processed and oocytes counted with methods adapted from Hunter and Macewicz (1985). Annual fecundity was calculated as 10 × batch fecundity, as Gray Triggerfish spawn 9–11 times throughout the duration of the spawning season (Lang and Fitzhugh 2015). Length at maturity and age at maturity were calculated for only females with methods adapted from Kacem et al. (2015).

Statistical analyses were performed with SPSS v. 23 (IBM 2013), with $\alpha = 0.05$. Mor-

Table 1. Histological characterization of the stages of the ovaries and testes of Gray Triggerfish. Modified from Kacem et al. (2015). Abbreviations: CA = cortical alveolar, Vtg1 = primary vitellogenic oocytes, Vtg2 = secondary vitellogenic oocytes, Vtg3 = tertiary vitellogenic oocytes, POFs = postovulatory follicles, GE = germinal epithelium, Sc1 = primary spermatocyte, Sc2 = secondary spermatocyte, Sg1 = primary spermatogonia, Sg2 = secondary spermatogonia, St = Spermatid, and Sz = spermatozoa.

Maturity (females)	Microscopic characteristics
I. Immature	Well-defined cellular organization with the presence of several unyolked oocytes in perinuclear regions. Only oogonia and primary growth.
II. Developing	Oogonia, CA, Vtg1, and Vtg2. No evidence of Vtg3.
III. Spawning capable	All oocyte stages present and abundance of Vtg3.
IV. Actively spawning	Hydrated oocytes (translucent) in large quantities and with few POFs.
V. Regressing	Few POFs, atresia (any stage), some residual CA, and Vtg1–Vtg2 oocytes.
VI. Resting	Reorganization of unyolked oocytes to the beginning of a new reproductive cycle, and atresia.

Maturity (males)	Microscopic characteristics
I. Immature	Small testes, often clear and threadlike. Only Sg1 present; no lumen in lobules.
II. Developing	Small testes but easily identified. Spermatocysts evident along lobules. Sg2, Sc1, Sc2, St, and Sz can be present in spermatocysts. GE continuous throughout.
III. Spawning capable	Large and firm testes. Sz in lumen of lobules and/or sperm ducts. All stages of spermatogenesis can be present. GE can be continuous or discontinuous.
IV. Regressing	Small and flaccid testes, no milt release with pressure. Residual Sz present in lumen of lobules and in sperm ducts. Widely scattered spermatocysts near periphery containing Sc2, St, and Sz. Spermatogonial proliferation and regeneration of GE common in periphery of testes.
V. Resting	Small testes, often threadlike. No spermatocysts. Lumen of lobule often absent. Proliferation of spermatogonia throughout testes. GE continuous throughout.

phometric and reproductive statistics were subject to tests of normality with visual Q–Q plots and by Kolmogorov–Smirnov test prior to running analyses on FL, *W*, and age. Levene's test confirmed data's homoscedasticity (Sokal and Rohlf 2012). Nested analysis of variance was used to compare morphometric and reproductive dependent variables with location (inshore versus offshore), with yearly quarter (Q) of collection (Q1 = January–March, Q2 = April–June, Q3 = July–September, Q4 = October–December), and with sex subsequently nested (Sokal and Rohlf 2012). This was deemed necessary because previous investigations of reef fish at the same sites were larger and older with an increased capture distance from shore (Alexander 2015). Linear regressions for batch fecundity with FL and *W*, with age, were performed to determine if any morphometric characteristics were significant predictors of batch fecundity. Fork length at maturity and *W* at maturity were calculated using a binary logistic regression analysis adapted from Kacem et al. (2015). Maturity, FL, and *W* did not violated assumptions prior to binary logistic regression testing.

A one-way analysis of variance (ANOVA) was performed to determine if lobe or sectioning region per lobe were associated with differences in oocyte yield. Compliance to the assumptions of normality and homoscedasticity was tested as before.

Ingram (2001) and Lang and Fitzhugh (2015) used linear and power functions, respectively, to represent their batch fecundity to FL relationships. For accurate comparison, the batch fecundity to FL relationships from this study are represented in both a linear and power format. In addition, Ingram (2001) provides a batch fecundity to *W* relationship, which is compared to findings from this study.

Female Gray Triggerfish collected from artificial reefs in the northwest Gulf of Mexico represented a smaller size range (FL and *W*) than was found in other studies of the same species elsewhere in the Gulf (Ingram 2001; Lang and Fitzhugh 2015). In order to standardize comparisons of batch fecundity across studies, 100 FLs and *W*s were randomly generated from the observed ranges for Gray Triggerfish, as reported in SEDAR (2015). The randomly generated values were used to calculate batch fecundity from relationships published among studies. Batch fecundity was then compared among studies using a one-way ANOVA. This approach was used to assess the capacity for this study's batch fecundity to FL and batch fecundity to *W* relationships to accurately estimate reproductive output of females, irrespective of size.

Sexual dimorphism has been reported in Gray Triggerfish throughout their natural range (Ingram 2001; Bernardes 2002; Burton et al. 2014; Allman et al. 2017). Subsequently, estimations of life history or reproductive history are represented as females only, males only, or a combination of both sexes (total) when appropriate.

Results

Significant differences in FL were observed between males and females ($F_{6,111} = 2.864$), and no significant differences were observed owing to location, site, or season sampled; FL ranged from 232 to 432 with a mean of 315 ± 39 mm (details in Table 2). Significant differences were found in *W* between males and females ($F_{6,111} = 3.435$), with no significant differences observed owing to other factors (*W* findings summarized in Table 3). While no significant differences were attributed to factors other than sex, smaller and lighter fish occurred at Inshore A, followed by intermediate sized/weighted fish from Inshore B, and fish from Offshore C were largest and heaviest. Thus, as distance from shore increased, FL and *W* of fish increased.

No significant differences in the age of Gray Triggerfish were attributable to any differences among treatments, with a summary of descriptive statistics for age presented in Table 4.

Sex ratios were significantly different at Inshore A ($\chi^2 = 5.0$) and Inshore B ($\chi^2 = 6.081$), skewed towards females at Inshore A and males at Inshore B, while the offshore site (Offshore C) exhibited a 1:1 sex ratio. Overall, the sex ratio for the catch year was observed at 1:1 ($\chi^2 = 0.571$).

All ovarian and testicular reproductive stages were present in both female (Figure 1)

Table 2. Fork length (mm) of Gray Triggerfish collected from three artificial reefs in the northwest Gulf of Mexico. All lengths reported in mm; *n* = 112; Max = maximum, Min = minimum, SD = standard deviation.

Site	Sex	Mean	Max	Min	SD
Inshore A	Female	284	375	235	29
	Male	320	374	262	37
	Total	296	375	235	36
Inshore B	Female	307	345	271	22
	Male	324	380	244	30
	Total	319	380	244	29
Offshore C	Female	328	432	232	44
	Male	355	405	297	30
	Total	338	432	232	41
All sites	Female	302	432	232	39
	Male	329	405	244	34
	Total	315	432	232	39

and male individuals (Figure 2), respectively, with peak spawning period in June to August and a peak spawn month in July (Figure 3). Female GSIs were pooled and averaged monthly. Female fish sampled in July had the highest mean GSI at 8.4. Fish caught in June, July, and August had a mean GSI ≥3.3, higher than all other months sampled. This indicated that the peak spawning season was June to August. During this period (June to August), 91.7% of the ovaries from all females sampled were identified as developing, spawning capable, actively spawning, or regressing.

Proportional FL and *W* at maturity were calculated for females. Females exhibited 50% maturity (FL_{50}) at 230 mm FL (Figure 4) and at 325 g (W_{50}).

Table 3. Weight of Gray Triggerfish collected from three artificial reefs in the northwest Gulf of Mexico. All weights reported in g; *n* = 112; Max = maximum, Min = minimum, SD = standard deviation.

Site	Sex	Mean	Max	Min	SD
Inshore A	Female	572.1	1,122.1	322.1	167.8
	Male	829.9	1,265.2	443.3	270.9
	Total	658.1	1,265.2	322.1	238.8
Inshore B	Female	684.5	1,033.8	465.0	155.8
	Male	797.4	1,258.5	349.0	194.1
	Total	763.8	1,258.5	349.0	188.8
Offshore C	Female	878.7	1,564.8	302.9	292.6
	Male	1,078.1	1,501.8	586.4	261.4
	Total	951.8	1,564.8	302.9	293.7
All sites	Female	689.8	1,564.8	302.9	250.2
	Male	866.2	1,501.8	349.0	254.0
	Total	771.9	1,564.8	302.9	265.9

Table 4. Age (years) of Gray Triggerfish collected from three artificial reefs in the northwest Gulf of Mexico. Abbreviations and sample size as in Table 2.

Site	Sex	Mean	Max	Min	SD
Inshore A	Female	2.6	5.2	0.7	1.1
	Male	2.4	4.2	1.0	0.8
	Total	2.5	5.2	0.7	1.0
Inshore B	Female	2.3	4.3	1.1	1.0
	Male	2.2	5.1	0.7	1.0
	Total	2.2	5.1	0.7	1.0
Offshore C	Female	2.8	4.2	1.1	1.1
	Male	3.0	4.2	2.1	0.8
	Total	2.9	4.2	1.1	1.0
All sites	Female	2.6	5.2	0.7	1.1
	Male	2.4	5.1	0.7	0.9
	Total	2.5	5.2	0.7	1.0

Comparisons among batch fecundity as a function of morphometric characteristics found log-transformed W with a power function to have the best fit (Figure 5, with alternative fits described in Figure 6 and Table 5). Oocyte distribution was not significantly different with regard to lobe (right or left) or lobe region (top, middle, lower, or bottom) (lobe: $F_{1, 15}$ = 1.439, region: $F_{3, 15}$ = 1.129). Age was a poor predictor of batch fecundity (data not shown). Batch fecundity ranges and summary statistics for collected females above FL_{50} and W_{50} are presented in Table 6.

Discussion

Red Snapper collected by Alexander (2015) exhibited a pattern similar to Gray Triggerfish in our study, with larger fish occurring at sites further from shore. However, unlike our study, a significant difference in the length and W of Red Snappers was observed among inshore and offshore sites. There is a significant difference in the FL and W of Gray Triggerfish, with males being larger than females. Female fish are dominant at Inshore A and males outnumber females at Inshore B. Fish from Offshore C maintain a perfect 1:1 ratio. Fish were larger at sites further from shore (i.e., Inshore A, smaller → Inshore B, intermediate → Offshore C, larger); however, this pattern may be the result of assortment by sex and not represent an increase of size with distance from shore, which has been observed for other species at these sites (Alexander 2015). Furthermore, it is unclear if the presence of larger fish occurring farther offshore is common among Gray Triggerfish populations elsewhere in the Gulf of Mexico.

Studies in the Gulf of Mexico and off Brazil have identified sexual dimorphism and differences in growth between males and females (Ingram 2001; Bernardes 2002; Burton et al. 2014; Allman et al. 2017). This is similar to the observed differences in FL and W between male and female Gray Triggerfish associated with artificial reefs in the northwest Gulf of Mexico.

Due to the lack of significant differences observed here and the impediment to identify homogeneous subsets of age among sites, the age structure of Gray Triggerfish can be considered similar across all sites in our

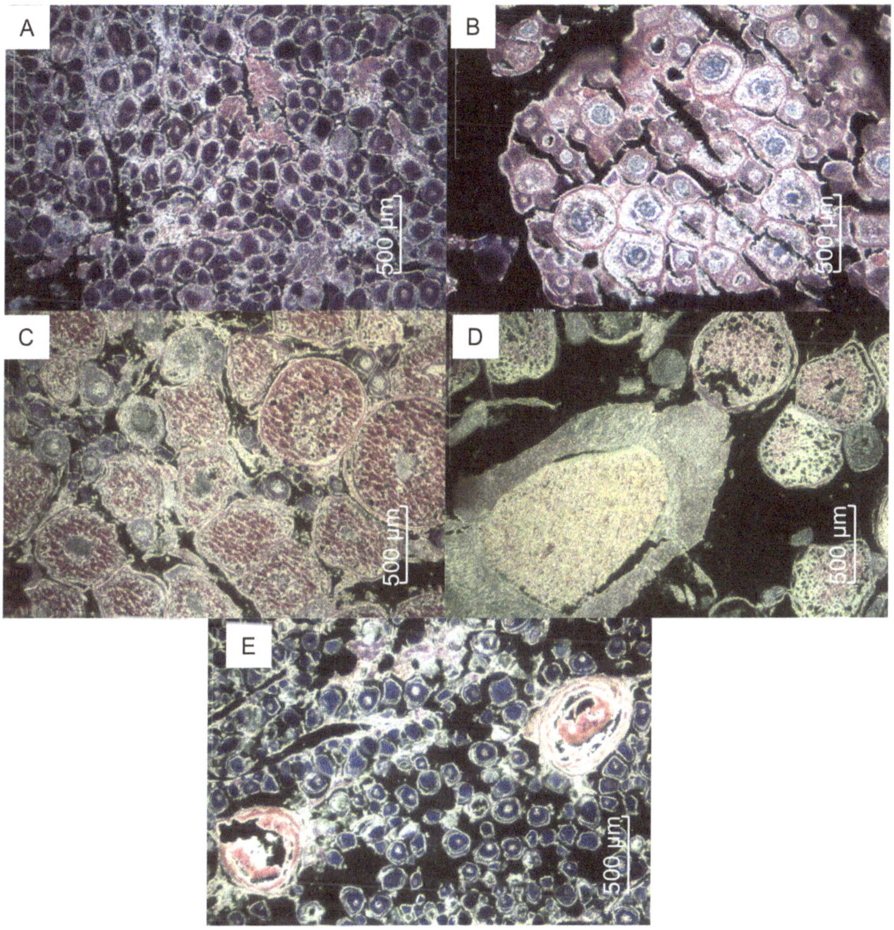

Figure 1. Ovarian stages of Gray Triggerfish collected from three artificial reefs in the northwest Gulf of Mexico. **(A)** immature or resting, **(B)** developing, **(C)** spawning capable, **(D)** active spawning, and **(E)** regressing. For descriptions of tissues, see Table 1.

study. Ages of fish sampled were generally restricted to the younger third (1–5 years) of the known age span (i.e., 0–15 years) and comprised only half of the typical ages (0–10 years) present at the artificial reefs sampled off south Texas (SEDAR 2015).

While the sex ratio was 1:1 for all the fish sampled here, apparently other unidentified factors skewed the sex ratio relative to site. Male Gray Triggerfish collect harems of up to five females. Members of each harem build a separate nest, which is well tended from fertilization until hatching that normally occurs about 48 h after fertilization (Lyczkowski-Shultz et al. 2003; MacKichan and Szedlmayer 2007; Lang and Fitzhugh 2015). Males actively guard the nest to exclude predators and competitors from the nest area. This behavior could be responsible for the apparent shift in the sex ratio towards females (observed in other studies, see SEDAR 2015), especially if some sites represent optimal nesting habitat and others do not. However, the assertion of male guarding behavior being responsible for persistently skewing the

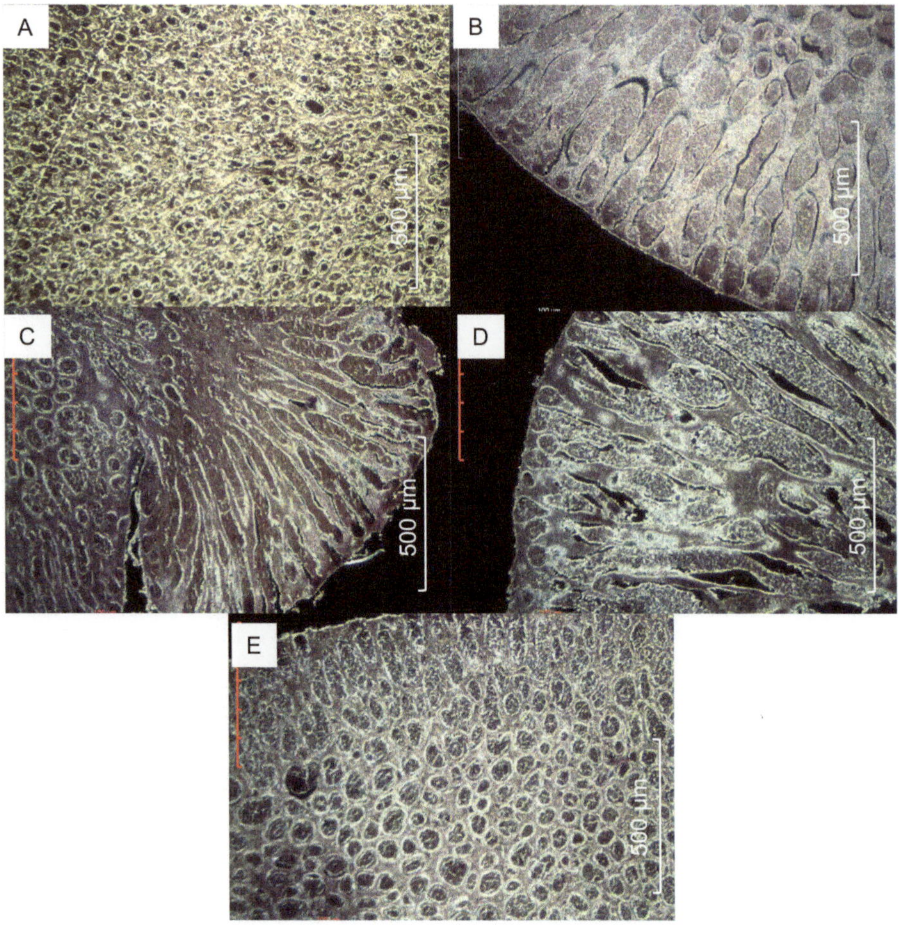

Figure 2. Testis stages of Gray Triggerfish collected from three artificial reefs in the northwest Gulf of Mexico. **(A)** immature or resting, **(B)** developing, **(C)** spawning capable, **(D)** active spawning, and **(E)** regressing. For descriptions of tissues, see Table 1.

sex ratios at sites remains speculative. An investigation into the monthly sex ratio in this study did not indicate a seasonal shift in sex ratio. Nevertheless, increases in the home range of Gray Triggerfish have been observed in months prior to and after the spawning season in the Gulf of Mexico (Herbig and Szedlmayer 2016). Home range extension could be attributed to searching for nesting habitat and mates. Dominant males could induce submissive males to move to less-suitable spawning habitat, resulting in sites dominated by a single sex (i.e., less-dominating males at some sites, few or single males and many females at harem sites). With suitable habitat at a premium, a lack of contiguous habitat (Mueller 2012), and site specificity and fidelity (Addis et al. 2007; Herbig and Szedlmayer 2016), it is feasible that aggregations of males and females may persist after the spawning season at south Texas artificial reefs. Two of three sites sampled had skewed sex ratios (Inshore A was dominated by females and Inshore B by males) while the third site (Inshore C) retained a 1:1 ratio. However, upon

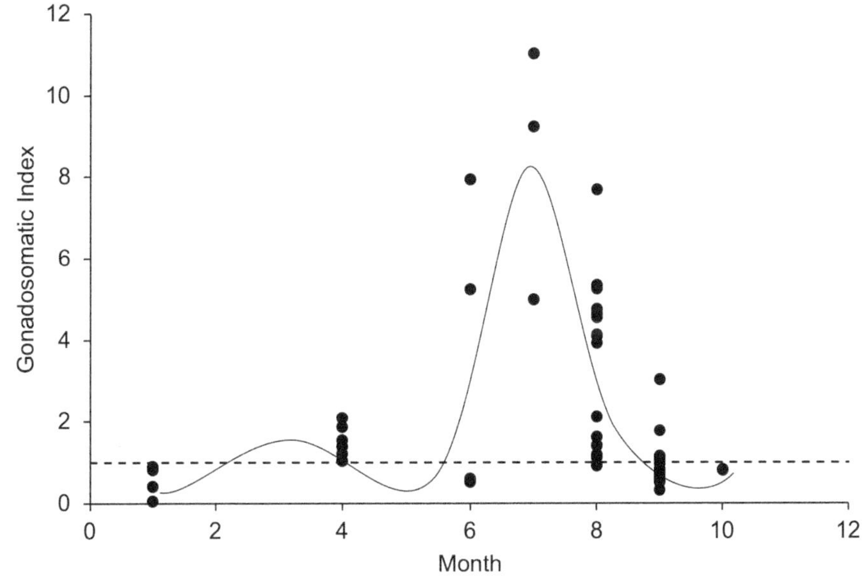

Figure 3. Gonadosomatic Index (GSI) of female Gray Triggerfish on artificial reefs in the northwest Gulf of Mexico by month captured throughout the sampled year. Solid line represents index path indicating the annual trend of the GSI value. Dashed line represents GSI = 1; individuals expressing index above 1 are considered spawn capable; $n = 60$.

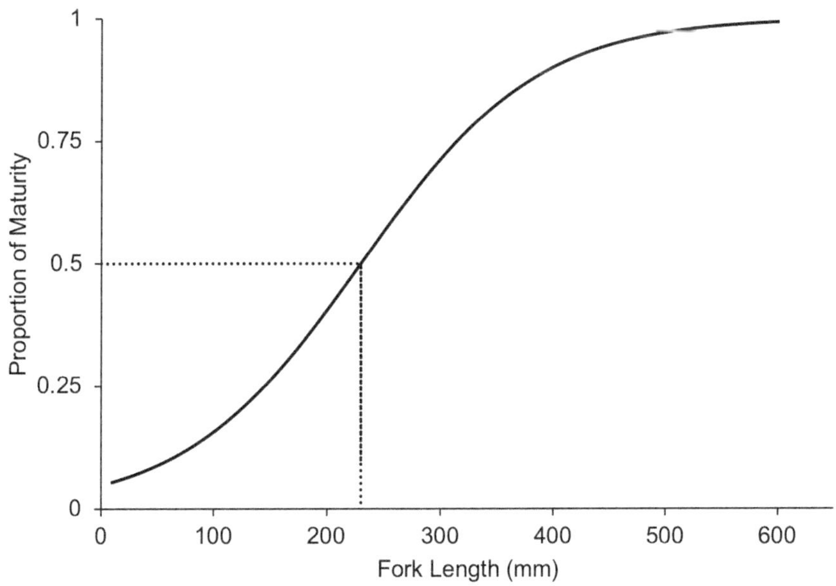

Figure 4. Proportional maturity ($n = 60$) at fork length (FL mm) for female Gray Triggerfish collected from three artificial reefs in the northwest Gulf of Mexico, where proportion mature = $1/1 + e^{(-2.999 + 0.013 * FL)}$.

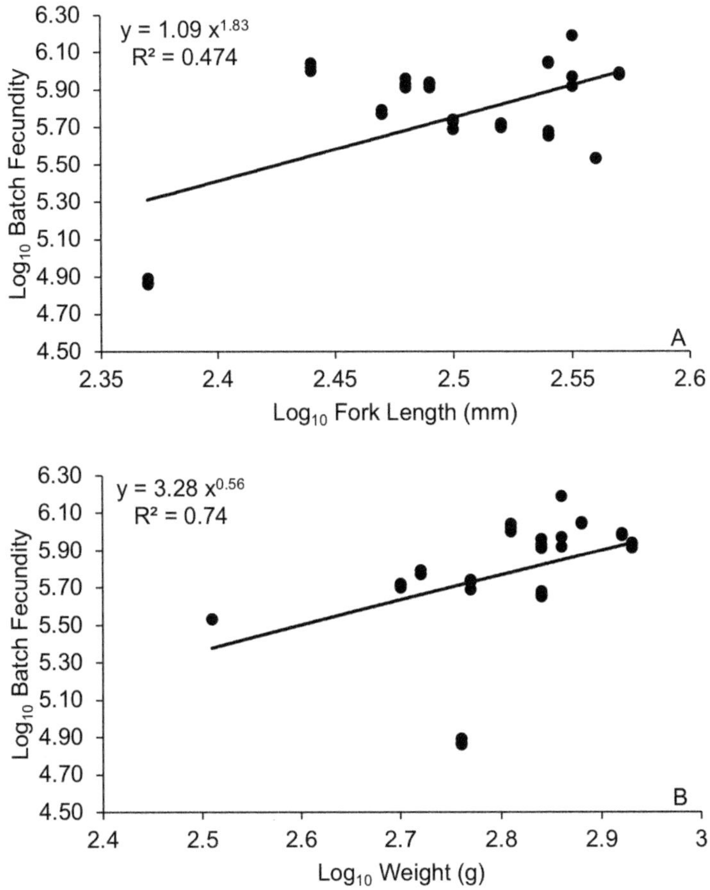

Figure 5. Log-transformed batch fecundity as a function of log-transformed **(A)** fork length (mm) and **(B)** weight (g) for Gray Triggerfish collected from three artificial reefs in the northwest Gulf of Mexico; *n* = 36.

further investigation, the number of males to females (11:19) more closely resembles a 1:2 annual average sex ratio. Further investigation is required before a persistent unequal sex ratio is adopted at these artificial reefs.

While artificial structures are occupied by Gray Triggerfish at all life stages throughout the Gulf of Mexico (MacKichan and Szedlmayer 2007; Simmons and Szedlmayer 2011; Allman et al. 2017), the identification of spawning-capable and spawning-active individuals (male and female) at all sites through histological characterization confirms that artificial reefs in south Texas are suitable habitat for Gray Triggerfish throughout their ontogeny.

Reproductive characteristics have been quantified for Gray Triggerfish on artificial reefs in the northwest Gulf of Mexico (Kjesbu 2009; Lang and Fitzhugh 2015). In this study, it was observed that female fish had a peak GSI just prior to the onset of spawning, as supported by histological observations. This July peak in the GSI of females mirrors other studies. Mean GSI for females throughout the spawning period in this study was >3, which is similar to values observed in Lang and Fitzhugh (2015).

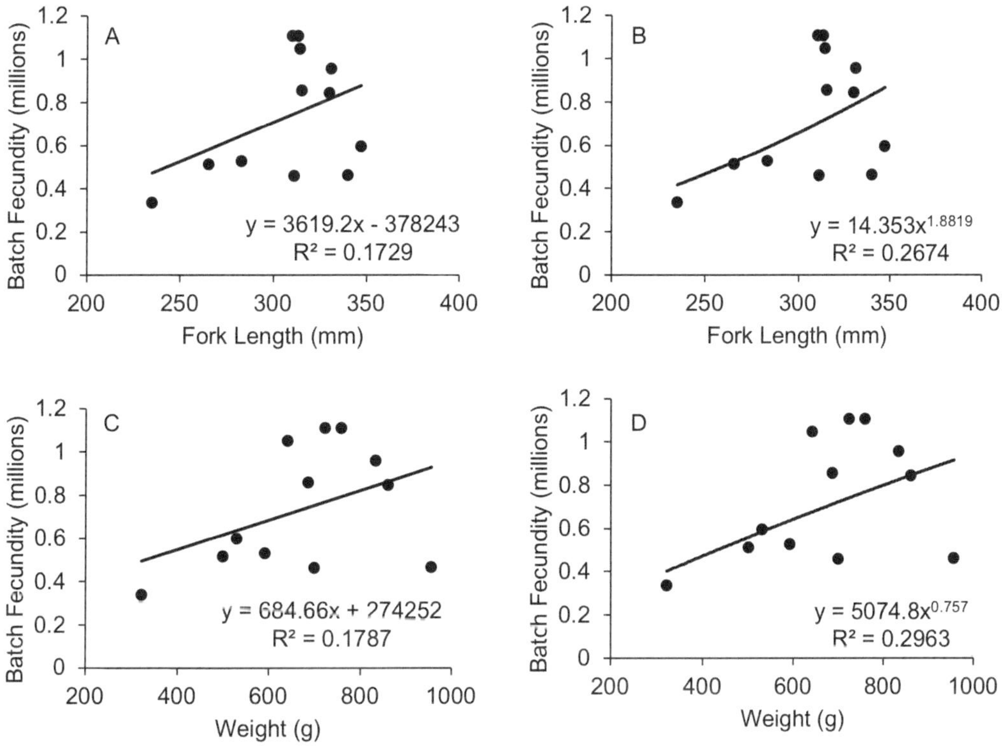

Figure 6. Batch fecundity as a function of **(A)** fork length (mm) linear fit, **(B)** fork length (mm) power fit, **(C)** weight (g) linear fit, and **(D)** weight (g) power fit for Gray Triggerfish collected from three artificial reefs in the northwest Gulf of Mexico; n = 12.

A peak GSI was also observed in male fish during July; however, the change in GSI for males is small (0.1–1.05) as compared to females (0.3–11). During July (as identified in this study and others), most females were identified as either spawning capable or actively spawning (Lang and Fitzhugh 2015).

Fork length at 50% maturity was similar for both males and females in this study (Δ = <5 mm). Kacem et al. (2015) observed that female size at maturity was about 200 mm FL and males at about 210 mm FL. Fitzhugh et al. (2015) observed 50% female maturity at 159 mm FL; this was markedly different from the size at maturity reported in other studies. Ismen et al. (2004) also calculated 50% maturity size as about 130 mm FL for both sexes. Differences in length at maturity may be attributed to a shortage of <200-mm-FL individuals in this study's samples, as well as geospatial differences in the areas studied. No females below 230 mm FL were collected in this study, while in all other studies, the sampled size range included individuals <200 mm FL. In addition, studies by Ismen et al. (2004) and Kacem et al. (2015) were performed in the Mediterranean Sea while Fitzhugh et al. (2015) and this study were performed in the northern Gulf of Mexico. Life history estimations from multiple studies suggest regional differences in the length at age among Gray Triggerfish (Allman et al. 2017).

Females used to calculate batch fecundity as a function of FL here were similar to sizes examined by Ingram (2001) and Lang and Fitzhugh (2015) (Figure 7). Batch fecundity

Table 5. Batch fecundity (BF) as a function of morphometrics from Gray Triggerfish from three studies conducted in the Gulf of Mexico.

Fork length (FL, mm)	Equation	R^2 value
This study	$BF = 14.353 \times FL^{1.88}$	0.27
	$BF = 3{,}691.2 \times FL - 378{,}243$	0.17
	$LogBF = 1.086 \times Log(FL^{1.83})$	0.47
Ingram (2001)	$BF = 13{,}573\ e^{[0.0109\ (FL)]}$	0.72
Lang and Fitzhugh (2015)	$BF = 8{,}703.69 \times FL - 1{,}776{,}482.66$	0.56

Total weight (*W*, g)	Equation	R^2 value
This study	$BF = 5{,}074 \times W^{0.757}$	0.30
	$BF = 684.66 \times W + 274{,}252$	0.18
	$LogBF = 3.285 \times Log(W^{0.56})$	0.74
Ingram (2001)	$BF = 836.415 \times W - 202{,}904$	0.83

estimates, as a function of FL, were statistically different among studies ($F_{3, 99} = 5.360$). However, post-hoc testing indicated that the power equation relationship between batch fecundity and FL from this study was not significantly different from the relationship described by Ingram (2001) and Lang and Fitzhugh (2015).

Batch fecundity as a function of *W* was significantly different among studies ($F_{2, 99} = 4.378$); post-hoc testing revealed that neither the power batch fecundity to *W* equation nor the linear batch fecundity to *W* relationship from this study was significantly different from Ingram (2001).

Longer and heavier fish produce more mature oocytes (Hixon et al. 2014). While Ingram (2001) found age to be a fair predictor of batch fecundity in Gray Triggerfish, it was a poor indicator of batch fecundity for individuals collected on the artificial reefs off the coast of south Texas. This is due in part to FL and *W* ranges, which overlapped several age bins. Additionally, individuals above 435 mm FL, 1,300 g *W*, and age 5.5 years were absent from our samples.

An absence of fish >435 mm FL and of an age >5.5 years suggests that artificial reefs off the south Texas coast either cannot support older, larger Gray Triggerfish or that an

Table 6. Batch fecundity of Gray Triggerfish: observed and calculated. Max = maximum; Min = minimum; SD = standard deviation; $n = 54$.

	Max	Min	Mean	SD
Observed[a]	1.11×10^6	3.39×10^5	7.36×10^5	2.69×10^5
Calculated[b]				
Fork length	1.31×10^6	4.06×10^5	6.77×10^5	2.27×10^4
Total weight	1.31×10^6	4.06×10^5	6.77×10^5	2.27×10^4

[a] $n = 12$.

[b] Calculated as a function of fork length (mm) or total weight (g) for fish collected from three artificial reefs in the northwest Gulf of Mexico.

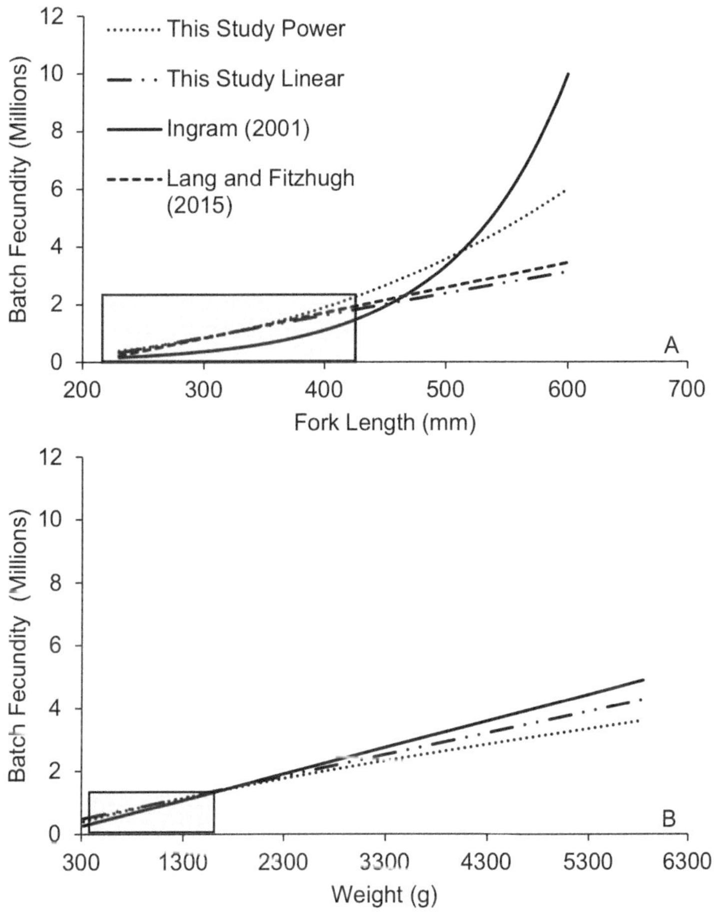

Figure 7. **(A)** Batch fecundity as a function of fork length (mm) from three studies of Gray Triggerfish in the Gulf of Mexico; gray box represents observed size range from this study. **(B)** Batch fecundity as a function of weight (g) from two studies of Gray Triggerfish in the Gulf of Mexico; gray box represents observed weight range from this study.

unknown variable is responsible for removing larger and older fish. Artificial structures and natural reefs throughout the Gulf of Mexico support populations of Gray Triggerfish with larger and older individuals (Johnson and Saloman 1984; Moore 1997; Ingram 2001; MacKichan and Szedlmayer 2007; Herbig and Szedlmayer 2016). This species exhibits a native range from Nova Scotia to Argentina, and temperate and subpolar populations are expressly exposed to a broader range of temperature than localized populations in the Gulf of Mexico. Thus, it is unlikely that environmental variables are exclusively responsible for the truncated age and size of Gray Triggerfish.

When abiotic factors are excluded, predation and fishing effort can potentially explain the lack of larger, older individuals. Large predators are common on artificial reefs in south Texas with Great Barracuda *Sphyraena barracuda* (Edwards, 1771), Greater Amberjack *Seriola dumerili* (Risso, 1810), Shark species (Shortfin Mako *Isurus oxyrinchus*, Rafinesque, 1810; Scalloped Hammerhead *Sphyrna lewini*, Griffith and Smith, 1834), and Goliath Grouper *Epinephelus itajara* (Lichtenstein, 1822) commonly recorded (Ajemian et al. 2015). Predation on Gray Triggerfish has yet to be quantified. However, it is difficult to speculate that predators observed at artificial reefs throughout the Gulf of Mexico would specifically target large (old) Gray Triggerfish in south Texas while ignoring other prey items (e.g., Red Snapper) that are in abundance.

Fishing effort could affect the size and age of Gray Triggerfish populations on artificial reefs in south Texas. The Gray Triggerfish recreational fishery in Texas State waters remains open year-round, with a bag limit of 20 individuals and a minimum size limit of 406 mm FL (TPWD 2017). Individuals <406 mm FL composed <1% of the sample. With other highly targeted reef-associated species being highly regulated, Gray Triggerfish offer an attractive fishing alternative to recreational anglers. However, a continuous loss of larger, older individuals could have significant management concerns for the Gray Triggerfish stock in the Gulf of Mexico (Hixon et al. 2014).

Genetic connectivity of Gray Triggerfish throughout their natural range suggests that the Gulf of Mexico populations can be considered as a single stock (Antoni et al. 2011; Antoni and Saillant 2017). Genetic connectivity also suggests that larval Gray Triggerfish may be transported up to 4,000 km (Antoni et al. 2011). While local retention of larval fish has been observed (Simmons and Szedlmayer 2011; Arney et al. 2017), it is difficult to assess the importance of local retention.

In the Gulf of Mexico, larval distribution roughly follows the Gulf Loop Current, which can extend either south into the Campeche Banks of Mexico or into the eastern Gulf of Mexico (Antoni et al. 2011). In either scenario, Texas waters can act as a source for the Gray Triggerfish population in the Gulf of Mexico, which is under constant fishing pressure. With the removal of larger, older individuals, it is suggested that the number of Gray Triggerfish larvae may be suboptimal, being produced by smaller adults with lower growth rates or with lower maximum size.

The U.S. federal fishery for Gray Triggerfish is in moratorium as of December 2017, and while this measure is intended to bolster the stock, the majority of artificial reefs in south Texas are in state waters (TPWD 2016) and open to fishing. While the impact of Texas fishery management of Gray Triggerfish is difficult to assess at this time, Gray Triggerfish recreational fisheries have been closed or highly regulated off all other U.S. states bordering the Gulf of Mexico with the exception of Mississippi (ADCNR 2017; FWC 2017; Gulf of Mexico Fishery Management Council 2017; LDWF 2017; MDMR 2017; TPWD 2017).

Conclusions

Gray Triggerfish on artificial reefs in the northwest Gulf of Mexico off the coast of south Texas exhibit reproductive characteristics similar to those found elsewhere in the Gulf of Mexico. The FL and *W* ranges of individuals collected were influenced by inconsistent sex ratios that varied among sample sites but were also different from the 1:1 ratios observed elsewhere (SEDAR 2015). The age (years) range of Gray Triggerfish was limited to the lower half of the commonly sampled age range reported in other studies, with individuals representing only a third of the known age range for the species (Allman et al. 2017).

Histological examination indicated a presence of all reproductive stages in both males and females, with a peak spawning season in June to August and a peak GSI in July that was similar for other Gray Triggerfish in the Gulf of Mexico. Fork length, at which 50% of individuals are mature, was about 230 mm, larger than reported by other studies in the Gulf of Mexico (e.g., Fitzhugh et al. 2015). Batch fecundity in female Gray Triggerfish, calculated as a function of both FL and *W*, separately, was not significantly different in comparison to other populations across the Gulf of Mexico (Ingram 2001; Lang and Fitzhugh 2015).

Fishing effort in Texas State waters should be further evaluated to identify the influence of present fishing practices on the population dynamics of Gray Triggerfish associated with artificial reefs in the northwest Gulf of Mexico. Notably, there was an absence of fish >435 mm FL and older than age 5.5 years sampled here. If this is a reflection of the population structure in the study area, this may have implications on the reproductive capacity of the population as older (larger) fish produce more mature oocytes than younger (smaller) fish. With the absence of older fish, stock replenishment is dependent on a smaller and less fecund proportion of the Gray Triggerfish population.

Acknowledgments

The authors thank the University of Texas Rio Grande Valley dive team, comprised of volunteers, faculty, academic staff, and graduate students, for help collecting samples and enduring rough weather conditions. The Texas Parks and Wildlife Department Artificial Reef Program provided funding for this study and granted Scientific Permit SPR-0314–050 to collect Gray Triggerfish in Texas waters. In addition, the National Oceanic and Atmospheric Administration–National Marine Fisheries Service provided an essential letter of acknowledgment, which allowed for collection outside the boundaries of fishing regulations.

References

Addis, D. T., W. F. Patterson, and M. A. Dance. 2007. Site fidelity and movement of reef fishes tagged at unreported artificial reef sites off NW Florida. Proceedings of the Gulf and Caribbean Fisheries Institute 60:297–304.

Ajemian, M. J., J. J. Wetz, B. Shipley-Lozano, J. D. Shively, and G. W. Stunz. 2015. An analysis of artificial reef fish community structure along the northwestern Gulf of Mexico shelf: potential impacts of "Rigs-to-Reefs" Programs. PLOS (Public Library of Science) ONE [online serial] 10(5):e0126354.

ADCNR (Alabama Department of Conservation and Natural Resources). 2017. Recreational size, daily creel and possession limits for state waters—revised May 2017. ADCNR, Public Summary, Montgomery.

Alexander, R. J. 2015. Comparing reproductive capacity of inshore and offshore Red Snapper, *Lutjanus campechanus*, on artificial reefs in the western Gulf of Mexico. Master's thesis. University of Texas Brownsville, Brownsville.

Allman, R. J., C. L. Fioramonti, W. F. Patterson, III, and A. E. Pacicco. 2015. Validation of annual growth zone formation in Gray Triggerfish (*Balistes capriscus*) dorsal spines, fin rays, and vertebrae. SEDAR (Southeast Data Assessment and Review), SEDAR 43-WP-01, North Charleston, South Carolina.

Allman, R. J., W. F. Patterson, C. L. Fioramonti, and A. E. Pacicco. 2017. Factors affecting estimates of size at age and growth in Gray Triggerfish *Balistes capriscus* from the northern Gulf of Mexico. Journal of Fish Biology 92:386–389.

Antoni, L., N. Emerick, and E. Saillant. 2011. Genetic variation of Gray Triggerfish in U.S. waters of the Gulf of Mexico and western Atlantic Ocean as inferred from mitochondrial DNA sequences. North American Journal of Fisheries Management 31:714–721.

Antoni, L., and E. Saillant. 2017. Spatial connectivity in an adult-sedentary reef fish with extended pelagic larval phase. Molecular Ecology 26:4955–4965.

Arney, R. N., C. Y. M. Froehlich, and R. J. Kline. 2017. Recruitment patterns of juvenile fish at an artificial reef area in the Gulf of Mexico. Marine and Coastal Fisheries: Dynamics, Management, and Ecosystem Science [online serial] 9:79–92.

Bernardes, R. A. 2002. Age, growth and longevity of the Gray Triggerfish *Balistes capriscus* (Tetradontiformes: Balistidae), from the southeastern Brazilian coast. Scientia Marina 66:167–173.

Bollinger, M. A., and R. J. Kline. 2017. Validating sidescan sonar as a fish survey tool over artificial reefs. Journal of Coastal Research 33:1397–1407.

Branco, I. S. L., D. L. Viana, R. T. S. Félix, D. P. Véras, and F. H. V. Hazin. 2013. Oocyte development and ovarian maturation of the Black Triggerfish, *Melichthys niger* (Actinopterygii: Balistidae) in São Pedro e São Paulo Archipelago, Brazil. Neotropical Ichthyology 11:597–606.

Brown-Peterson, N. J., D. M. Wyanski, F. Saborido-Rey, B. J. Macewicz, and S. K. Lowerre-Barbieri. 2011. A standardized terminology for describing reproductive development in fishes. Marine and Coastal Fisheries: Dynamics, Management, and Ecosystem Science [online serial] 3:52–70.

Burns, K. M., and J. T. Froeschke. 2012. Survival of Red Grouper (*Epinephelus morio*) and Red Snapper (*Lutjanus campechanus*) caught on J-hooks and circle hooks in the Florida recreational and recreational-for-hire fisheries. Bulletin of Marine Science 88:633–646.

Burton, M. L., J. C. Potts, and D. R. Carr. 2014. Age, growth, and mortality of Gray Triggerfish (*Balistes capriscus*) from the southeastern United States. U.S. National Marine Fisheries Service Fishery Bulletin 113:27–39.

Dickens, L. C., C. H. R. Goatley, J. K. Tanner, and D. R. Bellwood. 2011. Quantifying relative diver effects in underwater visual censuses. PLOS (Public Library of Science) ONE [online serial] 6(4):e18965.

Escorriola, J. I. 1991. Age and growth of the Gray Triggerfish, *Balistes capriscus*, from the southeastern United States. Master's thesis. University of North Carolina–Wilmington, Wilmington.

Fitzhugh, G. R., H. M. Lyon, and B. K. Barnett. 2015. Reproductive parameters of Gray Triggerfish (*Balistes capriscus*) from the Gulf of Mexico: sex ratio, maturity and spawning fraction. SEDAR (Southeast Data Assessment and Review), SEDAR 43-WP03, North Charleston, South Carolina.

Froehlich, C. Y. M., A. M. Lee, R. Oquita, C. E. Cintra-Buenrostro, J. D. Shively, and J. B. Shipley. 2018. A comparison of population dynamics from Red Snapper associated with inshore and offshore artificial reefs in the northwestern Gulf of Mexico. Pages 81–95 *in* S. A. Bortone, editor. Marine artificial reef research and development: integrating fisheries management objectives. American Fisheries Society, Symposium 86, Bethesda, Maryland.

FWC (Florida Fish and Wildlife Conservation Commission). 2017. Florida saltwater recreational fishing regulations 2017. FWC, Public Summary, Tallahassee.

Gulf of Mexico Fishery Management Council. 2014. Red Grouper recreational management measures. Gulf of Mexico Fishery Management Council, Tampa, Florida.

Gulf of Mexico Fishery Management Council. 2017. Recreational fishing regulations for the Gulf of Mexico federal waters: for species managed by the Gulf of Mexico Fishery Management Council. Gulf of Mexico Fishery Management Council, Tampa, Florida.

Herbig, J. L., and S. T. Szedlmayer. 2016. Movement patterns of Gray Triggerfish, *Balistes capriscus*, around artificial reefs in the northern Gulf of Mexico. Fisheries Management and Ecology 23:418–427.

Hixon, M., D. Johnson, and S. Sogard. 2014. BOFFFFs: on the importance of conserving old-growth age structure in fishery populations. ICES Journal of Marine Science 71:2171–2185.

Hood, P. B., and A. K. Johnson. 1997. A study of the age structure, growth, maturity schedules and fecundity of Gray Triggerfish (*Balistes capriscus*), Red Porgy (*Pagrus pagrus*), and Vermilion Snapper (*Rhomboplites aurorubens*) from the eastern Gulf of Mexico. MARFIN, Final Report FO499-95-F, St. Petersburg, Florida.

Hunter, J. R., and B. J. Macewicz. 1985. Measurement of spawning frequency in multiple spawning fishes. Pages 67–77 *in* R. Lasker, editor. An egg production method for estimating spawning biomass of pelagic fish: application to the Northern Anchovy, *Engraulis mordax*. NOAA Technical Report NMFS 36.

IBM. 2013. IBM SPSS statistics for Macintosh, version 23.0. IBM, Armonk, New York.

Ingram, G. W. 2001. Stock structure of the Gray Triggerfish, *Balistes capriscus*, on multiple spatial scales in the Gulf of Mexico. Doctoral dissertation. University of South Alabama, Mobile.

Ismen, A., M. Turkoglu, and C. C. Yigin. 2004. The age, growth and reproduction of Gray Triggerfish (*Balistes capriscus*, Gmelin 1789) in Skenderun Bay. Pakistan Journal of Biological Science 7:2135–2138.

Johnson, A. G., and C. H. Saloman. 1984. Age, growth, and mortality of the Gray Triggerfish, *Balistes capriscus*, from the northeastern Gulf of Mexico. U.S. National Marine Fisheries Service Fishery Bulletin 113:485–492.

Kacem, H., L. Boudaya, and L. Neifar. 2015. Gonadal maturation and histological observations of the Grey Triggerfish *Balistes capriscus* Gmelin, 1789 (Teleostei: Balistidae) in the Gulf of Gabes, Tunisia. Journal of Coastal Life Medicine 3:341–347.

Kjesbu, O. S. 2009. Applied fisheries reproductive biology: contribution of individual reproductive potential to recruitment and fisheries management. Pages 321–366 *in* T. Jakobsen, M. J. Fogarty, B. A. Megrey, and E. Moksness, editors. Fish reproductive biology: implications for assessment and management, 2nd edition. Wiley, Oxford, UK.

Lang, T. E., and G. R. Fitzhugh. 2015. Oogenesis and fecundity type of Gray Triggerfish in the Gulf of Mexico. Marine and Coastal Fisheries: Dynamics, Management, and Ecosystem Science [online serial] 7:338–348.

LDWF (Louisiana Department of Wildlife and Fisheries). 2017. Louisiana 2017 fishing regulations. LDWF, Baton Rouge.

Lyczkowski-Shultz, J., D. S. Hanisko, and G. A. Zapfe. 2003. Review of the early life history of Gray Triggerfish, *Balistes capriscus*, with a summary of data from SEAMAP plankton surveys in the Gulf of Mexico: 1982, 1984–2002. SEDAR (Southeast Data Assessment and Review), SEDAR 9-DW25, New Orleans, Louisiana.

MacKichan, C. A., and S. T. Szedlmayer. 2007. Reproductive behavior of the Gray Triggerfish, *Balistes capriscus*, in the northeastern Gulf of Mexico. Proceedings of the Gulf and Caribbean Fisheries Institute 59:231–235.

MDMR (Mississippi Department of Marine Resources). 2017. Guide to Mississippi saltwater fishing, rules and regulations. MDMR, Biloxi.

Moore, J. L. 1997. Age, growth and reproductive biology of the Gray Triggerfish (*Balistes capriscus*) from the southeastern United States. Master's thesis. College of Charleston, Charleston, South Carolina.

Mueller, M. 2012. Artificial structure and hardbottom spatial coverage in the Gulf of Mexico. SEDAR (Southeast Data Assessment and Review), SEDAR 31-DW29, North Charleston, South Carolina.

Nelson, B. D., and S. A. Bortone. 1996. Feeding guilds among artificial reef fishes in the northern Gulf of Mexico. Gulf of Mexico Science 14(2):66–80.

Randall, J. E. 1967. Food habitats of the West Indies. Studies in Tropical Oceanography 5:655–847.

SEAMAP (Southeast Monitoring and Assessment Program). 2012. Southeast Monitoring and Assessment Program vertical line survey protocol. Gulf States Marine Fisheries Commission, Ocean Springs, Mississippi.

SEDAR (Southeast Data Assessment and Review). 2015. Stock assessment for Gulf of Mexico Gray Triggerfish. SEDAR, SEDAR 43, Technical Report, North Charleston, South Carolina.

Simmons, C. M. 2001. Gray Triggerfish, *Balistes capriscus*, reproductive behavior, early life history, and competitive interactions between Red Snapper, *Lutjanus campechanus*, in the northern Gulf of Mexico. Doctoral dissertation. Auburn University, Auburn, Alabama.

Simmons, C. M., and S. T. Szedlmayer. 2011. Recruitment of age-0 Gray Triggerfish to benthic structured habitat in the northern Gulf of Mexico. Transactions of the American Fisheries Society 140:14–20.

Smith, M. W., D. Goethel, A. Rios, and J. Isley. 2015. Standardized catch rate indices for Gulf of Mexico Gray Triggerfish (*Balistes capriscus*) landed during 1986–2013 by the headboat fishery. SEDAR (Southeast Data Assessment and Review), SEDAR 43-WP-06, North Charleston, South Carolina.

Sokal, R. R., and F. J. Rohlf. 2012. Biometry, 4th edition. Freeman, London.

TPWD (Texas Parks and Wildlife Department). 2016. Texas artificial reef program material summary, cycle 21. TPWD, Dickson.

TPWD (Texas Parks and Wildlife Department). 2017. Texas parks and wildlife: outdoor annual, hunting and fishing regulations. TPWD, Austin.

Wilson, C. A., D. L. Nieland, and A. L. Stanley. 1995. Age, growth, and reproductive biology of Gray Triggerfish (*Balistes capriscus*) from the northern Gulf of Mexico commercial harvest. MARFIN, Final Report 8, St. Petersburg, Florida.

American Fisheries Society Symposium 86:117–140, 2018

Reef Fish Depend on Corals and Natural Outcrops: Are Predation and Migration Linked to the Fractal Characteristics of Habitats?

John F. Caddy*
Via Cervialto 3, Aprilia, Latina 04011, Italy

Abstract.—This paper focuses on interactions between reefs, fisheries, and their management and discusses the relationship between reef fish and their habitat. Hard bottom is much less extensive than fine sediments on most continental shelves but supports epifauna/flora as cover for reef fish and serves as habitat for their prey species. Corals, epifauna, and macroflora on natural reefs function as feeding, shelter, and aggregation sites and often have fractal characteristics. Size frequencies of catches from unfished banks may reflect the aggregate of crevice size spectra present, and this influences growth and natural mortality rates at size. It is suggested that migration is initiated when juvenile fish outgrow cover in nursery areas, which only offered predator protection at smaller sizes. Juveniles are then forced to move further offshore where cover for protecting larger reef fishes is often more available. Large cavities are rare on most natural outcrops or reefs, but these attract large fish and artificial reefs are often designed to have large cavities needed by mature reef fishes. Once artificial reefs are located by fishermen, large, resident fishes are readily captured. Unless some offshore reefs or refugia are protected, spawning potential could be affected.

Although natural mortality (M_t) declines with age for many free-living species (see Caddy 1991), information on reef-associated species suggests that natural mortality rises when potential bottlenecks occur; thus, M at size or age can increase for older individuals due to a shortage of suitable cover. Bottlenecks in recruitment may be due to a limited size-specific holding capacity of the present habitat or excessive distances of migration to habitats needed to survive to the successive life stage. Anthropogenic impacts on epifauna could also reduce holding capacity of habitats by eutrophication, hypoxia, or impacts from excessive use of bottom gear. A conceptual model suggests how increments in habitat complexity and proximity of habitats between sequential life history stages may reduce population bottlenecks and improve productivity of reef resources. It also suggests that the trophic development of reef ecosystems was linked to their fractal surface characteristics.

* Corresponding author: jfcaddy1@yahoo.co.uk

Introduction

The idea that artificial structures have a primary role in protecting reef-dependent fishes from predation seems commonly accepted, as shown by experimental studies and observations. Some conclusions drawn by experts are that predation success is inversely related to the physical complexity of an experimental habitat (Coen et al. 1981) and that fish abundance is correlated with substrate complexity (Carpenter et al. 1981). Beck 1995 noted that habitat structure can facilitate species persistence while the lack of availability of refuges may control population size.

Life-stage-specific bottlenecks in habitats have been described for many marine species (Table 1), and one principle when studying life histories of reef fishes is to avoid the dynamic pool assumption, namely that the risks their habitat exposes them to are uniform throughout the population range and life history. Various studies illustrate that

Table 1. Some habitat-mediated bottlenecks from the literature (after Caddy 2007).

Species/area	Bottleneck description	Age/life history stage	Author
American lobster *Homarus americanus*	Rare cobble substrate only 11% of the shoreline)	Cryptic stage between 5 and 40 mm carapace length	Wahle and Steneck 1991
Caribbean spiny lobster *Panulirus argus*	Turtle grass and coralline algae protect from snappers, moray eels, octopus, Nurse Sharks *Ginglymostoma cirratum*, etc.	Juvenile move from coralline algae to sea grass with increasing size. Plant cover is an ineffective protection over 55–75 mm carapace length.	Lipcius et al. 1998
Florida stone crab *Menippe mercenaria*	Size of available rocks determines crab size frequency	In bays with small rocks, crabs are stunted	Beck 1997
Atlantic Cod *Gadus morhua*	Prefers presence of sea grass, cobbles, and rocky reefs	Predation on 0+ juveniles is lower in these areas	Tupper and Boutlier 1995
Pomacentridae	Spawning and/or planktonic processes determine at least the early stages of recruitment events	Very early recruits of small reef fish may not be shelter-limited, but shelter limitation may increase in importance for larger sizes	Milicich et al. 1992
Coral reef fish	Recruitment of coral reef fish higher in grass beds near empty conch shells.	Juvenile coral reef fish may settle in sea grass beds and seek shelter in empty conch shells.	Shulman 1985
Iceland scallops	Juveniles often found inside dead shells of a large clam, *Spisula polynyma*	Juveniles	Arsenault and Himmelman 1998

species inhabiting or closely associated with reefs (e.g., Walters and Juanes 1993; Lindberg et al. 2006) depend on reefs for protection from predators and are obliged to concentrate most daytime feeding activities in the foraging arenas surrounding reefs or epifaunal/epifloral patches. Since most patches of hard bottom habitat are separated by expanses of flat sediment, migration between hard-bottom habitats suitable for successive life history stages of fish is inevitable but makes them predator-vulnerable during the migration. Turgeon et al. (2010) observed that reef fishes behave in relation to cover by using "functional connectivity," notably by making long detours to avoid swimming over open-sand bottom. This behavior suggests that migrations may not be self-initiated but may be sequels to unsuccessful predator attacks. This raises the possibility that bottlenecks affecting abundance later in life are related to increased vulnerability to predators. This could result if a critical habitat is reduced in size, if it is inefficient in protecting larger organisms, or if habitats for sequential life history stages are widely separated. A related observation is that a restriction in the foraging range is linked to stunting in growth in some reef populations (e.g., Lindberg et al. 2006). This presumably is because available food resources near limited cover patches are inadequate. Shortage of food, however, does not seem an adequate stimulus for migration, which is primarily a reaction to increased vulnerability to predation due to a shortage of cover.

The Refugium Concept

During a Canadian–U.S. workshop on lobster assessment (Anthony and Caddy 1986), the concept of an ecological refugium (from Latin: refugere—to escape) was found helpful. The *Webster's New Collegiate Dictionary* definition is "A refuge: a shelter or protection from danger or distress." This definition does not address the issue that a refugium may also be contributed to by other than spatial factors. A working definition of a refugium for fishery applications was proposed:

> A spatial or temporal enclave within the normal range of distribution of a species (elsewhere subject to potentially catastrophic mortality prior to maturity), or a characteristic of the scheduling of harvesting mortality, which allow a proportion of the stock to survive to maturity and reproduce the population as a whole.

In other words, the refugium could entail a combination of factors: a protected area, a seasonal closure, and/or a selective characteristic of the fishing gear. An example for offshore Canadian lobsters is a restricted size of trap opening or a regulation requiring the return of berried lobsters to the sea after tail clipping, contributing to a spawning refugium, which conserves large female lobsters. A perception of the importance of the refugium concept for fisheries management has been growing in recent years, with positive opinions expressed on its value as a safety margin or alternative to conventional management measures (e.g., Orensanz and Jamieson 1998; Caddy 2014a). In the Mediterranean, offshore refugia are beginning to be seen as logical counterparts to inshore nursery areas given that many inshore areas are predominantly inhabited by juveniles. Conserving nursery areas in some Mediterranean coastal waters has required a ban on bottom gear in shallow water and over sea grass beds. The impact of towed bottom gear on rocky reefs has been extensively documented (e.g., see Rooper et al. 2016), and as suspected, many reef areas are festooned by lost gear, which leads to ghost fishing (Macfadyen et al. 2009).

Fractal Theory and Habitats

Williamson and Lawton (1991) observed that much of nature looks like artificial fractal curves. Heck and Crowder (1991) saw the importance of the structural complexity of habitats, and Gee and Warwick (1994) observed that the fractal dimension of a habitat is a numerical expression of its complexity. They added that fractal characteristics are easy to calculate, independent of the type of habitat, and are related to the size hierarchies of animals occupying it. In Caddy and Stamatopoulos (1990), fractal theory (Figure 1, left) was applied to plot the number of holes at size predicted to occur on a surface of a given fractal dimension to explore how this would influence the natural mortality at age of cover-dependent species, assuming that the potential for occupancy by organisms that just fit within crevices was a requirement for survival (see Sale 1978). Thus, according to Hixon and Beets (1993), prey species should be most abundant on those reefs providing hole sizes near their body diameter.

Morse et al. (1985) showed that the number of cover units on a fractal surface is approximated by

$$N_L = k/L^{D+1}, \tag{1}$$

where N_L is the number of cover units at size L found in the surface, D is its fractal dimension, and k is a constant. Investigating a specific relationship between the geometry of a coral reef and predation probability led to a simulation of a fractal environment using a ratio based on this relationship, namely

$$N_{L2}/N_{L1} = L_1^{D+1}/L_2^{D+1}, \tag{2}$$

where N_{L1} and N_{L2} are the number of crevices of dimension L_1 and larger L_2 in a surface with a linear fractal coefficient D. Since factors other than fractality could lead to declines in numbers at size, $D + 1$ was substituted for by a slope parameter Q, which may be larger than a fractal dimension.

As applied to analyzing size-frequency data, the relationship used was

$$N_{L2}/N_{L1} = W_1^Q/W_2^Q, \tag{3}$$

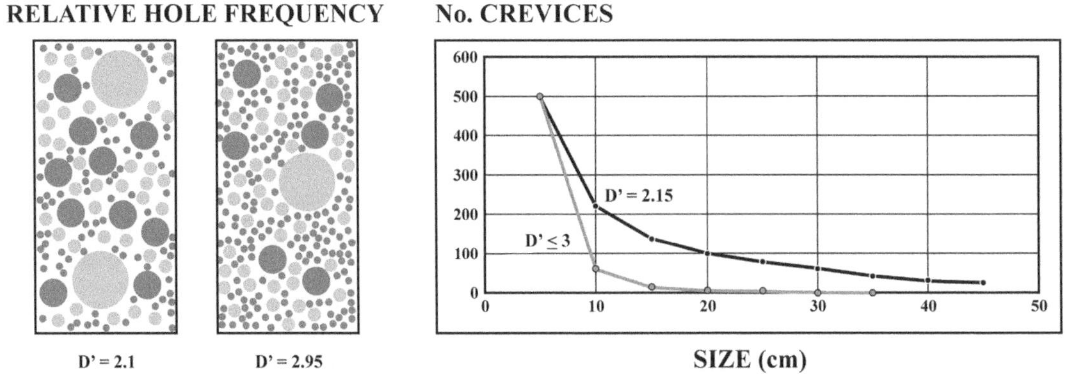

Figure 1. (Left): Illustrating the frequency distribution of hole diameters on surfaces corresponding to surface fractal coefficients D' = 2.1 and 2.95. (Right): the number of crevices with a high fractal coefficient tends to have many semi-enclosed cavities, with large cavities greatly outnumbered by small ones. (The number of cavities at size on a fractal surface must determine the limiting mortality rate at size, as shown in Figure 8, if a net migration is not involved).

where N_{L1} and N_{L2} are the number of crevices of size W_1 and larger W_2, within a surface having a surface fractal coefficient $D' = D + 1$, which was measured by the value of the parameter Q fitted to the observed data. It was assumed that all crevices were occupied by crevice-dwelling fish, thus also providing limiting natural mortalities at size, assuming that all fish excluded from crevices were predated.

Although Beck (1995) showed that stone crabs have distinct preferences for size of residence, seeking field evidence that a fractal surface selects for the size frequency of fish is problematic since other factors, such as fishing mortality and migration, affect the numbers present at size. However, an earlier study (Caddy 2011) focused on reanalysis of the invaluable surveys by Munro (1983) on then unfished reefs on Pedro Bank, Jamaica. Re-examining these data suggested that the fractal characteristics of the natural reef may be determined from the size frequency of resident species, and the evidence for this is revisited later in the chapter.

The life history of reef fishes often involves discontinuities between successive stages in the life history, when fishes are obliged to change their habitat preferences. This may occur once they outgrow their current habitat and it ceases to protect them from predators. The limited range of fish sizes protected by a given complex structure starts to apply early in life, but with growth in size, inshore cover provided to juveniles by algae, mangroves, or sponge colonies ceases to be effective for predator avoidance. Evidently, relationships between essential fish habitats such as nurseries and spawning areas cannot be investigated using survey strategies and conventional models based on biomass, which usually assume a statistical or dynamic pool configuration for fish in their overall habitat. Using standard stock assessment approaches to assess cover-dependent species often does not reflect the available biological information (e.g., Figure 2).

To illustrate the role of cover for reef fishes, Figure 2 details the likely life history events not shown in a conventional spawner–recruit relationship. The sequence of events begins with eggs and larvae but may lead to prerecruits being located in nursery areas having a particular structural complexity (e.g., Gotceitas and Colgan 1989; Hixon and Beets 1993). Migration may then ensue in one or more stages to eventually attain the habitat of older or mature fish (Figure 2). It is also necessary to discuss how the predation rate, and hence the natural mortality rate, changes with age and how this rate is dependent on the structural complexity of the habitat. Although this approach requires different data or assumptions than spawner–recruit relationships, it is important at least to outline the necessary framework.

The discovery by Mandelbrot (1983) of a geometrical category of natural structures referred to as fractal was followed by a rapid proliferation of investigations into their occurrence in natural habitats by marine and terrestrial ecologists (see Table 1). These studies revealed that many geological structures, plants, and colonial animals in aquatic ecosystems show broad similarities to the predictions of fractal theory, notably a broad self-similarity at different scales. Thus, while some natural environments may not be strictly fractal in the sense of not being mathematical identical over the whole size range of the habitat, the topography of complex natural surfaces is often adequately described by the fractal concept (Figure 3). Hacker and Steneck (1990) and Gee and Warwick (1994) asserted that faunal abundance at size often reflects the fractal geometry of habitats.

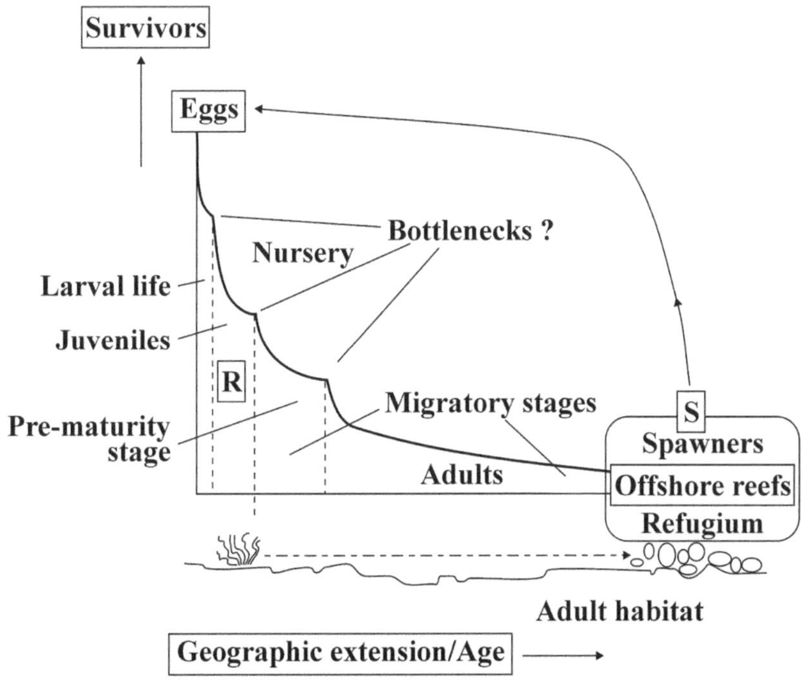

Figure 2. Modified from Caddy (2014b) to illustrate how the boundaries between successive stages in the life history of many reef fishes may be at points in time/space when the reduced cover efficiency of local habitats lead to increases in the overall natural mortality rate. Bottlenecks caused by a limited supply of habitat and subsequent migrations between different habitats must cause peaks in the predation rate. (These factors cannot be identified from a spawner–recruit relationship).

		DIMENSION:
Transect/cross section of irregular surface		>1 but < 2
Plane surface	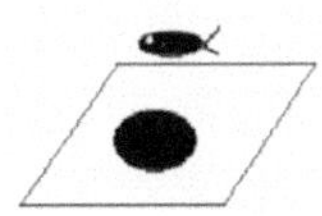	2
Convoluted surface		> 2 but < 3
Volume		3

Figure 3. Illustrating the fractal dimensions of lines, surfaces, and semi-enclosed volumes in the marine environment (modified from Caddy 2007). (In the top figure, suitable cover units at size are illustrated by gray balls).

In the following discussion, the emphasis is not on how specific fractal coefficients encountered vary between habitat types, but, as indicated by Caddy and Stamatopoulos (1990), that a common property of most natural surfaces is their scale independence. One consequence of scale independence is that the abundance of crevices or gaps in the surface falls off dramatically with increasing size. Consequently, the graph of cover abundance at size (Figure 1) should broadly resemble, or limit, the abundance at size of fish dependent on shelter there during a stage in their life history.

Field studies have suggested that many complex biological and mineral structures are fractal in configuration (see Table 2), and marine motile organisms often use fractal habitats for cover in their early life histories. A corollary to this is that migration to adult habitats becomes inevitable as they outgrow available small-scale cover and become progressively more visible to predators (i.e., specific types of cover decline in effectiveness with increases in fish size). Field data for older reef fishes suggest that higher natural mortality rates then occur (see Table 3), even in the absence of fishing (see data in Munro 1983 and the natural mortality rates for adults he estimated in Table 3). The extreme consequence of adapting to fractal surfaces is illustrated in Figure 4 for crevice-dependent species of reef fishes. These show stunted growth with abnormally high values of von Bertalanffy K and low values of L_∞ (e.g., Choat and Robertson 2002). This phenomenon is compatible with their evolution on fractal surfaces, as discussed above. This feature of von Bertalanffy parameters of small fishes, especially in the families Gobiidae, Blennidae, Apogonidae, and Pomacentridae, is discussed in Caddy (2013).

Biologically meaningful time intervals in a life history are not necessarily annual, especially if habitat changes with life history stage. Gillanders et al. (2003) demonstrated that many motile marine organisms change their preference for specific habitat types at least once during their life history. The juvenile stages of marine fish and invertebrates often occupy spatially restricted habitats early in life, expanding into larger bathymetric areas later in ontogeny, within which a preferred habitat may be sparsely distributed. Thus, the resulting spawning stock size for most mature reef fishes depends on the spatial extent of the habitat

Table 2. A few examples of physical and biological environments that broadly correspond to fractal theory (modified after Caddy 2007).

Environment	Range of fractal coefficients	Mean value	Author
Mussel bed surface	1.03–1.31	1.25	Commito and Rusingnuolo 2000
Red and chlorophyte algae	1.0–1.59	1.3	Gee and Warwick 1994
Forest patches	1.2–1.52	1.36	Krummel et al. 1987 (in Sugihara and May 1990)
Twigs from bushes (Ivy, yew, birch)	1.39, 1.47, 1.4	1.42	Burrough 1981
Terrestrial vegetation		1.6	Burrough 1981
Dendritic corals	1.5–1.72	1.68	Nakamura (1988).
Caves (surface measure)	2.04–3.0	95% of 3.0	Tercafs (1997)

Table 3. Munro's estimates of natural mortality rate (*M*) by species.

No.	Species	*M*
1	Red Hind *Epinephalus guttatus*	0.68
2	Coney *Cephalopholis fulva*	1.95
3	Graysby *Cephalopholis cruentata* (also known as *Petrometopon cruentatum*)	1.6
4	Bar Jack *Caranx ruber*	1.51
5	Blackfin Snapper *Lutjanus buccanella* (males)	2.24
6	Black Snapper *Apsilus dentatus*	1.9
7	Yellowtail Snapper *Ocyurus chrysurus*	0.6
8	White Grunt *Haemulon plumierii*	1.77
10	Margate *Haemulon album*	0.79
15	Queen Triggerfish *Balistes vetula*	2.6

most suitable for mature stages in the life history.

Structurally complex habitats are fragile and vulnerable to destruction, as occurs when epifauna is subject to fishing with bottom gear (e.g., Caddy and Seijo 2013) or as a result of natural or anthropogenic effects other than fishing, such as smothering by the fine sediments associated with river runoff documented in hypoxic environments in the Mediterranean Sea and Gulf of Mexico (Rabalais and Turner 2001). Specialized literature for many species shows that different life history stages can be distinguished by characteristics such as habitat type, depth range and food preferences, schooling, or solitary behavior. Eggleston et al. (1997) and Frazer and Lindberg (1984) noticed that growth of predators is linked to shelter-size effects and that refuge spacing similarly affected species from three phyla.

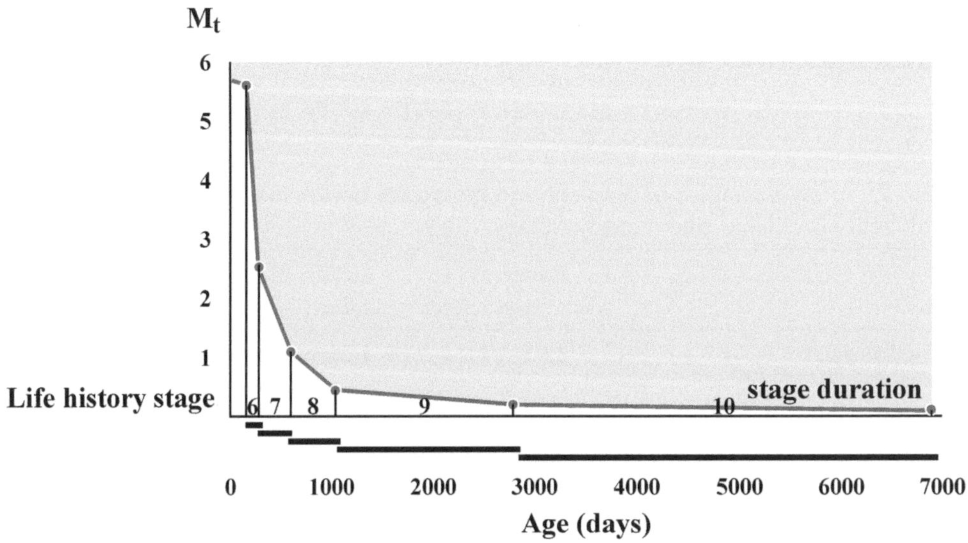

Figure 4. Reconstruction of natural mortality rates in gnomonic life history stages (data from Giménez-Hurtado et al. [2005] for Red Grouper): assuming here a constant natural mortality (*M*) for fish within each progressively longer gnomonic interval.

Natural Mortality Rates at Age (M_t), Life History Stages, and the Theory of Gnomonic Intervals

From limited data available on the natural mortality rate at age, the risk of predation for free-living fishes appears to fall from high levels for juveniles to a close to constant level as maturity is approached. This approach yielded reasonable mortality vectors for free-living species when the reciprocal model $M_t = (A + B)/t$ was fitted to juveniles of North Sea gadoids, where A and B are constants (Caddy 1991). This simple model provided adequate vectors of M at age (Abella et al. 1997) that were used for assessing harvests of juvenile gadoids in the Mediterranean Sea (Caddy 2018), where constant M-at-age assumptions are clearly inappropriate in assessments due to fishermen targeting juvenile fish. A question debated in the early literature on population dynamics must be raised, however, notably whether M at age continues to decline with age for older reef fishes.

In seeking to define so-called gnomonic intervals, the focus was on identifying temporal intervals in the life history, within all of which a more or less constant risk of predation occurs (Caddy 1996). Defining gnomonic time intervals would allow modeling of a progressive decline in natural mortality rates of highly fecund marine organisms throughout life (e.g., Figure 4), given that the although the net probability of predation in each interval Δ_i is the same, the intervals are progressively longer (i.e., a reasonable assumption for juveniles is that the mortality rate per unit time declines such that $G = M_i\Delta_i$ = k (i.e., roughly constant) for all gnomonic intervals in a life history.

The duration of the life span is modeled starting with a short "seed" interval a, which could be the egg and larval duration or the time to hatching. This is multiplied by a constant b to obtain the second gnomonic interval Δ_2, which is added to a. Subsequent intervals are generated by again multiplying b by the time elapsed from birth up to the end of the last interval. Consequently, the mean lifespan $T_n = a + \sum\Delta_i$ (from $i = 2 \rightarrow n$ stages) is made up of these summed intervals (Figure 4). If information is available on the number n of stages in the life history and the mean lifespan T, reasonable approximations for Δ_i can be found by iteration. This approach has been used to model life histories of short-lived invertebrates such as loliginid squids (e.g., Martínez-Aguilar et al. 2010), penaeid shrimps (Ramírez-Rodríguez and Arreguín-Sánchez 2003), sardines (Martínez-Aguilar et al. 2005), and multiannual life spans in species such as Red Grouper *Epinephelus morio* (Giménez-Hurtado et al. 2008). Gnomonic intervals can also be approximated if at least two time periods in a life history have been estimated and the number of life history stages are known.

The gnomonic strategy of life history division corresponds to an increase in stage duration with age, and similarly, an empirical relationship was found to provide reasonable estimates when modeling the mean time elapsed between molts of macrocrustaceans (Caddy 2003). An analogous approach was used for modeling life history stages of marine fishes under natural conditions; the assumption was then that the risk of predation was close to constant within each of these progressively longer time intervals. When Giménez-Hurtado et al. (2005) used this approach for Red Grouper on Campeche Bank to obtain estimates of M_i by life history stage i, the results of the gnomonic assumption were then compared against independent information on the duration of life history stages. Gnomonic intervals were found to be similar to those reported independently for life history stages in the literature. The gnomonic interval method methodology of the last-cited authors was used to estimate a vector of M at age from the duration of development stages and the mean annual fecundity and was calibrated using independent data from Colin et al. (1996).

A Return to Munro's Pedro Bank data

The method used to estimate surface fractal coefficients (Caddy 2011) from reef fish sample data from Munro (1983) was to plot the log frequencies of resident fish sampled against size. Plots of the aggregated slopes of the logarithm-transformed abundance at size for 15 species of reef fishes analyzed in Caddy (2011) show common features: two types of straight-line segments with similar slopes appear. These are mostly separated by what is referred to as a bottleneck, where the slope increases dramatically after a given size (Figure 5; Table 4). The similar slopes for species with different growth and natural mortality rates are strongly suggestive of an underlying feature of their common environment and/or behavior. Straight-line components were used to generate predicted lines at size for different components of the log size frequency, giving different values for parameter Q. For smaller fishes, the values of Q mostly ranged between 2.5 and 3. These values correspond well to physical measurements of the fractality of reef surfaces made by Purkis and Kohler (2008), who found a fractal dimension for reef surfaces closely approaching 3. They had suggested that similar values should be found in the log frequencies of resident fish populations. Using Munro's data, log-linear values of Q declined more rapidly with size for larger fish, implying that either migration, predation, or both were operational. Obviously, slopes exceeding 3 do not represent dimensionality.

Caddy (2011) showed plots of log numbers of 15 species sampled by Munro (1983), illustrating that two rates of decline in num-

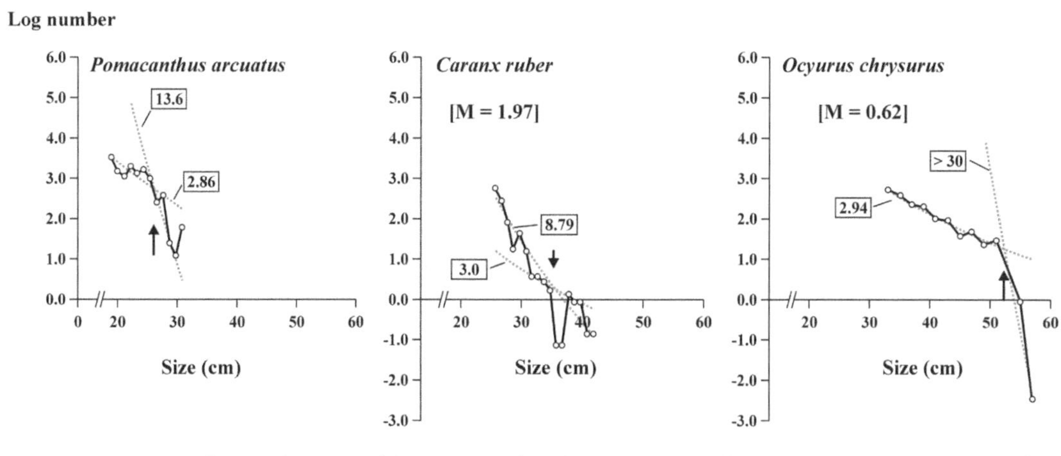

Figure 5. Redrawn from Caddy (2011) for three species showing trends in values of Q (dotted lines) superimposed on log values of Munro's log-frequency data (empty circles). (The vertical arrow is where two predicted lines appear to intersect—these are called "bottlenecks" in the text).

Table 4. Summary for nine species from Munro (1983) showing values of the parameter Q for two sections of their log size frequencies. An estimate of the intersection between size ranges where different values of Q seem to apply is given, implying perhaps that a bottleneck is in operation there? The maximum body depth at this size was measured from the FAO species identification sheets for Caribbean species (Carpenter 2002, Figure 8).

Species	First linear component of *Q*	Bottleneck fork length (cm)	Bottleneck max. body height (cm)	Second linear component of *Q*
Bar Jack *Caranx ruber*	3	30	9.1	7
Blackfin Snapper *Lutjanus buccanella*	1–2	41	13.4	15
Black Snapper *Apsilus dentatus*	3	51	16.7	30
Yellowtail Snapper *Ocyurus chrysurus*	3	52	15.8	>30
White Grunt *Haemulon plumierii*	1–2	31	14.0	>30
Cottonwick *Haemulon melanurum*	6	25	15.4	>30
Gray Angelfish *Pomacanthus arcuatus*	3	27	16.7	15
Acanthurus bahianus	1–2	19	7.8	24
Coney *Cephalopholis fulva*	5	28	8.2	>30

bers predominate: a moderate rate of decline common for many smaller fish and a much steeper rate of decline for older fish of most species, which did not reflect the limiting size of the opening of the fish trap used.

For reef-dwelling fish greater than 8–17 cm in total length on Pedro Bank (roughly 7.5–15.25 cm [3 6 in] body depth, as estimated by the method illustrated in Figure 6), there was often an abrupt changeover from a low rate of decline in numbers at size (perhaps due a decline in available crevice sizes) to a much steeper slope on the log frequency curves for larger fish (Caddy 2011). This leads to the hypothesis that crevices wider or deeper than 3–6 in (7.6–15.2 cm) were uncommon on the reef, with increasing risk of predation on larger fish (Figure 7), just as was predicted from fractal considerations (Figure 8). This last feature requires direct confirmation of reef configuration, but if Pedro Bank reefs resemble those studied by Purkis and Kohler (2008), this could explain the high natural mortality rates found by Munro (1983) as a function of the physical exclusion of larger fish from reef habitats due to their lack of shelter, hence higher vulnerability to predation. Fish are then obliged to migrate, be predated, or become stunted before reproduction is possible. The very steep slopes of the size frequency for larger reef fishes on Pedro Bank was noted by Munro as implying intensive shark predation, suggesting that if a fish is excluded from a crevice, it will most likely suffer a predator attack. If surviving, it might be forced to migrate to areas with larger crevices, a migration itself accompanied by a high probability of predation.

The Nature of the Seafloor Globally

Especially near the point of discharge of rivers, the seafloor is often dominated by soft sediments. Patches of live bottom with epifauna and epiflora such as coral reefs or macroalgae on rocky surfaces protruding through this sediment are relatively uncommon. Thorson (1950) suggested that this is a characteristic of most marine environments and how they differ from more com-

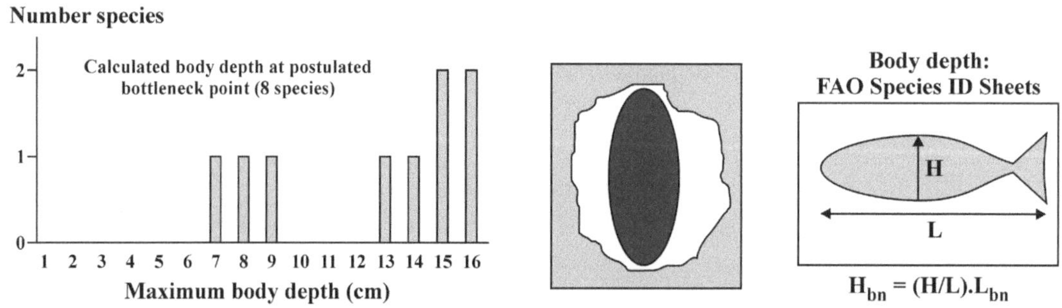

Figure 6. Body depth by species (left) where a bottleneck is postulated and (center; right) the means used to calculate effective body depth from data in Munro (1983) using body outlines of fishes shown in the Food and Agriculture Organization of the United Nations species identification sheets (Carpenter 2002).

plex vegetated environments on land. Fine sediment coverage increases near estuaries with concomitant increases in pollution and hypoxia (e.g., Caddy 2000; Rabelais and Turner 2001; Breitburg et al. 2009). Another human influence negatively impacting epifauna is bottom fishing gear, especially its effects on habitats used by juveniles of fish and crustaceans. Bottom trawling or dredging reduces the complexity and extent of existing outcrops and biological cover (see Caddy and Seijo 2013 for a summary of the impacts of towed bottom gears). These factors can also be responsible for the "gap and corridor" configurations noted by Kelton et al. (2012).

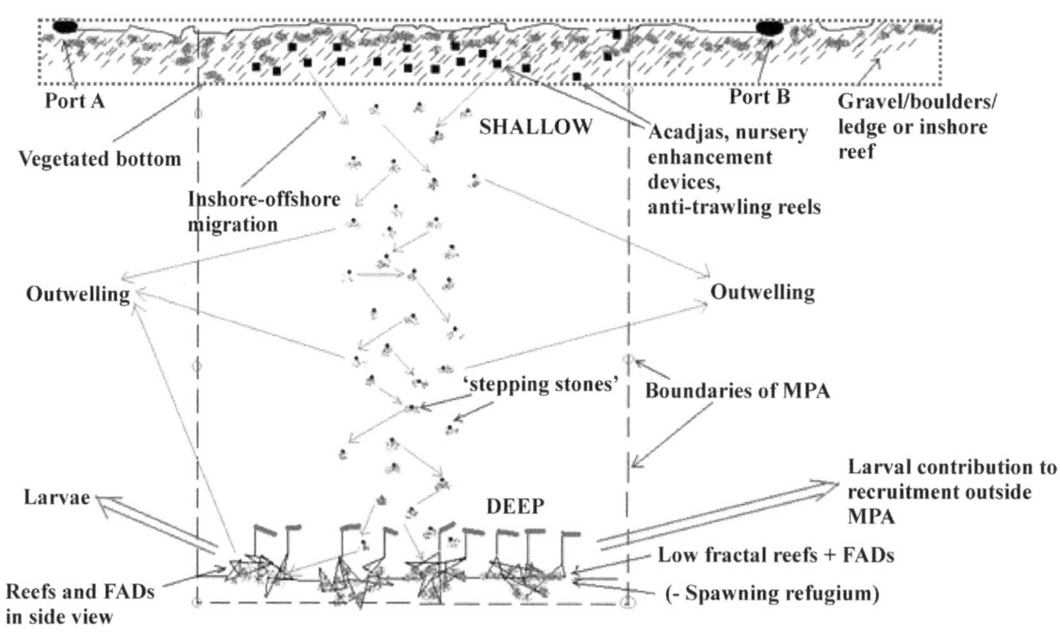

Figure 7. Habitat modification mechanisms for improving survival in life cycle migrations are the "migration corridor" or "stepping stones" (e.g., Tilman and Kareiva 1997; Gillanders et al. 2003; figure redrawn from Caddy 2007).

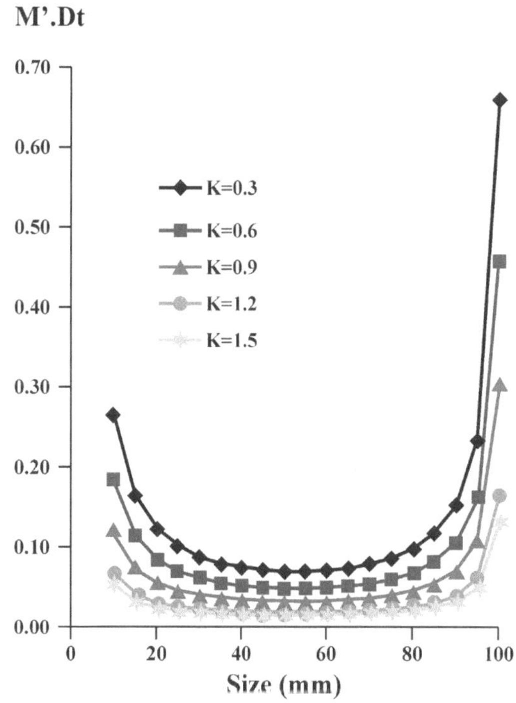

Figure 8. Simulated *Ms* at age for reef fishes resident in a fractal environment, under a range of assumptions for von Bertalanffy *K*. It is assumed that residents only survive if a suitably-sized crevice is available. Fish with high *K* values (stunted growth) survive best, but faster growth increases competition between the larger survivors for the few large crevices available (Caddy 2013), hence high *K* values and stunting in size is favorable for species needing to stay close to a perforated surface if migration is to be avoided.

Potential Impacts of Bottom Gear on Stage-Specific Habitats

The prohibition of trawling or dredging near pipelines or marine cables or the use of bottom gear where sediments have been extracted leaving uneven, untrawlable bottom implies that manipulation of the seafloor to leave structural irregularities may, on occasion, increase holding capacity and protect complex habitats and species. For example, rare species of skate are confined to such hilly, untrawlable bottom in the Irish Sea. The "sleeping policemen" concept (Guillén et al. 1994) is an artificial device used on some Mediterranean shelves to protect nursery areas from trawling. Such mechanisms potentially allow experiments of long duration, such as those on the utility of natural or artificial reefs.

The suggestion has been made that "migration corridors" or "stepping stones" (Tilman and Kareiva 1997; Gillanders et al. 2003) would help protect migrants from predation while they are migrating between age- or size-specific habitats. Figure 7 is a visualization of such a hypothetical scheme. It is envisaged to consist of rows of small artificial reefs or casitas (i.e., wooden, low-relief structures built to provide shelter for spiny lobsters), stretching between nursery areas and deepwater areas offshore, where artificial reefs with attached fish aggregating devices, as envisaged by Kellison and Sedberry 1998, would help to aggregate pelagic food fishes on the associated reefs. The cost of construction and maintenance of such a migration corridor would not be negligible, but the structure would hopefully persist for years if trawling or dredging were prohibited in the area. The high abundance of stunted cichlids achieved with brush parks (acadjas) in West Africa (e.g., Welcomme 1972; Hem and Avit 1994) is worth noting here since the acadja technology could be useful as a nursery area for juvenile fishes in this migration configuration.

Population Bottlenecks

A simple definition of the impact of a population bottleneck can be proposed: a sudden decrease in population size due to adverse factors in the environment or habitat. Although there are other causes of population bottlenecks, a rapid decline in population size at a given life history stage may be a function of the decline in quality and/or quantity of life-stage-specific habitat, perhaps as a result of natural or anthropogenic effects. Viewed in its broadest context, the generally limited population sizes of reef or outcrop-associated fish can be compared with the much larger stock sizes achieved by small pelagic fish and free-living demersal fishes (e.g., arctoboreal species such as many gadoids). That a specific habitat could be the cause of a population bottleneck becomes more evident when the life cycle is modeled by life history stage, even though quantitative stage estimates may be less precise.

Unlike the unit stock concept used in conventional assessments, here we are discussing juvenile to adult stages as subsets or stages in a local population occupying a coastal and shelf area. This approach differs from the annual time scale and population units used in stock assessment. Although timing by life history stage rather than annual time intervals is less precise, there are several advantages in such a qualitative description of population life histories. For example, when discussing migration and the concept of population bottlenecks, these may not be identified as such in annual data sets for the whole stock.

For many species fished as adults on artificial reefs, field observations suggest that in their early life histories, they often occupied different habitats, both bathymetrically and in terms of the geological or biological elements of their habitat. Gillanders et al. (2003) concluded that nursery areas are smaller and more densely occupied than the stage-specific habitats of older individuals. A habitat change in the life history for the Nassau Grouper *Epinephelus striatus* is that after inhabiting macroalgal meadows at 30–120 mm total length (TL), they migrate to patch reefs at 120–300 mm TL, then to offshore reefs at ages 3–4 years (Eggleston 1995). Another species typically showing multiple life history stages is the Yellowtail Snapper *Ocyurus chrysurus*. Its juveniles live in inshore sea grass nurseries; small adults congregate over hard bottom while adults live over sandy areas near deep reefs (10–70 m).

Formulating Hypotheses for Bottlenecks between Life History Stages

One cause of a population bottleneck for reef fish is implied when crevice abundance at size resembles the fractal distribution. With growth in size, only a fraction of small fish recruiting to the reef can be accommodated in the less-abundant larger crevices. Surplus fish have to either migrate, be predated, or remain stunted. As a consequence of the restriction in crevice numbers with size, a reduced curvature of the von Bertalanffy curve on approaching maturity means that older fish occupy larger holes for longer, increasing competition between larger fish for these few large, protective spaces (Figure 9).

It is interesting that small, resident reef fish have strongly inflected growth curves that show little increase in size as maturity approaches (see, e.g., Choat and Robertson 2002). The predicted natural mortality rate of reef fish increases due to the processes described in Figure 10 but at a much reduced rate compared with fish with low values of the von Bertalanffy *K* (Figure 9).

Figure 9 is based on data from Choat and Robertson 2002 and expresses how small species remaining on the reef (called "crevice-dependent" here) do so by drastic stunting, such that the values of their von Bertalanffy *K*s are extremely high. Larger schooling fish, referred to here as crevice-associated, cruise around

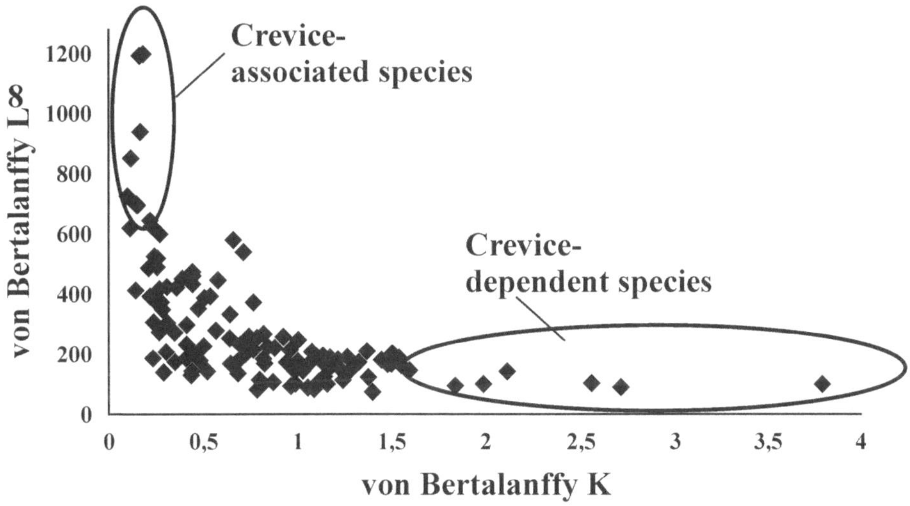

Figure 9. Summary of von Bertalanffy growth parameters for small, crevice-dependent fish that stay in, or close to, the reef surface and show a stunted growth form (Caddy 2013), in contrast to species that are reef-associated (i.e., often occur around reefs but only search for cover when attacked by predators).

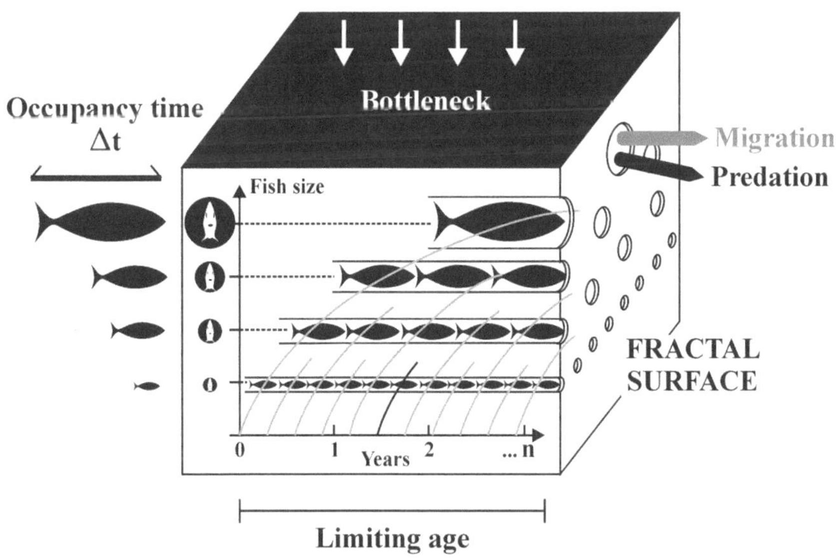

Figure 10. Illustrating factors that influence the holding capacity at size of a fractal surface and how as the growth rate of fish slows, their occupancy rate at size must increase. The mortality rate is determined by the spectrum of crevices, which is, at the limit, given by the surface fractal coefficient, where $2 < D' < 3$. Bottlenecks in reef fish production may be related to this process.

obstructions on the seafloor, diving for cover only when predators appear. The tendency of some reef species to remain within existing cover has been referred to as the functional connectivity of population subunits that are associated with patches of complex habitat (Turgeon et al. 2010). At this point, Sale's 1978 analogy between habitat restriction and the children's game "musical chairs" becomes evident, in that there are unlikely to be sufficient large chairs (crevices) to accommodate the whole school throughout life. Together with growth inhibition when areas of complex habitat are limited, functional connectivity appears to be a consequence of a predator avoidance response.

Figure 11 shows that it would be important to establish whether emigration is initiated after an increase in size makes fish more vulnerable to predation. However, a population in flux from habitat 1 to habitat 2 may not all be accommodated there if the limiting fish density D_2 there multiplied by the area of the second habitat, A_2, is $N_2 = A_2D_2$. A bottleneck will occur if $A_1D_1 >> A_2D_2$. Such a restriction in the holding capacity of habitat 2 may be either natural or anthropic and will occur if the extent and/or the cover density of habitat 2 has been reduced. If migrants cannot all be accommodated under cover in area A_2, high predation mortality on the survivors left without cover $(N_1 - A_2D_2)$ is inevitable. Such an effect could result if fish density D_2, and/or holding capacity A_2D_2 were reduced by an extrinsic activity such as trawling (see Caddy and Seijo 2013), which has been documented to erase natural cover.

A supplementary increase in vulnerability to predation (M_v) outside cover, is multiplied by the time (Δ_t) spent searching for alternative shelter. $M_v\Delta_t$ will increase with distance and time spent searching between habitats 1 and 2. Following the same argument through sequential stages to maturity, in each case, if successive bottlenecks exist in the extent and holding capacity of sequential habitats, this would mean that numbers reaching maturity are much lower than the number of recruits. Thus, habitat type and availability and spatial discontinuities between habitat types may lead to extensive migrations. These factors play a major role in increasing vulnerability

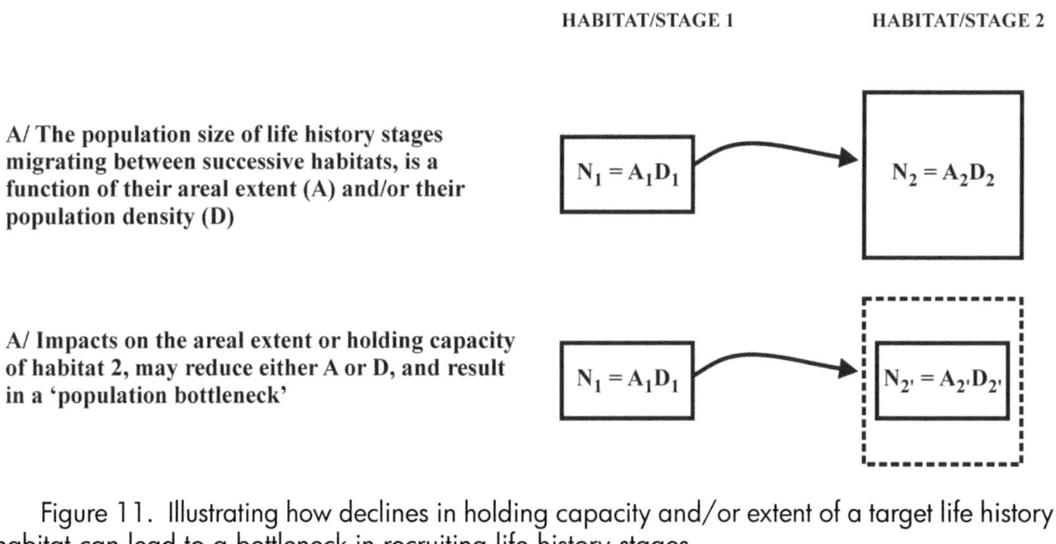

Figure 11. Illustrating how declines in holding capacity and/or extent of a target life history habitat can lead to a bottleneck in recruiting life history stages.

and hence will have demographic effects. The distance between successive artificial habitats that are important to successive life history stages should be as short as possible, and their areal extent and quality as antipredator devices should be restored or enhanced.

Some Thoughts on Reef Productivity

Although a detailed treatment of the food requirements of reef fishes is not attempted here, it is likely that local food shortages often lead to risk-taking and movements away from cover. It is also worth considering the implications of fractal theory for the biological productivity of a perforated surface. To illustrate the discussion, growth and natural mortality parameters for different life stages of the Red Grouper were modeled, but the general argument is intended to be applied for other inhabitants of a perforated surface, which in this calculation are expressed as spherical bodies of volume $(4/3)\pi(L/2)^3$, just fitting within crevices of diameter L (see Figure 12).

For Red Grouper, Stiles and Burton (1994) gave von Bertalanffy parameters (Table 5), which were used to calculate the equivalent age at size in order to apply a reciprocal relationship for this species between M at age and age. This was given by Giménez-Hurtado et al. (2005) as $M_t = 0.8976t^{-0.93}$, where M_t = natural mortality at age and t = age (years).

Figure 12 and Table 5 illustrate the sequence of calculated values leading to an indicative description of productivity at size on a fractal surface, assuming that different stages of Red Grouper are the occupants. This hypothetical example is intended to represent a more general case where spherical volumes of protoplasmic material just fit within crevice diameters. If summed over crevice sizes, their combined volume represents the biomass of living organisms held within a fully occupied set of reef crevices. This biomass indicator is multiplied by the natural mortality rate at age for each size of crevice to provide an index of their potential productivity. When plotted in Figure 13 as a pyramid, it is compared with Elton's (1927) conception of a "pyramid of numbers."

Creating a pyramid of crevice numbers at size starting with simple fractal theory resembles in its outcome the suggestion by Elton (1927). That is, one description of an ecosystem is as a pyramid of numbers of organisms at size. Since fractal surfaces on exposed outcrops presumably existed on the seafloor before evolution began, could it be that such a primitive habitat structure guided evolution of early ecosystems towards numerical pyramids at size, then onwards to trophic pyramids?

Although this abstract idea incorporates several major assumptions, it seems reasonable to suggest from Figure 13 that the productivity of small organisms on the reef greatly exceeds that of larger animals, including reef fishes. A further philosophical implication is pointed to. The spectrum of fractal crevice sizes, which already existed on mineral outcrops on the seafloor before evolution began, provided the structural framework for the complex trophic pyramids that have since developed on coral and natural reef outcrops.

Discussion

One reason for giving more priority to habitat structure in the management of fish populations is that where this has been studied (especially in inland waters), an enhancement of the physical complexity of a habitat surely increases the productivity of food resources, as well as assisting in predator avoidance by providing size-dependent refuges with predefined properties. Some of the ideas in this

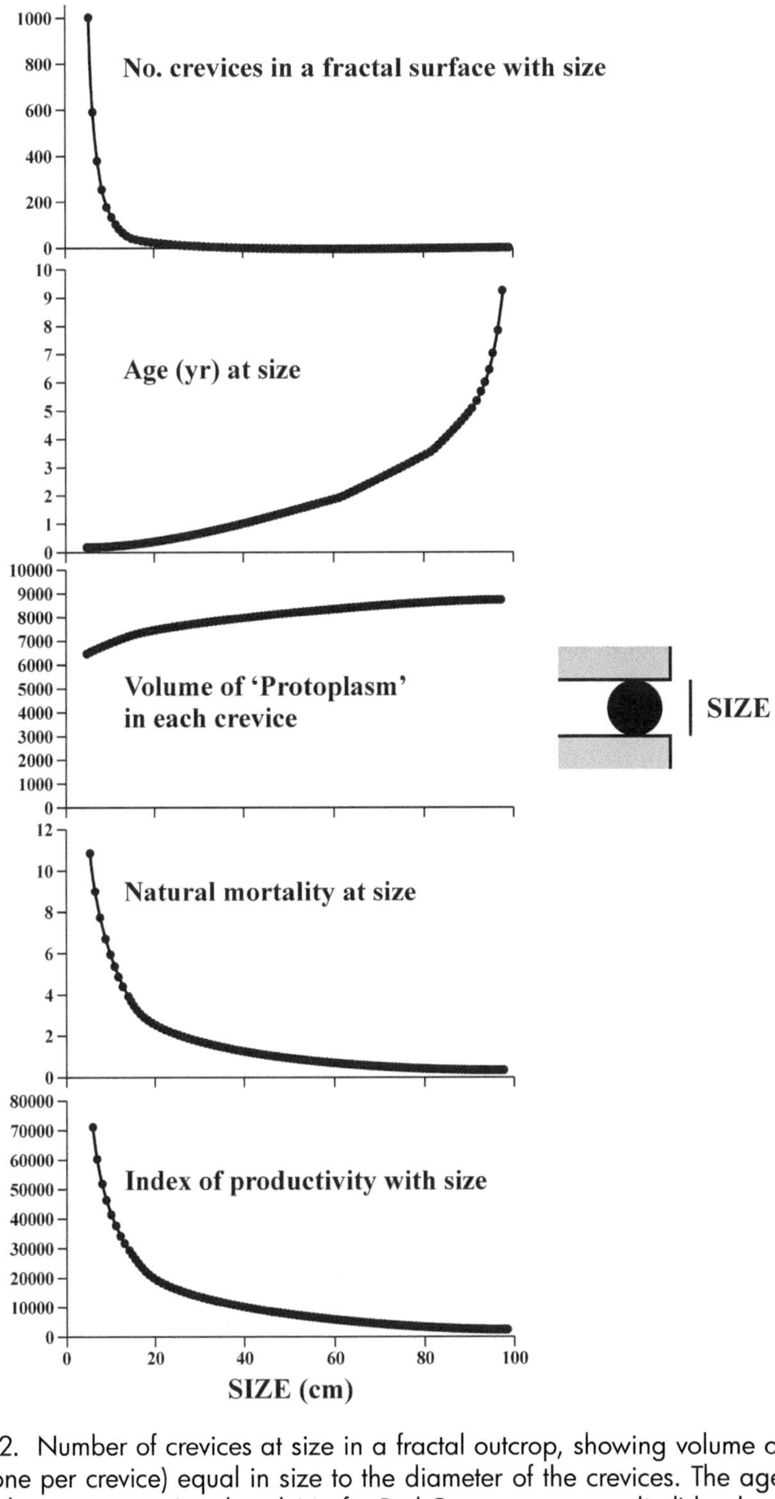

Figure 12. Number of crevices at size in a fractal outcrop, showing volume of protoplasm of spheres (one per crevice) equal in size to the diameter of the crevices. The age at size and natural mortality rate at size (predicted M_is for Red Grouper were applied) lead to a size-based index of productivity derived from the equations in Table 6.

Table 5. Hypothetical calculation assessing order of magnitude of production by crevice-dwelling organisms of different sizes represented as spheres just fitting within the crevice diameters. A common growth rate and natural mortality at size are assumed, and age/size related natural mortality rates are assumed to apply, using life history data for the Red Grouper. (The calculations underlie Figure 12).

Variable	Equation	Parameter value
No. crevices at size L (N_L) with surface fractal coefficient (D').	$N_{L+1} = N_L{*}[L^{D'}/(L+1)^{D'}]$	$D' = 2.9$
Approximating to age t at size L of crevice dwellers.	$t(L) = -(1/K){*}\ln[(L_\infty - L)/L_\infty]$	$K = 0.163$ $L_\infty = 93.8$ (cm)
Individual volumes (V) of crevice dwellers assumed spherical.	$V_L = (4/3)\pi(L/2)^3$	
M at age (M_t) (then converted to size with von Bertalanffy)	$M_t = (A + B)/t \quad (\geq M_L)$	$A = 0.898$ $B = 0.93$
Production index (P_L)	$P_L = M_L{*}V_L$	

paper could provide the basis for an experimental simulation of reef fish life histories to explore the relative importance of the different factors discussed.

A valid concern of ecologists is the possible adverse effects of constructing a few small, artificial reefs with large crevice sizes to attract mature fish. As noted by Polovina (1991), the high attraction of larger fish to large crevice sizes is surely based on the rarity of large crevices in most marine environments. Such a rare habitat as that required by large, spawning fish must limit spawning stock size and, if easily located by fishermen, will further increase catchability and reduce spawning potential. This is a risk to consider if conserving spawning potential is a key concern. One solution is that reefs for mature fish may also be constructed within no-fishing areas, or a rotational harvesting scheme for access to existing reefs be initiated, so as to protect a proportion of the mature popula-

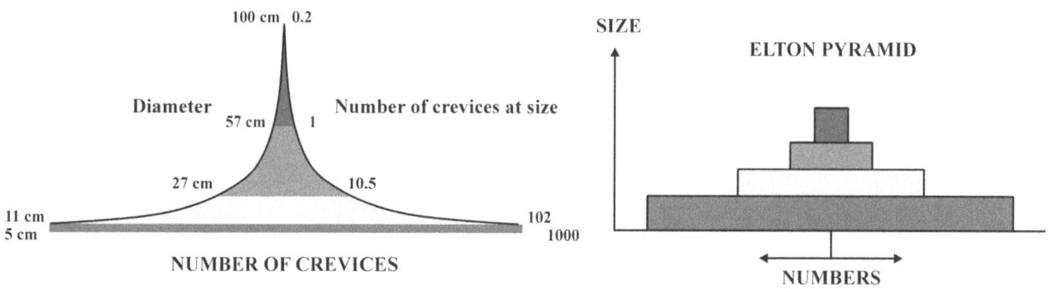

Figure 13. An index of the productivity of crevice-dwelling organisms of increasing size on a fractal outcrop, plotted (left), is seen as analogous to the conventional numerical pyramid of Elton (1927) (right). The productivity index so derived declines abruptly with size (see text for underlying assumptions).

tion in any given year. Such a scheme might involve harvesting such reefs every third or fourth year and could encourage spawning on the unfished reefs. Russ and Alcala (1996) showed that the closure of a spawning refugium will promote outwelling of gametes and juveniles to adjacent fished areas. The approach of establishing refugia for mature fish is being considered along the shelf edge in the Mediterranean (Caddy 2014a).

A field experiment ongoing in the Gulf of Mexico involves the use of abandoned oil rigs as artificial reefs and the potential negative implications for fisheries of deciding to deconstruct and remove abandoned rigs (Shipp and Bortone 2009). The potential financial return to constructors of reefs if charter boats pay a fee for access suggests that recovery of regulatory costs with local regulation or privatization of such structures would be possible.

Some Conclusions and Debating Points

Enhancing fisheries production by adding artificial complexity could be a source of recruits (nurseries) and enhance spawning aggregations (refugia). If natural reefs are the source of recruits to artificial reefs, Munro's data suggest that some natural reefs have restricted numbers of crevice sizes beginning at 8–10 cm diameter upwards. This suggests that reefs should be designed to accommodate larger girth sizes. Below are suggestions for environmental managers who opt for artificial reef deployments to enhance fisheries:

- Modeling natural mortality rate as a constant and ignoring how M_t varies with life history stage is unrealistic, especially for reef fishes. It appears likely that the vector of M at size for reef fishes, after experiencing a decline from a high natural mortality rate for juveniles, increases for larger fish due to restrictions in the availability of large crevice habitats.
- Migration is a high-risk activity and may be induced by predator attacks while fishes are outgrowing local cover. It seems likely that natural mortality increases during migrations; hence, there is an advantage in locating artificial reefs not too far offshore from natural outcrops or coral reefs, or linking them by "stepping stones."
- How can we best use the concept of "cover" in reef fish management? Removing epifauna by dragging (i.e., trawling) can eliminate suitable cover. It is suggested that this reduces holding capacity and or cover efficiency and potentially cause bottlenecks in the production of reef fish and will increase their natural mortality rate. Is there a regulatory mechanism in place for enhancing cover or protecting existing cover from erosion by towed bottom gears?
- Larger fish seek large crevices for protection. Have geographic information system methodologies been established to measure the natural abundance and location of natural reef habitats?
- Does the absence of adequate habitat for older stages of reef fishes oblige juveniles to remain in their original cover unit leading to stunting and hence maturity and breeding at a smaller size?

References

Abella, A. J., J. F. Caddy, and F. Serena. 1997. Do natural mortality and availability decline with age? An alternative yield paradigm for juvenile fisheries, illustrated by the Hake *Merluccius merluccius* fishery in the Mediterranean. Aquatic Living Resources 10:257–69.

Anthony, V. C., and J. F. Caddy, editors. 1986. Proceedings of the Canada–US workshop on status of assessment science for N.W. Atlantic lobster (*Homarus americanus*) stocks, St Andrews, N.B., Oct. 24–26, 1978. Canadian Technical Report of Fish and Aquatic Sciences 932.

Arsenault, D. J., and J. H. Himmelman. 1998. Size-related decrease in spatial refuge use by Iceland scallops (*Chlamys islandica*): ontogenetic behavioral changes or decreasing refuge availability? Marine Ecology Progress Series 162:153–61.

Beck, M. W. 1995. Size-specific shelter limitation in stone crabs: a test of the demographic bottleneck hypothesis. Ecology 76:968–980.

Beck, M. W. 1997. A test of the generality of shelter bottlenecks in four stone crab populations. Ecology 78:2487–2503.

Breitburg, D. L., D. W. Hondorp, L. Davias, and R. J. Diaz. 2009. Hypoxia, nitrogen and fisheries: integrating effects across local and global landscapes. Annual Review of Marine Science 1:329–49.

Burrough, P. A. 1981. Fractal dimensions of landscapes and other environmental data. Nature (London) 394:240–242.

Caddy, J. F. 1991. Death rates and time intervals: is there an alternative to the constant natural mortality axiom? Reviews in Fish Biology and Fisheries 1:109–138.

Caddy, J. F. 1996. Modeling natural mortality with age in short-lived invertebrate populations: definition of a strategy of gnomonic time division. Aquatic Living Resources 9:197–207.

Caddy, J. F. 2000. Marine catchment basin effects versus impacts of fisheries on semi-enclosed seas. ICES Journal of Marine Science 57:628–40.

Caddy, J. F. 2003. Scaling elapsed time: an alternative approach to modeling crustacean molting schedules? Fisheries Research 63:73–84.

Caddy, J. F. 2007. Marine habitat and cover: their importance for productive coastal fishery resources. UNESCO Publishing, Oceanographic Methodology Series, Paris.

Caddy, J. F. 2011. How artificial reefs could reduce the impacts of bottlenecks in reef fish productivity within natural fractal habitats. Pages 45–64 *in* S. A. Bortone, F. Pereira Brandini, G. Fabi, and S. Otake, editors. Artificial reefs in fisheries management. CRC Press, Boca Raton, Florida.

Caddy, J. F. 2013. Fractal environments select for high von Bertalanffy k's in crevice-dwelling fishes. Ciencia Pesquera 21(2):49–56.

Caddy, J. F. 2014a. The advantages offered by Mediterranean shelf habitats for the conservation of demersal resources using the refugium concept. Biologia Marina Mediterranea 21.192–208.

Caddy, J. F. 2014b. Why do assessments of demersal stocks largely ignore habitat? ICES Journal of Marine Science 10:1093.

Caddy, J. F. 2018. Conserving spawners and harvesting juveniles: is this better than postponing capture until sexual maturity? Chapter 4 *in* J. C. Seijo and J. G. Sutinen, editors. Advances in fisheries bioeconomics: theory and policy. Taylor and Francis, Milton Park, Abingdon, UK.

Caddy, J. F., and J.-C. Seijo. 2013. Destructive fishing practices by bottom trawls: a broad review of research and practice. Ciencia Pesquera 19(Special Issue):1–58.

Caddy, J. F., and C. Stamatopoulos. 1990. Mapping growth and mortality rates of crevice-dwelling organisms onto a perforated surface: the relevance of cover to the carrying capacity of natural and artificial habitats. Estuarine, Coastal and Shelf Science 31:87–106.

Carpenter, K., editor. 2002. The living marine resources of the western central Atlantic. Food and Agriculture Organization of the United Nations, FAO Species Identification Guide, Rome and American Society of Ichthyologists and Herpetologists, Special Publication No. 5, Lawrence, Kansas.

Carpenter, K. E., R. I. Miclat, V. D. Albaladejo, and V. T. Corpus. 1981. The influence of substrate structure on the local abundance and diversity of Philippine reef fishes. Pages 497–502 *in* E. D. Gomez, C. E. Birkeland, R. W. Buddemeier, R. E. Johannes, J. A. Marsh, Jr., and R. T. Tsuda, editors. Proceedings of the 4th international coral reef symposium, volume 2. Marine Science Center, University of the Philippines, Manila, Philippines.

Coen, L. D., K. L .Heck, Jr., and L. G. Abele. 1981. Experiments on competition among shrimps of seagrass meadows. Ecology 62:1484–1493.

Choat, J. H., and D. R. Robertson. 2002. Age-based studies on coral reef fishes. Pages 57–80 *in* P. F. Sale, editor. Coral reef fishes: dynamics and diversity in a complex ecosystem. Academic Press, San Diego, California.

Colin, P. L., C. C. Koenig, and W. A. Laroche. 1996. Development from egg to juvenile of the Red Grouper (*Epinephelus morio* Pisces: Serranidae) in the laboratory. Pages 399–414 *in* F. Arreguín-Sanchez, J. L. Munro, M. C. Balgos, and D. Pauly, editors. Biology, fisheries and culture of tropical groupers and snapper. ICLARM (International Center for Living Aquatic Resources Management) Contribution 1323.

Commito, J. A., and B. R. Rusignuolo. 2000. Structural complexity in mussel beds: the fractal geometry of surface topography. Journal of Experimental Marine Biology and Ecology 255(2):132–152.

Eggleston, D. B. 1995. Recruitment in Nassau Grouper, *Epinephelus striatus*: post-settlement abundance, microhabitat features and ontogenetic shifts. Marine Ecology Progress Series 124:9–22.

Eggleston, D. B., R. N. Lipcius, and J. J. Grover. 1997. Predator and shelter-size effects on coral reef fish and spiny lobster prey. Marine Ecology Progress Series 149:43–59.

Elton, C. 1927. Animal ecology. Macmillan, New York.

Frazer, T. K., and W. J. Lindberg. 1984. Refuge spacing similarly affects reef-associated species from three phyla. Bulletin of Marine Science 55:388–400.

Gee, J. M., and R. M. Warwick. 1994. Body size distribution in a marine metazoan community and the fractal dimensions of macroalgae. Journal of Experimental Marine Biology and Ecology 178:247–249.

Gillanders, B. M., K. W. Able, J. A. Brown, D. B. Eggleston, and P.F. Sheridan. 2003. Evidence of connectivity between juvenile and adult habitats for mobile marine fauna: an important component of nurseries. Marine Ecology Progress Series 247:281–95.

Giménez-Hurtado, R., E. R. Coyula-Pérez-Puelles, S. E. Lluch-Cotac, A. A. González-Yañez, V. Moreno-García, and R. Burgos-de-la-Rosa. 2005. Historical biomass, fishing mortality, and recruitment trends of the Campeche Bank Red Grouper (*Epinephelus morio*). Fisheries Research 71:267–277.

Gotceitas, V., and P. Colgan. 1989. Predator foraging success and habitat complexity: quantitative test of the threshold hypothesis. Oecologia 80:158–66.

Guillén, J. E., A. A. Ramos, L. Martinez, and J. L. Sanchez Lizaso. 1994. Antitrawling reefs and the protection of *Posidonia oceanica* (L.) Delile, meadows in the western Mediterranean Sea: demand and aims. Bulletin of Marine Science 55:645–650.

Hacker, S., and R. Steneck, 1990. Habitat architecture and the abundance and body-size-dependent habitat selection of a phytal amphipod. Ecology 71:2269–85.

Heck, K. L., Jr., and L. B. Crowder 1991. Habitat structure and predator–prey interactions in vegetated aquatic systems. Pages 281–299 *in* S. S. Bell, E. D. McCoy, and H. R. Mushinsky, editors. Habitat structure:

the physical arrangement of objects in space. Chapman and Hall, New York.

Hem, S., and J. L. B. Avit. 1994. First results on "acadja-enclos" as an extensive aquaculture system (West Africa). Bulletin of Marine Science 55:1038–1049.

Hixon, M. A., and J. P. Beets. 1993. Predation, prey refuges and the structure of coral-reef assemblages. Ecological Monographs 63:77–101.

Kellison, G. T., and G. R. Sedberry. 1998. The effects of artificial reef vertical profile and hole diameter on fishes off South Carolina. Bulletin of Marine Science 62:763–780.

Kelton, W., K. W. McMahon, M. L. Berumen, and S. R. Thorrold. 2012. Linking habitat mosaics and connectivity in a coral reef seascape. Proceedings of the National Academy of Sciences of the United States of America 109:15372–15376.

Krummel, J. R., R. H. Gardner, G. Sugahara, R. V. O'Neill, and P. R. Colman 1987. Landscape patterns in a disturbed environment. Oikos 48:321–324.

Lindberg, W. J., T. K. Frazer, K. M. Portier, F. Vose, J. Loftin, D.J. Murie, D. M. Mason, B. Nagy, and M. K. Hart. 2006. Density-dependent habitat selection and performance by a large mobile reef fish. Ecological Applications 16:731–746.

Macfadyen, G. T. Huntington, and R. Cappell. 2009. Abandoned, lost or otherwise discarded fishing gear. Food and Agriculture Organization of the United Nations, Fisheries and Aquaculture Technical Paper 523, Rome.

Mandelbrot, B. B. 1983. The fractal geometry of nature. W. H. Freeman and Company, New York.

Martínez-Aguilar, S., F. Arreguín-Sánchez, and E. Morales-Bojórquez. 2005. Natural mortality and life history stage duration of Pacific Sardine (*Sardinops caeruleus*) based on gnomonic time divisions. Fisheries Research 71:103.

Martínez-Aguilar, S., J. G. Diez-Uribe, S. J. De Anda-Montanez, F. Arreguín-Sánchez, and M. A. Cisneros-Mata. 2010. Natural mortality and life stage duration for the jumbo squid (*Dosidicus gigas*) in the Gulf of California, Mexico, using the gnomonic time division. Ciencia Pesquera 18:31–42.

Milicich, M. J., M. G. Meekan, and P. J. Doherty. 1992. Larval supply: a good predictor of recruitment of three species of reef fish (Pomacentridae). Marine Ecology Progress Series 86(2):153–166.

Morse, D. R., J. H. Lawton, M. M. Dodson, and M. H. Williamson. 1985. Fractal distribution of vegetation and the distribution of arthropod body lengths. Nature (London) 314:731–733.

Munro, J. L. 1983. Caribbean coral reef fishery resources. ICLARM (International Center for Living Aquatic Resources Management) Contribution 125.

Orensanz, J. M., and G. S. Jamieson. 1998. The assessment and management of spatially-structured stocks: an overview of the North Pacific symposium on invertebrate stock assessment and management. Pages 441–459 *in* G. S. Jamieson and A. Campbell, editors. Proceedings of the North Pacific symposium on invertebrate stock assessment and management. Canadian Special Publication of Fisheries and Aquatic Sciences 125.

Polovina, J. J. 1991. Ecological considerations on the applications of artificial reefs in the management of artisanal fisheries. Tropical Coastal Area Management 6:1–4.

Purkis, S. J., and K. E. Kohler. 2008. The role of topography in promoting fractal patchiness in a carbonate shelf landscape. Coral Reefs 27:977–989.

Rabalais, N. N., and R. E. Turner, editors. 2001. Coastal hypoxia: consequences for living resources and eosystems. American Geophysical Union, Coastal and Estuarine Studies, Washington, D.C.

Ramírez-Rodríguez, M., and F. Arreguín-Sánchez. 2003. Life history stage duration and natural mortality for the pink shrimp *Farfantepenaeus duorarum* (Burkenroad 1939) in the southern Gulf of Mexico, using the gnomonic model for time division. Fisheries Research 60:45–51.

Rooper, C. N., P. J. Etnoyer, K. L. Stierhoff, and J. V. Olson. 2016. Effects of fishing gear on deep-sea corals and sponges in U.S. waters. Pages 93–111 *in* T. F. Hourigan, P. J. Etnoyer, and S. D. Cairns, editors. The state of deep-sea coral and sponge ecosystems of the United States. NOAA Technical Memorandum NMFS-OHC-4.

Russ, G. R., and A. C. Alcala. 1996. Do marine reserves export adult fish biomass? Evidence from Apo Island, central Philippines. Marine Ecology Progress Series 132:1–9.

Sale, P. F. 1978. Coexistence of coral reef fishes: a lottery for living space. Environmental Biology of Fishes 3:85–102.

Shipp, R. L., and S. A. Bortone. 2009. A perspective of the importance of artificial habitat on the management of Red Snapper in the Gulf of Mexico. Reviews in Fisheries Science 17:41–47.

Shulman, M. J. 1985. Recruitment of coral reef fishes: effects of distribution of predators and shelter. Ecology 66:1056–66.

Stiles, T. C., and M. L. Burton. 1994. Age, growth and mortality of Red Grouper (*Epinephelus morio*) from the southwestern US. Proceedings of the 43rd Gulf and Caribbean Fisheries Institute 43:123–147.

Sugihara, G., and R. M. 1990. Nonlinear forecasting as a way of distinguishing chaos from measurement error in time series. Nature 344:734–741.

Tercafs, R. 1997. A fractal model of cave ecological environment. Memoires de Biospéologie 24:1–8.

Thorson, G. 1950. Reproduction and larval ecology of marine bottom invertebrates. Biological Reviews 25:1–45.

Tilman, D., and P. Kareiva, editors. 1997. Spatial ecology: the role of space in population dynamics and interspecific interactions. Princeton University Press, Princeton, New Jersey.

Tupper, M., and R. G. Boutilier. 1995. Effects of habitat on settlement, growth and post-settlement survival of Atlantic Cod (*Gadus morhua*). Canadian Journal of Fisheries and Aquatic Sciences 52:183–84.

Turgeon, K. A. Robillard, J. Gregoire, V. Duclos, and D. L. Kramer. 2010. Functional connectivity from a reef fish perspective: behavioral tactics for moving in a fragmented landscape. Ecology 91:3332–3342.

Wahle, R. A., and R. S. Steneck. 1991. Recruitment habitats and nursery grounds of the American lobster *Homarus americanus*: a demographic bottleneck? Marine Ecology Progress Series 69:231–243.

Walters, C., and F. Juanes. 1993. Recruitment limitation as a consequence of natural selection for use of restricted feeding habitats and predation risk taking by juvenile fishes. Canadian Journal of Fisheries and Aquatic Sciences 50:2058–2070.

Welcomme, R. L. 1972. An evaluation of the acadja method of fishing as practised in the coastal lagoons of Dahomey (West Africa). Journal of Fish Biology 4:39–55.

Williamson, M. H, and J. H. Lawton 1991. Fractal geometry of ecological habitats. Pages 69–86 *in* S.S . Bell, E. D. McCoy, and H. R. Mushinsky, editors. Habitat structure: the physical arrangement of objects in space. Chapman and Hall, London.

American Fisheries Society Symposium 86:141–163, 2018

Comparison of Reef-Fish Assemblages between Artificial and Geologic Habitats in the Northeastern Gulf of Mexico: Implications for Fishery-Independent Surveys

SEAN F. KEENAN*, THEODORE S. SWITZER, KEVIN A. THOMPSON, AND AMANDA J. TYLER-JEDLUND

Florida Fish and Wildlife Conservation Commission, Fish and Wildlife Research Institute
100 8th Avenue SE, St. Petersburg, Florida 33701, USA

ANTHONY R. KNAPP

Florida Fish and Wildlife Conservation Commission, Fish and Wildlife Research Institute
Senator George Kirkpatrick Field Laboratory
11350 SW 153rd Court, Cedar Key, Florida 32625, USA

Abstract.—Reef-fish assemblage structure was compared among multiple artificial and geologic (i.e., naturally occurring hard bottom) habitats in the northeastern Gulf of Mexico during 2014–2016 as part of a larger fishery-independent survey. Baited remote underwater video systems equipped with stereo cameras were deployed (n = 348) on 11 habitat types, classified through interpretation of side-scan sonar imagery. In the video samples, 11,801 fish were enumerated. Nonparametric analysis of reef-fish assemblages detected four clusters related to habitat; assemblages associated with geologic habitats were distinct, whereas the remaining three clusters represented groupings of artificial habitats of different size, scale, and complexity. While many species, including Vermilion Snapper *Rhomboplites aurorubens* and Red Snapper *Lutjanus campechanus*, were observed in greater numbers on artificial reef habitats, most species were observed in all habitats sampled. Among artificial reef habitats, the habitat cluster consisting of unidentified depressions, unidentified artificial reefs, construction materials, and reef modules was similar to geologic habitats in supporting larger individuals, specifically Gray Triggerfish *Balistes capriscus* and Red Snapper. In contrast, the habitat cluster consisting of smaller, generally solitary chicken-transport cages was inhabited by smaller individuals, including smaller Red Snapper. Although geologic reefs are the predominant reef habitat throughout much of the eastern Gulf, artificial reefs are important locally, especially in the Florida Panhandle. Accordingly, continued incorporation of artificial reef

* Corresponding author: sean.keenan@myfwc.com

habitats within large-scale fishery-independent monitoring efforts is critical to the accurate assessment of the status of reef-fish stocks on broad spatial scales.

Introduction

Reef fishes in the southeastern United States have supported extensive commercial and recreational fisheries for decades, contributing US$30 million to $60 million in annual landings revenue and $\$3 \times 10^9$ to $\$8 \times 10^9$ in annual economic impact in the Gulf of Mexico region (Keithly and Roberts 2017). In particular, species such as Red Snapper *Lutjanus campechanus*, Gag *Mycteroperca microlepis*, Vermilion Snapper *Rhomboplites aurorubens*, Gray Triggerfish *Balistes capriscus*, and Greater Amberjack *Seriola dumerili* are targeted by both commercial and recreational fisheries (NMFS 2017). In recent decades, many reef fishes have been assessed as overfished or undergoing overfishing, necessitating increasingly restrictive management regulations to end overfishing and rebuild overfished stocks. Such measures include individual fishing quotas, changes in size or bag limits, and seasonal fishing closures. These management strategies often indirectly modify behavior in fishery sectors and make it more difficult to use fishery-dependent data to evaluate changes in relative abundance and size/age composition of managed fish populations. There is, therefore, an increasing need for fishery-independent data that accurately characterize the entire population of managed reef fishes, not just that portion sampled by the fishery (Rotherham et al. 2007).

Although expanding the fishery-independent survey effort has long been a high priority in the southeastern United States, implementing new or expanding existing surveys has been challenging. A significant impediment to broad-scale surveys of reef fishes, particularly in the Gulf of Mexico, is an incomplete understanding of the quantity and distribution of reef habitat upon which reef-fish populations rely. The identification and characterization of critical reef habitat in the eastern Gulf is complicated because (1) the area containing potential reef habitat is extremely large and (2) reef habitat is widely distributed throughout much of the region. Overall, the west Florida continental shelf extends for ~900 km along the 75-m isobath that varies between 25 and 250 km in width, encompassing approximately 125,000 km^2 (Hine and Locker 2011). In addition, the region's geologic history contributes to significant variability in the quantity, quality, and distribution of benthic habitat throughout the region. The west-central Florida shelf is characterized as a sediment-starved, gently sloping surface with a sediment veneer overlaying a lithified carbonate surface (Brooks et al. 2003; Obrochta et al. 2003). In contrast, the near-shelf off the Florida Panhandle (USA) contains relatively thick quartz sediments originating from the Apalachicola River and smaller rivers to the west (Davis 2017). Hard-bottom habitats on the west-central Florida shelf are widely distributed across the shelf and include ledges and large areas of low-relief hard-bottom habitat covered with encrusting infauna such as gorgonians and macroalgae. Drowned shorelines comprise the majority of hard-bottom habitats off the Florida Panhandle and are often concentrated between 90- and 120-m isobaths (Davis 2017). Because of its large areal extent, mapping the entire eastern Gulf in support of fishery-independent monitoring activities is cost-prohibitive. Moreover, focused mapping efforts within regions (e.g., Madison-Swanson, Florida Middle Grounds, and Pulley Ridge; Mallinson et al. 2014) of ecological or fisheries importance do not encompass the full range of habitats important to all species and life history stages of managed reef fishes. Accordingly, an alternative approach (e.g., stratified random, systematic) to mapping the eastern Gulf may be required to identify representative reef habitats in support of fishery-independent monitoring surveys.

Broad-scale surveys of reef-fish populations in the southeastern United States have relied on either diver-based, point-count methods in the Florida Keys reef tract (Ault et al. 2005) or baited remote underwater video (BRUV) systems off the southeastern United States in the Atlantic Ocean and the Gulf of Mexico (Campbell et al. 2015; Thompson et al. 2017a). Although these surveys provide valuable data for the assessment of managed reef fishes, most focus entirely on geologic reef habitats (defined here as naturally occurring hard-bottom habitat—not biogenic, reef-building coral) and exclude artificial reef structures. Over recent decades, deployments of artificial reefs have increased worldwide (Stone et al. 1991) with objectives as varied as the materials used to create them, including mitigating loss of bottom habitat (Dupont 2008), creating diving and fishing opportunities (Adams et al. 2011), and enhancing angling opportunities (Lindberg 1997; Baine 2001; Leeworthy et al. 2006). In general, the focus of scientific studies has evolved from the attraction-versus-production hypothesis to evaluating the ecological function of artificial reefs (Bohnsack and Sutherland 1985; Lindberg 1997). Improved technologies that foster research into specific aspects of artificial (and geologic) reef ecology include remotely operated vehicles (ROVs) and BRUVs (Rooker et al. 1997; Patterson et al. 2014; Streich et al. 2017), acoustic telemetry (Lindberg et al. 2006; Topping and Szedlmayer 2011; Biesinger et al. 2013), stable isotopes (Simonsen et al. 2015), and active and passive acoustics (Stanley and Wilson 2000; Wall 2012).

We conducted a 3-year pilot study to determine whether artificial reefs could appropriately be incorporated into ongoing surveys of geologic reefs in the eastern Gulf. Data on reef habitats and reef-associated fish populations were analyzed to (1) assess the distribution of different artificial and geologic reef habitats and (2) determine whether reef-fish assemblages differed between artificial and geologic reef habitats. Based on the outcomes of this study, we determined whether ongoing surveys of geologic reef habitats may be improved upon through the incorporation of artificial reef habitats and discussed implications that such improvements may have on the assessment and management of reef-fish populations.

Methods

Study Area

This study was conducted in nearshore (10–37 m) waters of the northeastern Gulf of Mexico off the Florida Panhandle (Figure 1), a region within which the state of Florida's fishery-independent survey of reef fishes is currently conducted (Thompson et al. 2017b). Numerous artificial reefs and geologic reef habitats exist within this region. This region is of particular interest because approximately 3,000 additional artificial reefs are planned to be deployed through Natural Resource Damage Assessment oil spill recovery efforts (Keith Mille, Florida Fish and Wildlife Conservation Commission, personal communication).

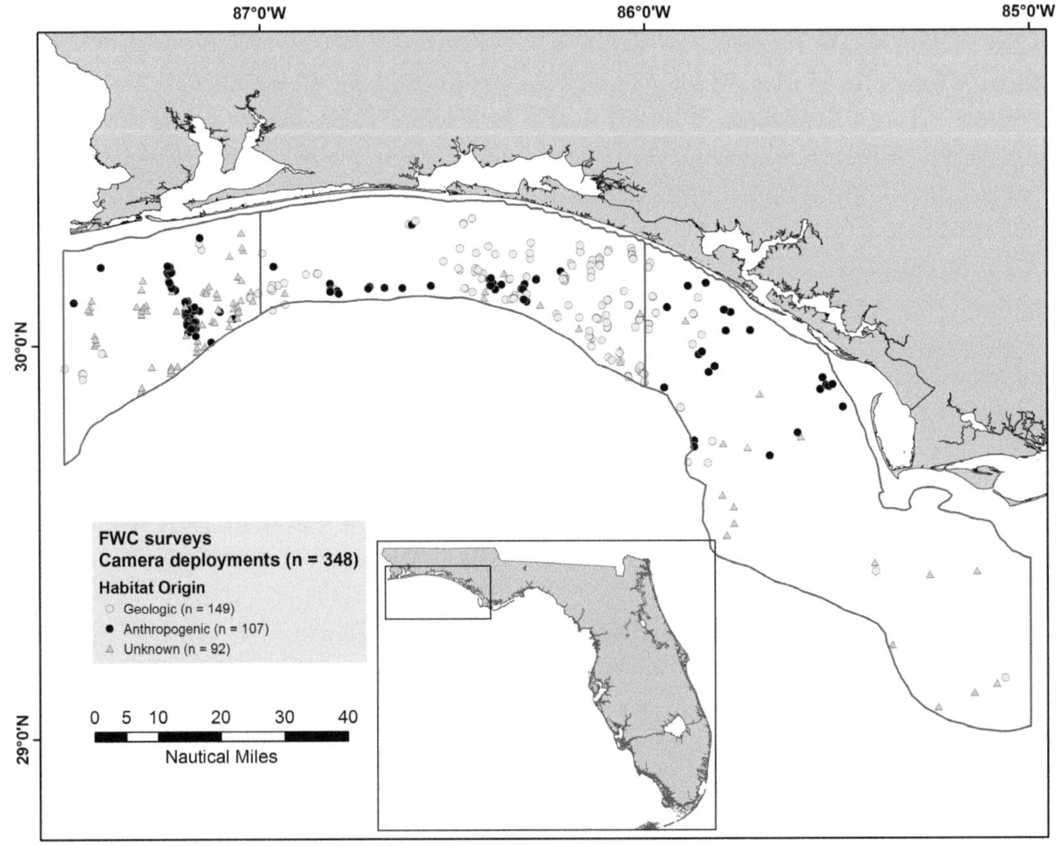

Figure 1. Sampling sites (2014–2016) within the Florida Panhandle study area in the Gulf of Mexico. The three National Marine Fisheries Service statistical zones are indicated within the 50-m isobaths, and sampling sites are characterized as geologic (gray circles), anthropogenic/artificial (black circles), or unidentified depressions (gray triangles).

Habitat Classification and Site Selection

Annual sampling (2014, 2015, and 2016) was allocated between both spatial (longitudinal grids corresponding to National Marine Fisheries Service [NMFS] statistical zones) and habitat strata (artificial reefs and geologic reefs). High-resolution, benthic-mapping data on the distribution of reef habitat are unavailable for much of the study area. To identify and delineate all geologic and artificial reef habitats prior to conducting fishery-independent surveys, spatially randomized habitat-mapping surveys (each covering an area of 2.1 km^2) were completed using a side-scan sonar (L3-Klein 3900) operating at 445 kHz. Each survey location was randomly selected from a gridded universe within the study area via ArcGIS 10.3 and Geospatial Modeling Environment procedures. A standardized survey for geologic habitats covered an area 0.55 km E–W and 3.89 km N–S, running perpendicular to the coast to increase the probability of locating habitats. A standardized survey for artificial reefs covered an area 1.30 km E–W and 1.67 km N–S, with the randomly selected artificial reef centered in the surveyed area. This format provided mapping data on the artificial reef (locations presumably known) and the surrounding area while covering the same area as the geologic surveys. Surveys were processed to correct for time-varied gain and navigational errors using Chesapeake Technologies SonarWiz software to produce a geotiff with 0.25 m resolution. Geotiffs were imported into an ArcGIS 10.3 project in which habitats were identified and manually delineated using a polygon-drawing tool and the Habitat Digitizer extension. The habitat classification scheme was a derivative of the geoform and surface geological component of the Coastal and Marine Ecological Classification Standard developed by the National Oceanic and Atmospheric Administration's Office for Coastal Management (www.fgdc.gov/standards/projects/cmecs-folder) and consisted of a series of distinct geologic and artificial reef habitats (Table 1). Another habitat identified and sampled was termed "unidentified depressions," which may represent either an artificial or a natural habitat.

A routine was created within ArcGIS 10.3 to randomly select specific deployment locations from identified habitats within the geologic and artificial reef surveys. For the geologic surveys, any identified geologic habitat was equally eligible to be sampled, and

Table 1. Habitats occurring in the study area as identified through interpretation of side-scan sonar imagery and criteria used to delineate habitat types.

Habitat class	Definition
Flat hard bottom (HB)	Flat or nearly flat areas (<0.1-m relief) of hard bottom generally colonized by benthic biota
Fragmented hard bottom (FB)	Areas dominated by exposed rock or coral that may be separated by narrow channels of finer sediment that has been eroded leaving the rock elevated above the seafloor with relief of >0.1 m
Mixed hard bottom (MB)	Mainly flat areas of hard bottom containing some features that have relief of >0.1 m
Ledge (LD)	A linear change in elevation of the seafloor that is associated with a rocky outcrop or underwater ridge of rocks. Ledges are defined spatially as the area within 5 m of the identified ledge
Unidentified artificial reef (UR)	Any artificial material >2 m in size that cannot be identified
Construction material (CM)	Any material deposited on the seafloor that was originally intended for construction purposes
Chicken-transport cages (CP)	A wire cage used for transporting poultry
Vessel (VL)	A vessel that was either intentionally or unintentionally sunk
Military tank (MT)	Decommissioned tanks that were deposited as a part of an artificial reef
Reef module (RM)	Prefabricated structures that have been constructed for the purpose of being deployed as reef-fish habitat (reef balls, pyramids, etc.)
Unidentified depression (PH)	Small (2–10 m in diameter) indentations or depressions lower than the surrounding surface usually occurring in unconsolidated sediments. This feature could be naturally made such as a pothole excavated by groupers or created by current scour.

one or two BRUVs were deployed in each geologic survey. Sites in artificial reef surveys focused on the initially selected central reef, but if no artificial reef habitat was identified, other artificial structures in a survey area were randomly selected. Any identified artificial habitat was eligible for inclusion as a sampling point, but geologic habitats identified in artificial reef surveys were not sampled; similarly, artificial habitats were not sampled from geologic reef surveys. All randomly selected sampling locations were spaced at least 100 m apart.

Collection and Processing of Video Data

Sampling was conducted using a BRUV consisting of two independent stereo-video recorders mounted opposite each other inside a 0.6 × 0.6 × 0.6 m aluminum frame. Each stereo-video recorder consisted of two digital video cameras set to record 10 images per second. Additionally, GoPro cameras were mounted orthogonal to the stereo-video recorders to provide habitat view in all directions. Baited remote underwater videos were baited with four cut mackerel *Scomber* spp. halves and deployed for 35 min.

For deployments for which at least 20 min of video was available, MaxN was determined for each species (i.e., the maximum number of individuals of the species observed in one of the screen shots for that deployment). Viewing of each video began at the point approximately 10 min after the gear reached the bottom. Only fishes identified to species were retained in the database for analyses. Because it is so difficult to count individuals in large schools, MaxN was assigned at 300 individuals in such cases. Stereo-video cameras allow measurement of individuals that are viewed by both lenses. Measurements were generated for selected species and examined as fork length (FL in mm) or total length (TL in mm) if the fish did not have a forked tail (e.g., Gray Triggerfish and lionfishes *Pterois* spp.). Only measurements from individuals less than 5 m from the camera were included in analyses.

Analytical Methods

Reef-fish assemblage structure in association with habitat type was compared using a Bray–Curtis similarity matrix calculated using fourth-root-transformed MaxN to reduce the influence of highly abundant taxa. Species-specific MaxN were first averaged by habitat type (flat hard bottom, ledge, fragmented hard bottom, mixed hard bottom, unknown artificial reef, construction materials, chicken-transport cages, vessels, military tanks, artificial reef modules, and unidentified depressions; see Table 1 for descriptions). Rare species (i.e., those occurring in five or fewer samples) were removed from the analyses. All analyses were conducted using the PRIMER v7 computer program (Clarke and Gorley 2015). To visualize the patterns of assemblage structure among habitats, we constructed a nonmetric multidimensional scaling (nMDS) ordination. To identify habitat clusters with statistically distinct fish assemblages, a similarity profile (SIMPROF) test was conducted using group averaging, and statistically significant clusters were then overlaid on the nMDS ordination. Species that contributed to 80% of the dissimilarity in assemblage structure between habitat clusters were identified via similarity percentages analyses (SIMPER). Length-frequency distribution of selected species contributing to the differences among habitat clusters were generated to qualitatively evaluate trends in size among clusters.

To explore the presumption of whether artificial reef habitats deployed within permit areas were properly documented by or reported to the Florida Fish and Wildlife Conservation Commission (FWC), the number of habitat polygons and area coverage of identified habitats were examined. Habitat data were evaluated within the FWC artificial reef permit areas, as well as the FWC data set of artificial reef coordinates. Area of habitat reported within a permit area was defined as any habitat located within 100 m of a documented artificial reef inside the permit area; unreported artificial reefs inside a permit area were more than 100 m from any coordinate in the database. Also summarized were artificial reef habitats identified outside of permit boundaries and interpreted as private reefs.

Results

Baited remote underwater videos were deployed 348 times in 2014–2016, 149 times on geologic habitats, 107 times on artificial habitats, and 92 times on unidentified depressions (Figure 1). Most deployments on geologic habitats occurred on flat hard bottom (n = 117); most deployments on artificial habitats occurred on unidentified artificial reefs (n = 34). Although not explicitly delineated as a sampling stratum within our design, many deployments were made on unidentified depressions (n = 92), which initially appeared similar to grouper potholes (sensu Coleman et al. 2010) when viewed using side-scan sonar imagery. Based on examination of video from earlier surveys and this study, we could not determine whether these depressions originated from active excavations by fish or from current-induced scouring. Because these features could include artificial or natural reef habitats, they were treated as a separate habitat class for analyses.

A total of 11,801 individual fish from 75 species were identified from BRUV deployments (Table 2), but 25 of the 75 species were represented by only one or two individuals. The most commonly observed species were Vermilion Snapper (n = 4,212), Tomtate *Haemulon aurolineatum* (n = 2,011), and Red Snapper (n = 1,058); more than 60% of observed individuals represented these valuable fishery and forage species (Table 2). The five most abundant species were observed in all 11 habitat classes. But individuals of 14 species were viewed solely from geologic habitats (e.g., Short Bigeye *Pristigenys alta* and Banded Rudderfish *Seriola zonata*), those of 10 species were viewed solely from artificial habitats (e.g., Hardhead Catfish *Ariopsis felis* and Bigeye *Priacanthus arenatus*), and those of 9 species were observed only on unidentified depressions (e.g., Blackwing Searobin [also known as Blackfin Searobin] *Prionotus rubio* and Goliath Grouper *Epinephelus itajara*), and in most instances (n = 30), only one individual of the species was seen.

Reef-fish assemblage structure differed significantly among habitat classes (Figure 2). The SIMPROF identified four significant habitat groups: group A (habitat classes: vessel, military tank) comprised artificial habitats with a high degree of vertical relief, group B (chicken-transport cages) represented chicken-transport cages, group C (flat hard bottom, ledge, mixed hard bottom, fragmented hard bottom) consisted of all geologic habitats, and group D (construction material, reef module, unidentified depression, unidentified artificial reef) comprised artificial habitats with less vertical relief and unidentified reefs or depressions. There were 20 species that, in combination, contributed nearly 80% of the dissimilarity between at least one pair of habitat groups (Figure 3). While many species were more abundant in artificial habitats, Red Porgy *Pagrus pagrus*, Blue Angelfish *Holacanthus bermudensis*, Bank Sea Bass *Centropristis ocyurus*, Spotfin Butterflyfish *Chaetodon ocellatus*, and Yellowtail

Table 2. Species observed in each habitat class within each reef type (i.e., geologic or artificial) where values represent the sum of MaxN for each identifiable species by habitat class. Species are ranked by increasing total number observed. The number of deployments is the number of camera deployments on each habitat. See Table 1 for habitat class abbreviations.

		Geologic				Artificial						Unknown	
		HB	FB	MB	LD	UR	CM	CP	VL	MT	RM	PH	Total
	Number of deployments	117	19	7	6	34	29	7	11	4	22	92	348
Scientific name	Common name												
Rhomboplites aurorubens	Vermilion Snapper	712	437	44	82	596	433	66	135	28	639	1,040	4,212
Haemulon aurolineatum	Tomtate	304	387	35	8	162	81	392	21	17	277	327	2,011
Lutjanus campechanus	Red Snapper	182	54	14	10	168	156	44	35	28	57	310	1,058
Pagrus pagrus	Red Porgy	384	50	30	26	24	23	11	2	4	32	162	748
Seriola dumerili	Greater Amberjack	96	11	7	2	163	30	5	21	8	31	31	405
Balistes capriscus	Gray Triggerfish	132	19	3	8	35	40	9	4	9	66	69	394
Caranx crysos	Blue Runner	206	1	3	49	13	2	0	0	89	9	20	392
Lutjanus griseus	Gray Snapper	26	13	2	2	95	43	0	4	30	63	66	344
Chromis enchrysura	Yellowtail Reeffish	258	5	20	14	0	0	0	0	0	8	0	305
Centropristis ocyurus	Bank Sea Bass	132	8	5	5	16	0	1	0	0	3	105	275
Lutjanus synagris	Lane Snapper	47	5	1	0	86	18	11	2	2	7	61	240
Pterois spp.	Lionfishes	33	15	4	5	26	12	9	0	0	32	73	209
Holacanthus bermudensis	Blue Angelfish	80	30	11	5	21	27	2	3	3	12	7	201
Equetus lanceolatus	Jackknife-fish	63	1	1	3	37	0	7	0	0	3	22	137
Chaetodon ocellatus	Spotfin Butterflyfish	52	8	1	3	4	13	1	2	0	10	6	100
Seriola rivoliana	Almaco Jack	7	5	2	0	24	11	5	4	4	14	10	86
Serranus phoebe	Tattler	55	2	5	3	2	1	0	1	0	1	11	81
Mycteroperca phenax	Scamp	13	16	3	2	7	10	0	2	2	0	5	60
Xyrichtys novacula	Pearly Razorfish	8	0	2	0	6	0	2	1	0	4	30	53
Chaetodon sedentarius	Reef Butterflyfish	22	11	4	4	1	0	0	0	0	0	0	42
Aluterus monoceros	Unicorn Filefish	4	0	0	0	17	0	0	0	0	10	8	39
Calamus proridens	Littlehead Porgy	21	3	0	0	5	4	0	0	0	3	2	38
Halichoeres bathyphilus	Greenband Wrasse	14	0	1	0	3	1	0	0	0	2	11	32

Table 2. Continued.

		Geologic				Artificial						Unknown	
Scientific name	Common name	HB	FB	MB	LD	UR	CM	CP	VL	MT	RM	PH	Total
Pristigenys alta	Short Bigeye	20	7	0	3	0	0	0	0	0	0	0	30
Sphoeroides spengleri	Bandtail Puffer	8	0	1	3	2	0	0	0	0	3	7	24
Canthigaster rostrata	Sharpnose Puffer	12	0	1	2	1	7	0	0	0	0	0	23
Gymnothorax saxicola	Honeycomb Moray	16	1	0	1	1	0	0	0	0	0	4	23
Epinephelus morio	Red Grouper	14	3	1	2	1	0	1	0	0	0	0	22
Mycteroperca microlepis	Gag	0	0	0	1	2	2	2	0	1	1	10	19
Rypticus maculatus	Whitespotted Soapfish	4	1	0	0	0	7	0	4	0	0	3	19
Seriola zonata	Banded Rudderfish	11	0	0	4	0	0	0	0	0	0	0	15
Archosargus probatocephalus	Sheepshead	1	0	0	0	1	3	0	5	2	2	0	14
Serranus subligarius	Belted Sandfish	7	1	3	0	1	1	0	0	0	0	0	13
Muraena retifera	Reticulate Moray	4	3	0	1	0	1	1	0	0	0	0	10
Ophichthus puncticeps	Palespotted Eel	2	0	1	0	0	1	0	0	0	0	6	10
Cephalopholis cruentata	Graysby	3	4	0	0	0	1	0	0	0	1	0	9
Priacanthus arenatus	Bigeye	0	0	0	0	0	0	0	8	0	0	0	8
Holacanthus ciliaris	Queen Angelfish	5	1	0	0	0	0	1	0	0	0	0	7
Aluterus schoepfii	Orange Filefish	3	1	0	0	0	0	0	1	0	0	1	6
Ariopsis felis	Hardhead Catfish	0	0	0	0	0	6	0	0	0	0	0	6
Lagocephalus laevigatus	Smooth Puffer	4	0	0	0	1	0	0	0	0	0	1	6
Stegastes partitus	Bicolor Damselfish	2	3	0	0	0	0	0	0	0	1	0	6
Bodianus pulchellus	Spotfin Hogfish	0	2	0	3	0	0	0	0	0	0	0	5
Sphyraena barracuda	Great Barracuda	0	0	0	0	1	1	1	1	0	0	1	5
Carcharhinus plumbeus	Sandbar Shark	1	0	0	0	0	0	0	0	1	0	2	4
Halichoeres bivittatus	Slippery Dick	4	0	0	0	0	0	0	0	0	0	0	4
Lagodon rhomboides	Pinfish	2	0	0	0	0	0	0	0	0	1	1	4
Stegastes variabilis	Cocoa Damselfish	3	0	0	0	0	1	0	0	0	0	0	4
Chaetodipterus faber	Atlantic Spadefish	0	0	0	0	0	0	0	0	0	3	0	3
Prognathodes aya	Bank Butterflyfish	1	0	0	2	0	0	0	0	0	0	0	3
Rachycentron canadum	Cobia	2	1	0	0	0	0	0	0	0	0	0	3

Table 2. Continued.

Scieintific name	Common name	Geologic				Artificial						Unknown	
		HB	FB	MB	LD	UR	CM	CP	VL	MT	RM	PH	Total
Bodianus rufus	Spanish Hogfish	0	0	0	0	0	1	0	0	1	0	0	2
Calamus bajonado	Jolthead Porgy	0	0	0	0	0	0	0	2	0	0	0	2
Canthigaster jamestyleri	Goldface Toby	0	0	0	0	0	1	0	1	0	0	0	2
Caranx hippos	Crevalle Jack	0	0	0	0	0	1	0	0	0	0	1	2
Carcharhinus leucas	Bull Shark	0	0	0	0	1	0	0	0	0	0	1	2
Ginglymostoma cirratum	Nurse Shark	0	0	0	0	0	1	1	0	0	0	0	2
Parablennius marmoreus	Seaweed Blenny	2	0	0	0	0	0	0	0	0	0	0	2
Paranthias furcifer	Atlatnic Creolefish	0	0	0	2	0	0	0	0	0	0	0	2
Prionotus rubio	Blackwing Searobin	0	0	0	0	0	0	0	0	0	0	2	2
Seriola fasciata	Lesser Amberjack	0	0	0	0	0	0	0	0	0	2	0	2
Acanthostracion polygonius	Honeycomb Cowfish	1	0	0	0	0	0	0	0	0	0	0	1
Epinephelus itajara	Goliath Grouper	0	0	0	0	0	0	0	0	0	0	1	1
Gonioplectrus hispanus	Spanish Flag	0	0	0	1	0	0	0	0	0	0	0	1
Gymnothorax moringa	Spotted Moray	0	0	0	1	0	0	0	0	0	0	0	1
Haemulon plumierii	White Grunt	0	0	0	0	0	0	0	0	0	1	0	1
Hyporthodus niveatus	Snowy Grouper	0	0	0	0	0	0	0	0	0	0	1	1
Liopropoma eukrines	Wrasse Basslet	0	0	0	1	0	0	0	0	0	0	0	1
Orthopristis chrysoptera	Pigfish	0	0	0	0	1	0	0	0	0	0	0	1
Paradiplogrammus bairdi	Lancer Dragonet	0	0	0	0	0	0	0	0	0	0	1	1
Pronotogrammus martinicensis	Roughtongue Bass	0	0	0	1	0	0	0	0	0	0	0	1
Serranus annularis	Orangeback Bass	1	0	0	0	0	0	0	0	0	0	0	1
Serranus notospilus	Saddle Bass	0	0	0	0	0	0	0	0	0	0	1	1
Sphyrna lewini	Scalloped Hammerhead	0	0	0	0	0	1	0	0	0	0	0	1
Sphyrna mokarran	Great Hammerhead	0	0	0	0	0	0	0	0	0	0	1	1
	Total	2,984	1,109	205	259	1,524	941	572	259	229	1,298	2,421	11,801

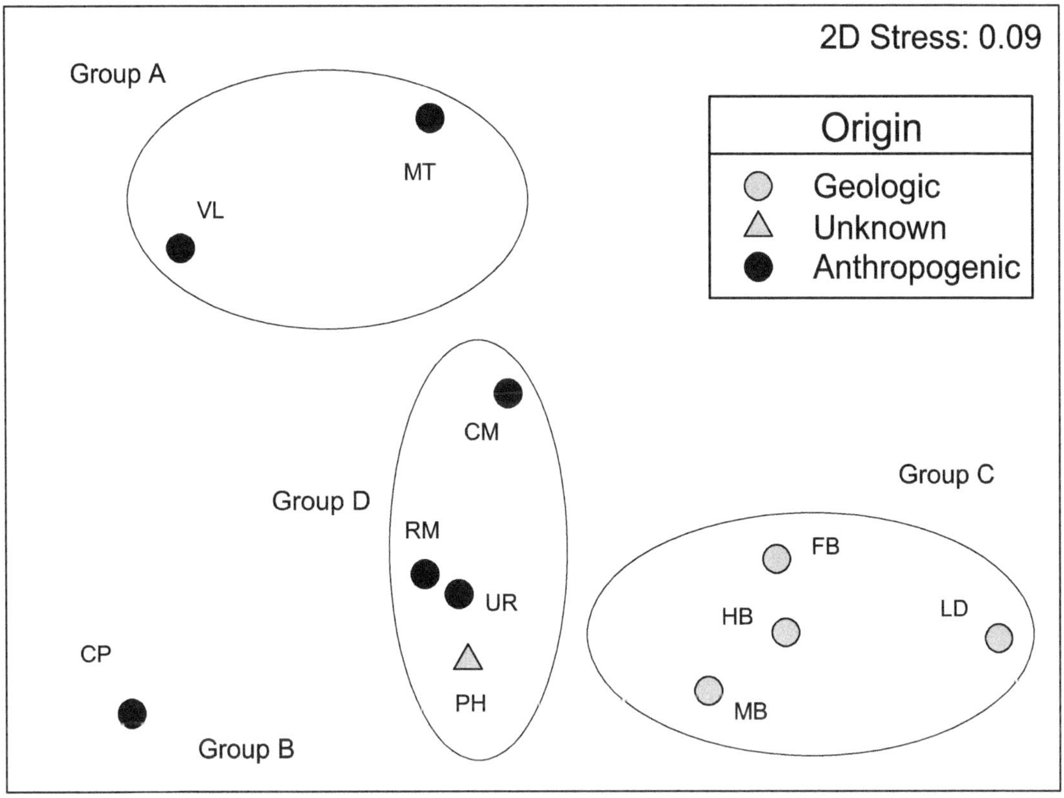

Figure 2. Nonmetric multidimensional ordination of reef-fish assemblage species abundance by habitat class representing geologic, anthropogenic, and unknown habitats. Lines represent significant clusters as identified via SIMPROF analyses. See Table 1 for habitat class abbreviations.

Reeffish *Chromis enchrysura* were more abundant in geologic habitats. Economically important Red Snapper and Vermilion Snapper were abundant in all habitats, with greatest abundance from chicken-transport cages, while Greater Amberjack and Gray Snapper *Lutjanus griseus* were more abundant from vessels and military tanks. Lionfishes were also observed in all habitats, with greatest abundances observed on chicken-transport cages.

While not quantitatively assessed, fishes contributing to differences in reef-fish community structure among habitat groups also exhibited marked variability in size distribution among habitat groups. These results should be interpreted cautiously due to substantially greater sampling effort and subsequently greater number of overall length measurements coming from group C (geologic habitats) and group D (reef modules, construction material, and unidentified depressions). The largest Red Snapper were observed in association with habitat groups C and D while the smallest Red Snapper were associated with group B (Figure 4). Habitat-related size composition of Gray Snapper was not as distinct, but larger individuals were observed in habitat group C (Figure 4). Vermilion Snapper length distributions overlapped among habitats, but small individuals (≤100 mm fork length) were observed only

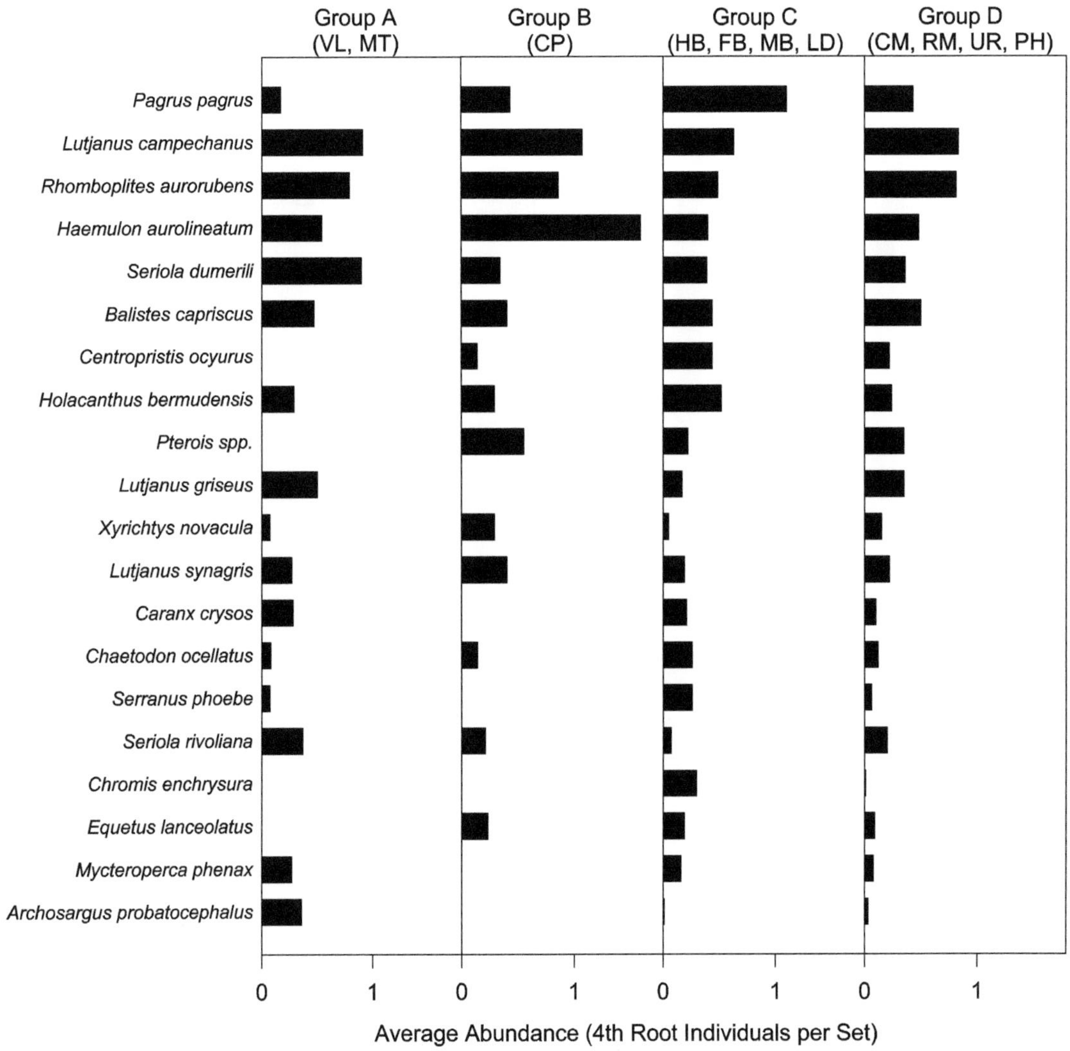

Figure 3. Fourth-root transformed abundance data for species that contributed up to 80% of differences among habitat groups. See Table 1 for habitat class abbreviations.

in association with habitat group D (Figure 5). This pattern was similar in Red Porgy, with the smaller individuals observed in habitat group D and larger individuals observed from geologic habitats in group C (Figure 5). The largest Gray Triggerfish were observed in group C and D habitats, whereas the smaller lionfishes were associated with habitat group D (Figure 6).

In the study region, 601.108 km^2 of the ocean bottom was scanned in support of reef-fish survey efforts, including areas in and outside Florida artificial reef permit areas. Of this total, approximately 1.2% of the area (7.155 km^2), represented by 5,755 individual polygons, was coded as reef habitat (Table 3). Flat hard bottom (5.577 km^2) comprised the greatest area of geologic habitat; construction materials (0.136 km^2) comprised the greatest area of artificial habitat. Unidentified depressions accounted for 373 polygons, totaling 0.025 km^2. In the scanned

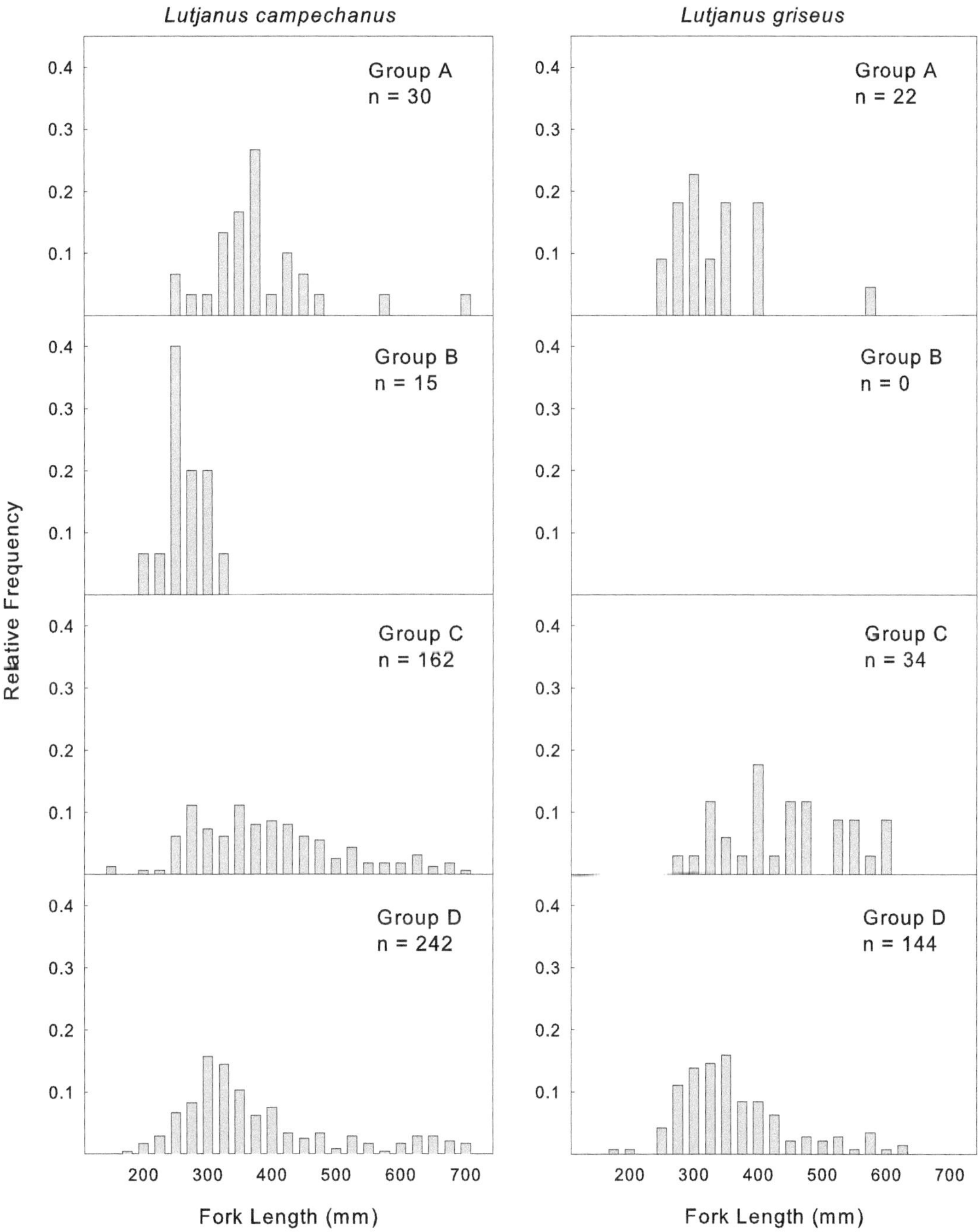

Figure 4. Length-frequency (mm fork length) histograms for Red Snapper and Gray Snapper separated by habitat groups. Habitat groups included group A—VL, MT; group B—CP; group C—HB, FB, MB, LD; and group D—CM, RM, PH, UR. See Table 1 for habitat class abbreviations.

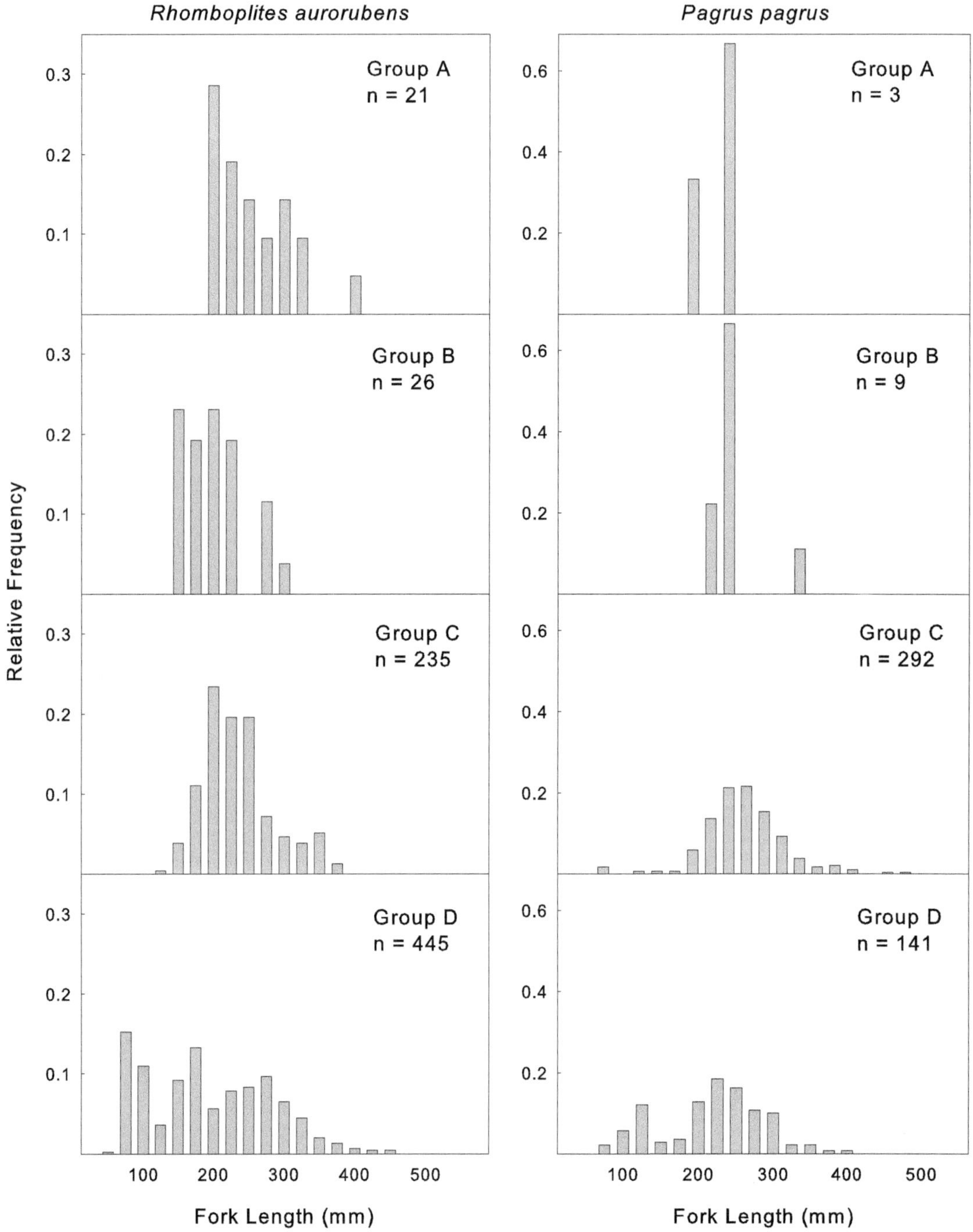

Figure 5. Length-frequency (mm fork length) histograms for Vermilion Snapper and Red Porgy separated by habitat groups. Habitat groups included group A—VL, MT; group B—CP; group C—HB, FB, MB, LD; and group D—CM, RM, PH, UR. See Table 1 for habitat class abbreviations.

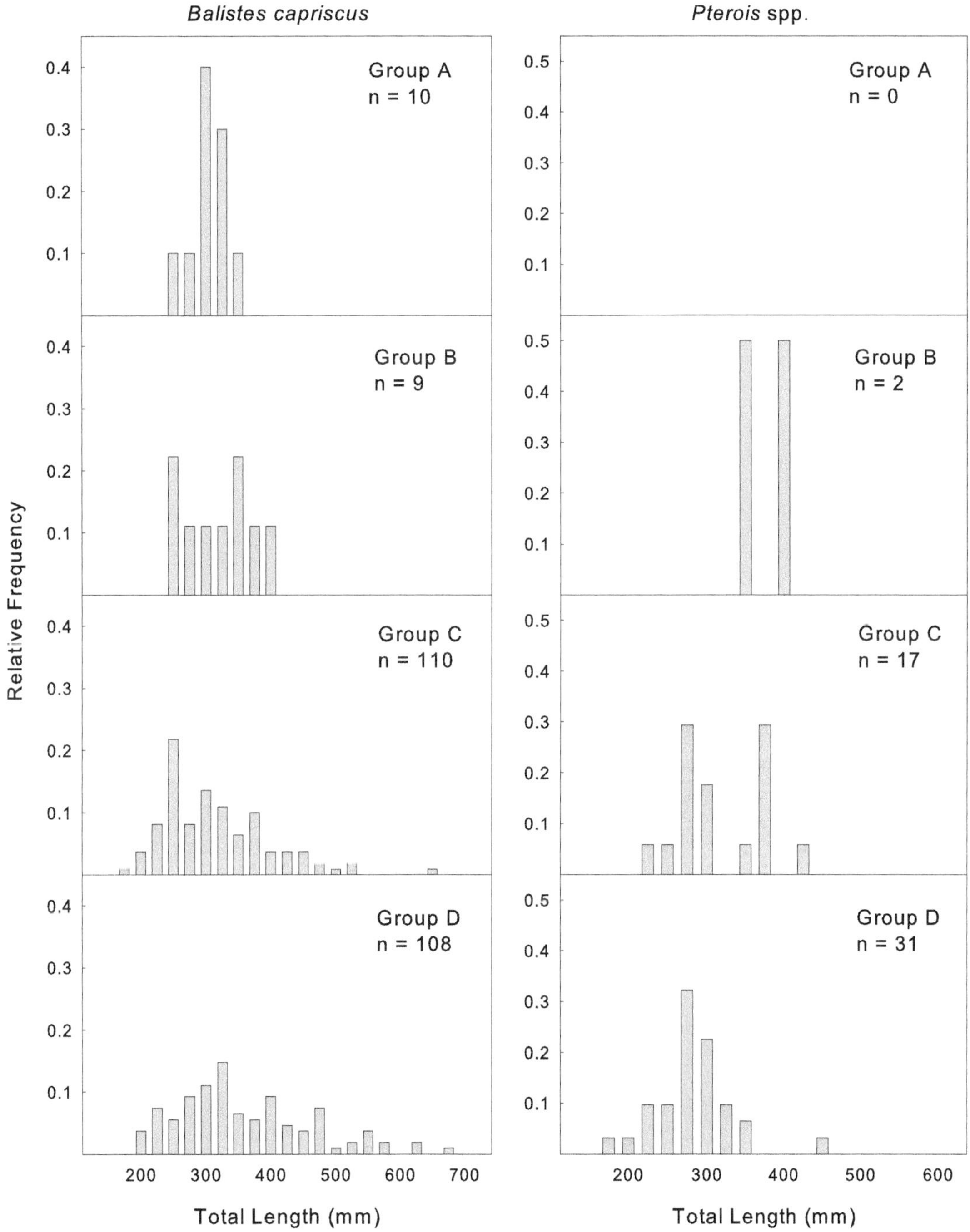

Figure 6. Length-frequency (mm total length) histograms for Gray Triggerfish and lionfishes separated by habitat groups. Habitat groups included group A—VL, MT; group B—CP; group C—HB, FB, MB, LD; and group D—CM, RM, PH, UR. See Table 1 for habitat class abbreviations.

Table 3. Habitat parameters from side-scan-sonar mapping by habitat class. The number of individual habitat features (See Table 1 for abbreviations) and their total area are summarized for all habitats. The area coverage of artificial habitats respective to published Florida artificial reef permit zones is also summarized. Area reported in permit area includes habitats within 100 m of known artificial reef deployments (points); unreported in the permit area are habitats identified in the permit area boundary but outside 100 m.

	Habitat class											
	Geologic				Artificial						Unknown	
	HB	FB	MB	LD	UR	CM	CP	VL	MT	RM	PH	Total
Number of polygons	1,892	692	37	86	1,225	389	236	30	13	782	373	5,755
Total area (km^2)	5.5769	1.0812	0.1327	0.0800	0.0618	0.1362	0.0077	0.0278	0.0010	0.0246	0.0248	7.1546
Area reported inside artificial reef permit zone					0.0081	0.0790	0.0001	0.0125	0.0006	0.0189		0.1192
Area unreported inside artificial reef permit zone					0.0270	0.0489	0.0028	0.0053	0.0004	0.0048		0.0892
Area outside artificial reef permit zone					0.0266	0.0083	0.0048	0.0100	–	0.0009		0.0506

areas of the FWC artificial reef permit areas, most of the mapped artificial habitat had previously been identified or reported, but we identified many unreported artificial reefs. Most artificial habitats exhibited greater coverage within permit areas (i.e., in the FWC database, within 100 m of a documented artificial reef), but unidentified artificial reefs (from side-scan sonar interpretation) and chicken-transport cages also occurred outside the FWC reef permit areas. Less total artificial habitat identified was outside of permit areas (0.051 km^2) than inside (0.208 km^2); however, there was greater coverage of chicken-transport cages outside permit areas (0.005 km^2 versus 0.003 km^2), indicating that these reefs were likely deployed predominantly as private reefs. Since many of these chicken-transport cages had a small footprint (approximately 30 m^2), our limited mapping effort identified hundreds of unreported individual reefs based off the comparative area identified.

Discussion

We demonstrated the utility of incorporating artificial reef surveys into broad-scale (>10,000 km^2), shelf-wide surveys of reef fishes focused primarily on geologic habitats. Most studies comparing reef-fish assemblages between artificial and geologic habitats in the Gulf of Mexico have been limited in spatial extent or have incorporated a limited number of habitat types (Rooker et al. 1997; Patterson et al. 2014; Streich et al. 2017). By examining similarities and differences in reef-fish assemblages between artificial and geologic features, these studies improve understanding of the influence of artificial reefs on marine fish populations (Rooker et al. 1997; Streich et al. 2017). Additionally, these baseline data can help in monitoring changes in fish assemblages associated with environmental perturbations, including invasive species introductions (e.g., lionfishes). Nevertheless, more comprehensive efforts on broad spatial scales are required to most effectively support fishery assessment and management. As did the above-mentioned studies, our study indicated significant differences in assemblage structure and, for certain taxa, size composition between artificial and geologic habitats. Although most species were observed in association with both artificial and geologic habitats, many species were more abundant in artificial habitats, so artificial reef habitats are likely to serve locally as important habitat for a significant proportion of managed reef-fish populations.

Results from this study indicate that artificial reefs can be sampled with BRUVs, although important questions remain as to the overall effectiveness of this approach, including evaluation of gear selectivity (Wells et al. 2008; Patterson et al. 2014). In general, stationary camera gear is more effective at sampling reef-associated species across a wide range of sizes in comparison to traditional hooked gears and traps (Bacheler et al. 2013). Nevertheless, there are concerns regarding deploying remote sampling gear near complex habitats such as sunken vessels, complex artificial reef modules, or naturally forming pinnacles. While most geologic habitats in the northern Gulf of Mexico consist of low-relief reefs covering large expanses, artificial reefs typically have a smaller footprint with higher relief, presenting a twofold challenge to field logistics. Chiefly, it may be difficult to effectively target small-scale reef features with stationary sampling gear deployed remotely from the ocean surface. Second, retrieval of sampling gear near high-relief reef habitats is also challenging. During this study, cameras were deployed 37 times over high-relief (>2 m) artificial habitats, including sunken vessels and tanks and reef modules, as well as six times on ledges (high-relief geologic habitats). All gear was recovered. We also successfully deployed and recovered BRUVs on small-scale geologic features such as depressions (n = 92), many of which were only 3–4 m across. Detectability of fish should also be considered. Many fish may occupy positions above the field of view of the cameras (i.e., higher in the water column) or reside within complex structure out of view of the BRUV (a characteristic of high-relief artificial and geologic reefs). Measures of selectivity and gear effectiveness on reef habitats may therefore benefit from the use of complementary sampling approaches, including active acoustics and ROVs (Dance et al. 2011; Streich et al. 2017).

Evidence suggested that overall fish assemblages differed between artificial and geologic habitats, but the breadth of our sampling effort revealed other differences among several artificial reef habitat groups. All four geologic habitats supported similar reef-fish assemblages that generally differed from those of artificial habitats by more abundant smaller-bodied species such as Blue Angelfish, Bank Sea Bass, Yellowtail Reeffish, and Tattler *Serranus phoebe*. These species are more likely to feed on benthic invertebrates than are the more pelagic and piscivorous species such as Red Snapper or Greater Amberjack (Nelson and Bortone 1996). In contrast, the planktivorous Tomtate (Norberg 2015) occurred in greater densities in artificial reef habitats. Among artificial reef habitats, several distinct habitat groupings were evident. Interestingly, chicken-transport cages supported a unique assemblage characterized by greater abundances of Tomtate, Red Snapper, and lionfishes. Chicken-transport cages, often featuring strong vertical relief (2–3 m; author's personal observation), were generally isolated and surrounded by sand. Because they were most often seen in areas devoid of other reef habitats, they are probably private reefs that few anglers could find. Therefore, these habitats may serve as refugia, experiencing lower fishing rates than other better known reefs, natural or artificial (Addis et al. 2013). Further information is needed to support or refute this assertion, however, as length frequency data from Red Snapper indicated that juveniles also may recruit to these types of habitats (Szedlmayer and Conti 1999). Unidentified depressions, initially considered similar in origin to potholes excavated and maintained by Red Grouper (Coleman et al. 2010; Harter et al. 2017), were found to be chiefly anthropogenic. Red Grouper were never observed in association with these unidentified depressions, which were more likely created by mechanisms such as scouring or excavation from other species. Similarity in assemblage structure among unidentified depressions, unidentified artificial reefs, reef modules, and construction material is notable as it may indicate depressions function similarly to artificial habitats. The ecological function of these habitats warrants further investigation because for several managed reef species (e.g., Red Snapper, Gray Snapper, Vermilion Snapper, and Gray Triggerfish), they typically supported larger individuals.

Habitat

In general, our understanding of the distribution of benthic habitats in the nearshore waters of the northeastern Gulf of Mexico is limited, although general descriptions from Davis (2017) and McBride and Byrnes (1995) summarize sediment composition and large-scale structure. Habitat mapping efforts in support of this project yielded valuable insights into the quantity and distribution of reef habitats in the region. Within the area described in this study (10,885 km^2), more than 600 km^2 has been surveyed and approximately 1% of the area we mapped contained reef habitat. The majority of the reef habitat identified (i.e., flat hard bottom [5.577 km^2]) has little vertical relief, although these habitats often support rich epifaunal assemblages (e.g., sponges, soft corals, etc.). Though not within the scope of this project, the spatial distribution and composition of reef habitat in the area as elucidated in this study does provide insight that will be helpful for planned restoration activities. The large areas of unconsolidated sediment we mapped could be considered for future artificial reef deployments or considered as potential sources for beach renourishment (i.e., the addition of sand to beaches) projects.

The state of Florida has one of the most active artificial reef programs in the United States (FWC 2003), with thousands of artificial reefs in its waters and 3,000 more planned in association with Natural Resource Damage Assessment oil spill recovery efforts (Mille, personal communication). Within the study area, more than 600 artificial reefs have been documented, representing more than 20% of all the artificial reefs known by FWC. Although most reefs occur in artificial reef permit zones established by the FWC, our mapping effort identified hundreds of additional, unreported private reefs on the shelf. When considering that only 5% of the study area has been mapped, well more than 3,000 such habitats may exist within the unmapped portions of the region. Individual private reefs are small in scale, but cumulatively, they may represent a significant component of reef habitat in the region.

Recommendations

This study is an important first step toward incorporating standardized methods for evaluating reef-fish communities across multiple habitats in a large region. Our results highlight the advantages of a randomized approach to habitat mapping in support of fishery-independent surveys. This approach provides reasonable estimates of habitat quantity and quality that are likely representative of those in unmapped areas; therefore, information from these surveys can reasonably be extrapolated to provide regionwide estimates of habitat availability. There is increased demand for finer-scale spatial data to further inform the assessment and management of reef-fishes, and the collection of concurrent fish-habitat data is critical for accurate regionwide estimates of population status (Campbell et al. 2015; Karnauskas et al. 2017; Streich et al. 2017). The sampling gear used in this study (i.e., stereo-video-equipped BRUVs) provided valuable data (counts and lengths) for multiple species across complex habitats in a standardized manner. Moreover, data from this survey have already been used to produce regionwide estimates of size composition of Red Snapper for artificial and geologic reef habitats (Walter et al. 2017). Moving forward, data from FWC and ongoing NMFS surveys will be modeled to generate relative abundance indices that more accurately reflect temporal changes in population abundance by including data from both geologic and artificial reef habitats.

Many unreported artificial habitats have been identified through habitat mapping, and considerations should be made as to how to appropriately incorporate these unreported artificial reefs into the survey design (Dance 2008; Addis et al. 2013). Fishing pressure and fishing mortality are known to differ between reported and unreported artificial reefs (Simard et al. 2016), but with improved navigational and acoustic sensing technology, anglers eventually discover most larger, private reefs. Many of the unreported reefs identified in this study are small and so less likely to be exploited. If these habitats were incorporated into ongoing surveys as a distinct habitat type, information would be generated to better explain their function compared with those of reported and geologic reefs in the region. Artificial reefs are used extensively as restoration tools to mitigate lost angling opportunities, so their function in the Gulf and their relative influence on managed reef species (e.g., Red Snapper, Gray Triggerfish, and Greater Amberjack) merits standardized and routine monitoring to improve stock assessment and management efforts. Our findings indicated that similar reef-fish communities exist on different artificial reef materials, which will contribute to evolving fishery-independent survey design, improve efficiency, and increase confidence in relative abundance estimates.

Acknowledgments

We gratefully acknowledge staff of the Florida Fish and Wildlife Conservation Commission's Fish and Wildlife Research Institute who put in countless hours collecting and processing side-scan sonar and video data, especially R. Caillouet, S. Parks, B. Pittinger, and N. Roman for their oversight of data-processing activities. We thank E. Matheson, M. Schrandt, and B. Crowder for thoughtful insight and comments on this manuscript. Thanks to captains and crew of R/V *Apalachee*, Tommy Munro, and Kimberly Dawn for assistance in collecting side-scan sonar and video data. Funding for this project was provided by the National Fish and Wildlife Foundation Gulf Environmental Benefit Fund (grant numbers FL 40624, FL 45766, FL 50347) and by proceeds of State of Florida saltwater fishing license sales. The statements, findings, views, conclusions, and recommendations contained in this document are those of the authors, do not necessarily reflect the views of the U.S. Department of Commerce or the National Fish and Wildlife Foundation, and should not be interpreted as representing the opinions or policies of the U.S. Government or the National Fish and Wildlife Foundation. Mention of trade names or commercial products does not constitute their endorsement by the U.S. Government or the National Fish and Wildlife Foundation.

References

Adams, C., W. J. Lindberg, and J. Stevely. 2011. The economic benefits associated with Florida artificial reefs. Florida Cooperative Extension Service, Institute of Food and Agricultural Sciences, University of Florida, Gainesville.

Addis, D. T., W. F. Patterson, III, M. A. Dance, and G. W. Ingram, Jr. 2013. Implications of reef fish movement from unreported artificial reef sites in the northern Gulf of Mexico. Fisheries Research 147:349–358.

Ault, J. S., J. A. Bohnsack, S. G. Smith, and J. Luo. 2005. Towards sustainable multispecies fisheries in the Florida, USA, coral reef ecosystem. Bulletin of Marine Science 76:595–622.

Bacheler, N. M., C. M. Schobernd, Z. H. Schobernd, W. A. Mitchell, D. J. Berrane, G. T. Kellison, and M. J. M. Reichert. 2013. Comparison of trap and underwater video gears for indexing reef fish presence and abundance in the southeast United States. Fisheries Research 143:81–88.

Baine, M. 2001. Artificial reefs: a review of their design, application, management, and performance. Ocean and Coastal Management 44:241–259.

Biesinger, Z., B. M. Bolker, D. Marcinek, and W. J. Lindberg. 2013. Gag (*Mycteroperca microlepis*) space-use correlations with landscape structure and environmental conditions. Journal of Experimental Marine Biology and Ecology 443:1–11.

Bohnsack, J. A., and D. L. Sutherland. 1985. Artificial reef research: a review with recommendations for future priorities. Bulletin of Marine Science 37:11–39.

Brooks G. R., L. J. Doyle, R. A. Davis, N. T. DeWitt, and B. C. Suthard. 2003. Patterns and controls of surface sediment distribution: west-central Florida inner shelf. Marine Geology 200:307–324.

Campbell, M. D., K. R. Rademacher, M. Hendon, P. Felts, B. Noble, J. Salisbury, and J. Moser. 2015. SEAMAP reef fish video survey: relative indices of abundance of Vermilion Snapper. SEDAR (Southeast Data, Assessment and Review), SEDAR45-WP-11, North Charleston, South Carolina.

Clarke, K. R., and R. N. Gorley. 2015. PRIMER v7: user manual/tutorial. PRIMER-E, Plymouth, UK.

Coleman, F. C., C. C. Koenig, K. M. Scanlon, S. Heppell, S. Heppell, and M. W. Miller. 2010. Benthic habitat modification through excavation by Red Grouper, *Epinephelus morio*, in the northeastern Gulf of Mexico. Open Fish Science Journal [online serial] 3:1–15.

Dance, M. A. 2008. Reef fish community structure at unpublished artificial reef sites off northwest Florida: implications for harvest refugia. Master's thesis. University of West Florida, Pensacola.

Dance, M. A., W. F. Patterson, III, and D. T. Addis. 2011. Fish community and trophic structure at artificial reef sites in the northwestern Gulf of Mexico. Bulletin of Marine Science 87:301–324.

Davis, R. A. Jr. 2017. Sediments of the Gulf of Mexico. Pages 165–215 in C. H. Ward, editor. Habitats and biota of the Gulf of Mexico: before the Deepwater Horizon oil spill, volume 1. Springer, New York.

Dupont, J. M. 2008. Artificial reefs as restoration tools: a case study on the west Florida shelf. Coastal Management 36:495–507.

FWC (Florida Fish and Wildlife Conservation Commission). 2003. State of Florida artificial reef strategic plan. FWC, Tallahassee, Florida.

Harter, S. L., H. Moe, J. K. Reed, and A. W. David. 2017. Fish assemblages associated with Red Grouper pits at Pulley Ridge, a mesophotic reef in the Gulf of Mexico. U.S. National Marine Fisheries Service Fishery Bulletin 115:419–432.

Hine, A. C., and S. D. Locker. 2011. Florida continental shelf: great contrasts and significant transitions. Pages 101–127 *in* N. A. Buster and C. W. Holmes, editors. Gulf of Mexico: origin, waters and biota. Volume 3: geology. Texas A&M University Press, College Station.

Karnauskas, M., J. R. Walter, III, M. D. Campbell, A. G. Pollack, J. M. Drymon, and S. Powers. 2017. Red Snapper distribution on natural habitats and artificial structures in the northern Gulf of Mexico. Marine and Coastal Fisheries: Dynamics, Management, and Ecosystem Science 9:50–67.

Keithly, W. R., Jr., and K. J. Roberts. 2017. Commercial and recreational fisheries of the Gulf of Mexico. Pages 1039–1188 *in* C. H. Ward, editor. Habitats and biota of the Gulf of Mexico: before the *Deepwater Horizon* oil spill. Springer, New York.

Leeworthy, V. R., T. Maher, and E. A. Stone. 2006. Can artificial reefs reduce or alter user pressure on adjacent natural reefs? Bulletin of Marine Science 78:29–37.

Lindberg, W. J. 1997. Can science solve the attraction versus production debate? Fisheries 22(4):10–13.

Lindberg, W. J., T. K. Frazer, K. M. Portier, F. Vose, J. Loftin, D. J. Murie, D. M. Mason, B. Nagy, and M. K. Hart. 2006. Density-dependent habitat selection and performance by a large mobile reef fish. Ecological Applications 16:731–746.

Mallinson, D., A. Hine, D. Naar, S. Locker, and B. Donahue. 2014. New perspectives on the geology and origin of the Florida Middle Ground carbonate banks, west Florida shelf, USA. Marine Geology 355:54–70.

McBride, R. A., and M. R. Byrnes. 1995. Surficial sediments and morphology of the southwestern Alabama/western Florida Panhandle coast and shelf. Gulf Coast Association of Geological Societies Transactions 45:392–404.

NMFS (National Marine Fisheries Service). 2017. Fisheries economics of the United States, 2015. NOAA Technical Memorandum NMFS-F/SPO-170.

Nelson, B. D., and S. A. Bortone. 1996. Feeding guilds among artificial-reef fishes in the northern Gulf of Mexico. Gulf of Mexico Science 14(2):66–80.

Norberg, M. J. 2015. The ecology of Tomtate, Haemulon aurolineatum, in the northern Gulf of Mexico and the effects of the *Deepwater Horizon* oil spill. Master's thesis. University of South Alabama, Mobile.

Obrochta, S. P., D. S. Duncan, and G. R. Brooks. 2003. Hardbottom development and significance to the sediment-starved west-central Florida inner continental shelf. Marine Geology 200:291–306.

Patterson, W. F., III, J. H. Tarnecki, D. T. Addis, and L. R. Barbieri. 2014. Reef fish community structure at natural versus artificial reefs in the Northern Gulf of Mexico. Proceedings of the Gulf and Caribbean Fisheries Institute 66:4–8.

Rotherham, D., A. J. Underwood, M. G. Chapman, and C. A. Gray. 2007. A strategy for developing scientific sampling tools for fishery-independent surveys of estuarine fish in New South Wales, Australia. ICES Journal of Marine Science 64:1512–1516.

Rooker, J. R., Q. R. Dokken, C. V. Pattengill, and G. J. Holt. 1997. Fish assemblages on artificial and natural reefs in the Flower Garden Banks National Marine Sanctuary, USA. Coral Reefs 16:83–92.

Simard, P., K. R. Wall, D. A. Mann, C. C. Wall, and C. D. Stallings. 2016. Quantification of boat visitation rates at artificial and natural reefs in the eastern Gulf of Mexico using acoustic recorders. PLOS (Public Library of Science) ONE [online serial] 11(8):e0160695.

Simonsen, K. A., J. H. Cowan, Jr., and K. M. Boswell. 2015. Habitat differences in the feeding ecology of Red Snapper (*Lutjanus campechanus*, Poey 1860): a comparison between artificial and natural reefs in the northern Gulf of Mexico. Environmental Biology of Fishes 98:811 – 824.

Stanley, D. R., and C. A. Wilson. 2000. Variation in the density and species composition of fishes associated with three petroleum platforms using dual beam hydroacoustics. Fisheries Research 47:161–172.

Stone, R. B., L. M. Sprague, J. M. McGurrin, and W. Seaman, Jr. 1991. Artificial habitats of the world: synopsis and major trends. Pages 31–57 *in* W. Seaman and L. Sprague, editors. Artificial habitats for marine and freshwater fisheries. Academic Press, San Diego, California.

Streich, M. K., M. J. Ajemian, J. J. Wetz, and G. W. Stunz. 2017. A comparison of fish community structure at mesophotic artificial reefs and natural banks in the western Gulf of Mexico. Marine and Coastal Fisheries: Dynamics, Management, and Ecosystem Science 9:170–189.

Szedlmayer, S. T., and J. Conti. 1999. Nursery habitats, growth rates, and seasonality of age-0 Red Snapper, *Lutjanus campechanus*, in the northeast Gulf of Mexico. U.S. National Marine Fisheries Service Fishery Bulletin 97:626–635.

Thompson, K. A., T. S. Switzer, M. C. Christman, S. F. Keenan, C. Gardner, M. Campbell, and A. Pollack. 2017a. Indices of abundance for Gray Snapper *Lutjanus griseus* using combined data from three independent video surveys. SEDAR (Southeast Data, Assessment and Review), SEDAR51-DW15, North Charleston, South Carolina.

Thompson, K. A., T. S. Switzer, and S. F. Keenan. 2017b. Indices of abundance for Red Snapper *Lutjanus campechanus* from the NFWF-expanded Florida Fish and Wildlife Research Institute (FWRI) video survey in the eastern Gulf of Mexico. SEDAR (Southeast Data, Assessment and Review), SEDAR52-WP-04, North Charleston, South Carolina.

Topping, D. T., and S. T. Szedlmayer. 2011. Site fidelity, residence time and movement of Red Snapper *Lutjanus campechanus* estimated with long-term acoustic monitoring. Marine Ecology Progress Series 437:183–200.

Wall, C. C., 2012. Shelf-scale mapping of fish distribution using active and passive acoustics. Doctoral dissertation. University of South Florida, Tampa.

Walter, J., M. Drymon, C. Gardner, C. Hightower, W. Patterson, S. Powers, K. Thompson and T. Switzer. 2017. A multinomial predictive model to incorporate visual surveys of red snapper lengths. SEDAR (Southeast Data, Assessment and Review), SEDAR52-WP-22, North Charleston, South Carolina.

Wells, R. J. D., K. M. Boswell, J. H. Cowan, Jr., and W. F. Patterson, III. 2008. Size selectivity of sampling gears targeting Red Snapper in the northern Gulf of Mexico. Fisheries Research 89:294–299.

American Fisheries Society Symposium 86:165–186, 2018

Restoring a Nearshore Rocky Reef Ecosystem in the Challenge of an Urban Setting

Daniel J. Pondella II*, Jonathan P. Williams, and Chelsea M. Williams
Vantuna Research Group, Occidental College
1600 Campus Road, Los Angeles, California 90041, USA

Jeremy T. Claisse
Biological Science Department, California State Polytechnic University–Pomona
3801 West Temple Avenue, Pomona, California 91768, USA

David Witting
National Oceanic and Atmospheric Administration, Montrose Settlements Restoration Program
501 West Ocean Boulevard, Suite 4470, Long Beach, California 90802, USA

Abstract.—The subtidal rocky reefs that surround the Palos Verdes Peninsula in Los Angeles County, California, USA are subject to multiple anthropogenic impacts, including many (at least partially) human-induced landslides over the past half century, which have resulted in chronic sedimentation (e.g., reef burial and scour) and associated turbidity effects along a major stretch of this coastline. The amount of rocky reef habitat has significantly decreased; as such, determining the optimum technique for restoring this lost habitat is the focus of this study. Over the past decade, we mapped and intensively surveyed the nearshore physical and biological characteristics both inside this impacted area and in surrounding reference areas that contain extensive rocky reef habitat with established kelp forests. Notable among all survey locations is a relatively high-relief (~5 m) area of reef within the sediment impacted area that consistently has the highest fish biomass density among anywhere on the peninsula. The high structural relief prevents sediment accumulation, scour, and subsequent reef burial, and this reef ultimately served as the example for the design of sets of quarry rock reef "blocks" that together form the proposed restoration reef. Our primary objective was to use the 63,500 metric tons of quarry rock the budget would allow us to create the most productive habitat by restoring the natural reef environment while balancing scientific study design considerations (i.e., replicated reef components at multiple spatial scales) with maximizing the potential for an effective restoration effort across the range of important species and overall kelp forest biodiversity. To meet this objective, we considered multiple criteria that incorporated engineering specifications and

* Corresponding author: pondella@oxy.edu

biological performance and were informed by the scientific literature and results of natural and artificial reef surveys in this region. Ultimately, the design incorporated heterogeneity at multiple spatial scales while attempting to maximize high relief components, surface area to volume ratio, perimeter, ecotones, and small-scale current flow features and nutrient flux and while being consistent with the size of natural reefs along the Palos Verdes Peninsula. Further, placement and spacing of individual reef blocks (i.e., 2 × 48 m heterogeneous quarry rock reefs) included space for sand channels between blocks to permit sediment transport and create sand/rock ecotone habitats while remaining close enough to each other and existing natural reefs to maintain biological connectivity. Reef blocks were also located at the depth (15–20 m) where the most productive reef habitat in the region was observed. Finally, we discuss a proposed pre- and post-construction monitoring program and additional studies that could be performed that would leverage the replicated elements in the restoration reef design to inform future reef restoration programs.

Setting

The nearshore environment of the Palos Verdes Peninsula in Los Angeles County, California, USA (Figure 1) has been intensively studied for decades. The subtidal rocky reefs that surround this headland have garnered attention due to multiple anthropogenic impacts (e.g., commercial and recreational fishing, the establishment of marine protected areas, giant kelp beds lost to pollution, and landslides; Foster and Schiel 2010; Pondella 2009; Stull et al. 1987). Historically, the greatest deleterious impact to the reefs at Palos Verdes was the loss of its kelp beds due to pollution from the Joint Water Pollution Control Plant's Whites Point outfall. By 1960, untreated sewage had reduced the kelp canopy on the peninsula to small patches at Abalone Cove and in Portuguese Bend (Figure 2; North 1964). To exacerbate the situation, road construction on Palos Verdes Drive triggered the Portuguese Bend landslide in 1956. From 1956 to 1999, approximately 5.7–9.4 million metric tons (mt) of sediment slid onto the inner shelf (Kayen et al. 2002). By 1999, due to infrastructure improvements, the landslide slowed appreciably with present sediment releases primarily associated with wave action. Nonetheless, the biological damage has been extreme (highlighted by the loss of the Portuguese Bend kelp bed), leaving only the Abalone Cove kelp bed by 1974. A series of infrastructure improvements have been made to the Whites Point outfall that helped remediate impacts to the local environment. Between 1937 and 1967, the three deep outfalls were built and the two deepest outfalls, ~2.4 km offshore in 61 m of water, are used. In 1984, partial secondary treatment of the flow (60:40 mix of secondary : primary) was implemented and this continued until late 2002 with the implementation of 100% secondary effluent treatment. In the early 1970s, giant kelp *Macrocystis pyrifera* restoration efforts at the Palos Verdes Peninsula were successful and kelp remains present to this time (Foster and Schiel 2010).

While these restoration and enhancement efforts ameliorated the historical consequences of the Whites Point outfall

Figure 1. Satellite image of the Palos Verdes Peninsula (image adapted from NASA Jet Propulsion Laboratory taken on October 18, 2016).

throughout the peninsula, sedimentation and associated turbidity continue to have chronic impacts. First, there is continued turbidity related to sediment transport and scour associated with the sediment deposited in Portuguese Bend from the landslide (Figure 3). Presently, turbidity is caused by wave action and is much reduced compared to historical values, as can be observed by the plume prior to stabilization of the landslide (Figure 3, left photo). Further exacerbating this influx of sediment was the 6.5-ha landslide on June 2, 1999 from the 18th hole of what is now the Trump National Golf Club, positioned above Bunker Point. Reef burial via sediments near Bunker Point was not observed in extensive surveys of this region in the 1990s (Stephens et al. 1996) but has been observed in more recent surveys (Pondella et al. 2012, 2015a). Proximity to the Trump National Golf Club

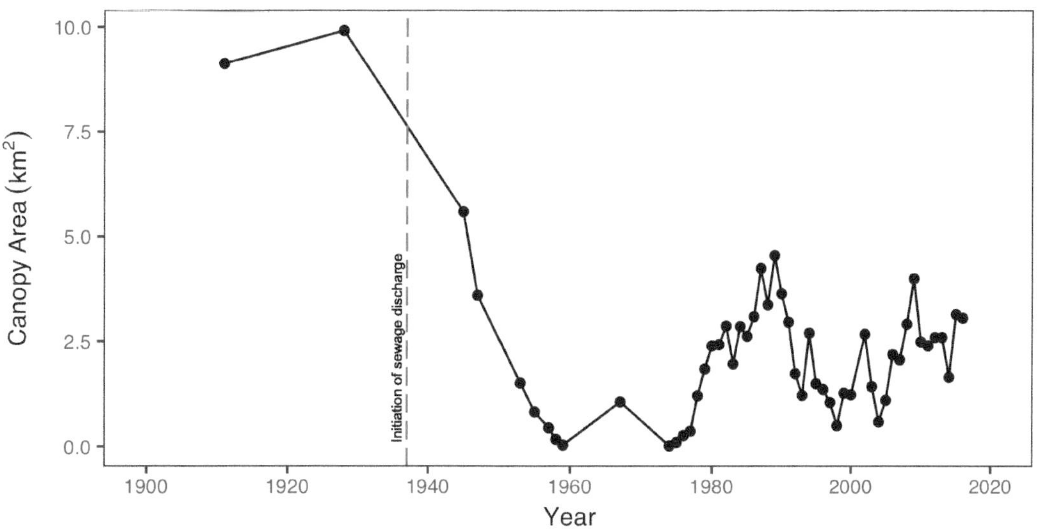

Figure 2. Kelp canopy coverage (km^2) from 1911 to 2016 on the Palos Verdes Peninsula (data from MBC Applied Environmental Sciences, 2016).

Figure 3. Turbidity plume from the Portuguese Bend landslide (left: circa 1980s; right: April 2016). White arrows indicate location of sediment source.

landslide suggests that the reef has likely been buried since 1999. A third point source of turbidity and sedimentation comes from the large storm drains that empty into this nearshore environment. With these various chronic stressors, there is continued deleterious impacts to the nearshore rocky environment, especially from Portuguese Bend (buried reef) to Point Fermin (Stephens et al. 1996; Pondella et al. 2012).

Mapping and Spatial Analysis

We mapped subtidal rocky reef habitat from the shoreline to the 30-m isobaths for the Palos Verdes coast from the Redondo Canyon to Point Fermin in a geographic information system. This area consisted of approximately 2.9 km^2 (2,899,280 m^2) of nearshore environment. The geographic extent and character of marine hard bottom/reef was determined by combining several different spatial data sets into a preliminary habitat data layer (Claisse et al. 2012). This layer was then ground-truthed and corrected using underwater field observations and analyses of aerial and satellite photography. All mapping and spatial analysis was done using ArcGIS (ArcMap v. 10.3). Spatial data layers were created and maintained in the shapefile format using the UTM Zone 11 North, WGS84 projection to minimize distortion in both area and length measurements for the region. Kelp canopy was a highly precise polygon spatial layer created by using a 2-m rectangular grid to classify georeferenced aerial photography (Kelner et al. 2005). In this layer, several years (2008 and 2011–2014) of data were used cumulatively. Kelp canopy is an indication of a rocky reef substratum, although reefs can be buried and/or suffer from scour that prohibits kelp growth (Pondella et al. 2012). Multibeam bathymetry and side-scan sonar data were obtained from the Sea Floor Mapping Laboratory at California State University–Monterey Bay. For this restoration effort, we focused our study on the south-facing reefs from Long Point to Point Fermin (Figure 4). The amount of exposed

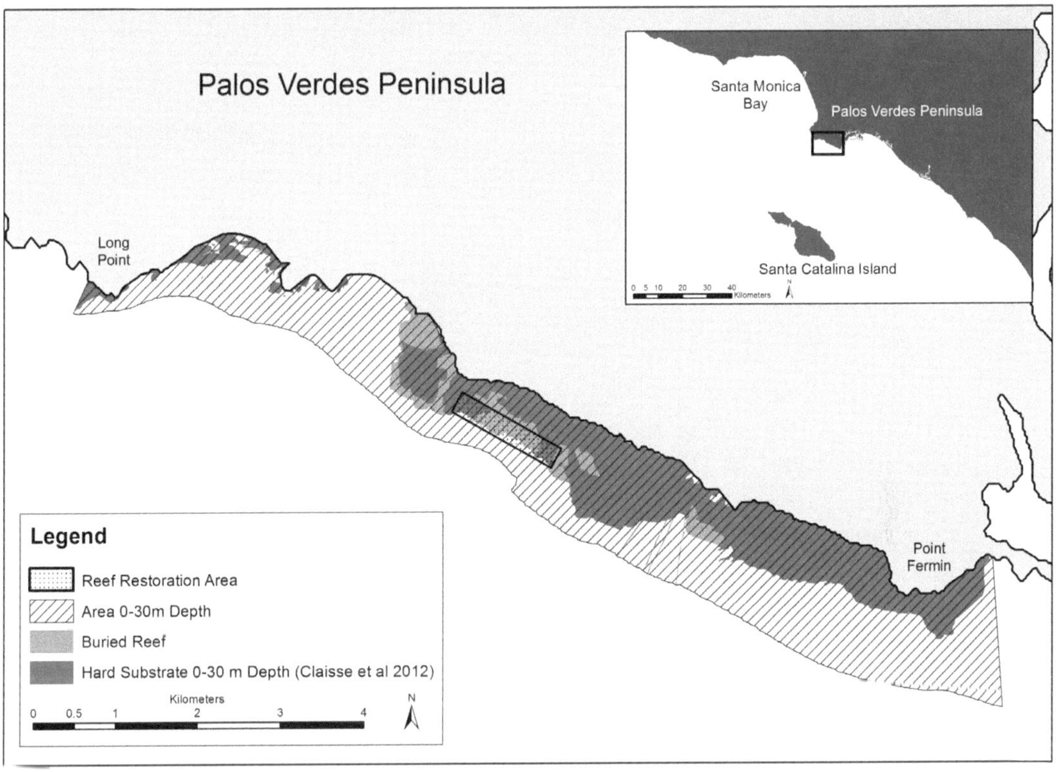

Figure 4. Hard substrate and buried reef within the 0–30-m isobaths from Long Point to Point Fermin, Palos Verdes.

(hard substrate) and buried reef by depth zone is shown in Table 1.

Field Surveys and Analysis

We examined the potential efficacy of restoration reefs in this region by conducting an intensive biological and physical sampling program throughout the subtidal areas of Palos Verdes Peninsula. As part of multiple kelp forest monitoring programs, we have conducted 752 surveys at 38 sites from 2004 to 2017 in this region (Figure 5) using the

Table 1. Area (km^2) of hard substrate and buried reef within the area between the 0–30-m isobaths from Long Point to Point Fermin, Palos Verdes.

Zone	Isobath	Area (km^2)	Percent of total area	Percent of hard substrate
Inner	2.5–7.5 m	1.58	12.2%	30.1%
Middle	7.5–12.5 m	1.45	11.1%	27.6%
Outer	12.5–17.5 m	1.16	8.9%	22.0%
Deep	17.5–30 m	1.07	8.2%	20.3%
Total hard substrate	0–30 m	5.26	40.5%	100.0%
Total area	0–30 m	13.00	100.0%	
Buried reef area	0–30 m	0.81	6.2%	

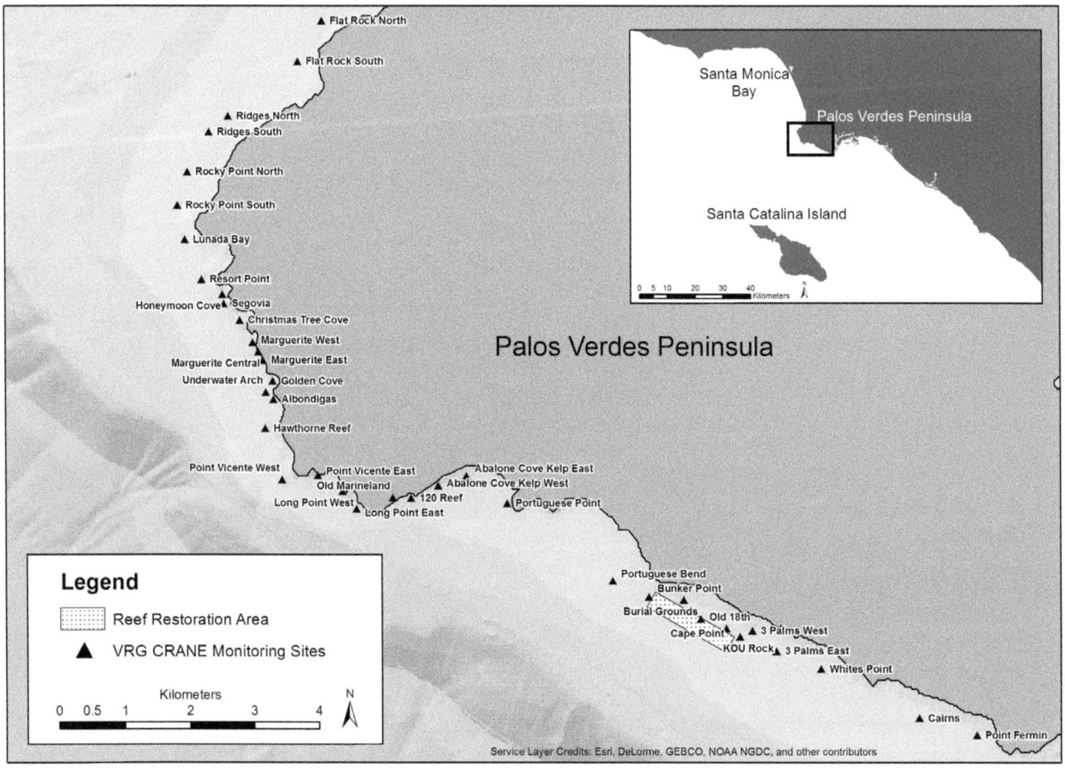

Figure 5. Ventuna (VRG) Cooperative Research and Assessment of Nearshore Ecosystems (CRANE) monitoring sites around the Palos Verdes peninsula with the restoration area (white).

Cooperative Research and Assessment of Nearshore Ecosystems (CRANE) protocol. This is a standardized comprehensive community monitoring survey method that quantifies fishes, invertebrates, algae, and habitat characteristics within multiple depth zones at each site (for more details on the protocol, see Claisse et al. 2012; Pondella et al. 2015a, 2015b; Zahn et al. 2016). This protocol primarily focuses on sampling rocky reef habitats. Consequently, areas that were primarily soft bottom, including the proposed locations for the restoration reef blocks, were sampled with additional CRANE surveys. To determine the effects of the sedimentation and turbidity on rocky reef habitats around the Palos Verdes Peninsula, we conducted a habitat characterization utilizing metrics generated from uniform point contact (UPC) data from the CRANE surveys. The physical substrate and relief of reefs varied throughout the peninsula. Most of the variation in substrate was associated with the ratio of sand versus bedrock, and most of the variation in physical relief was associated with the proportion of flat (0–0.1 m) reef versus moderate- (1–2 m) and high-relief (>2 m) reef. The restoration study area was characterized by flat to low-relief reef with larger proportions of sand, cobble, and boulders versus other areas of the peninsula where bedrock reefs were the dominant feature.

The complexity of this ecosystem between Bunker Point and Whites Point is illustrated by the presence of high fish biomass for commercial and recreational fish species where rocky reefs were present above the sediment despite the sedimentation and turbidity problems. Fish total length estimates were converted to biomass using standard species-specific length–weight conversions from the literature (e.g., Williams et al. 2013). Biomass density (g/m^2) was calculated by summing the total and species-specific biomass for each transect (note the survey protocol includes portions of the transect at multiple levels in the water column, including bottom, midwater, and canopy where present) and scaled to surface area; and this relative biomass estimate (hereafter, "biomass") was used throughout. The biomass density of Kelp Bass *Paralabrax clathratus*, the top recreationally fished species in the area, was higher at nearly every reef from Portuguese Bend to Cairns compared to other regions of the peninsula (Figure 6a). California Sheephead *Semicossyphus pulcher* exhibited a similar pattern, with higher biomass densities in the sediment-impacted area from Portuguese Bend to KOU Rock, as well as at sites within marine protected areas located along another section of the peninsula to the west (Long Point West, Long Point East, and Point Vicente West; Figure 6b). This increased biomass of fishery species may reflect differential fishing pressure around the peninsula, attraction, and/or elevated production along this section of coastline.

Most salient to the design of the restoration reef are the depth-zone-specific patterns of fish biomass density. A unique fish community resides in each depth zone at sites along the Palos Verdes Peninsula (ANOSIM, $p < 0.001$). A two-dimensional, nonmetric multidimensional scaling ordi-

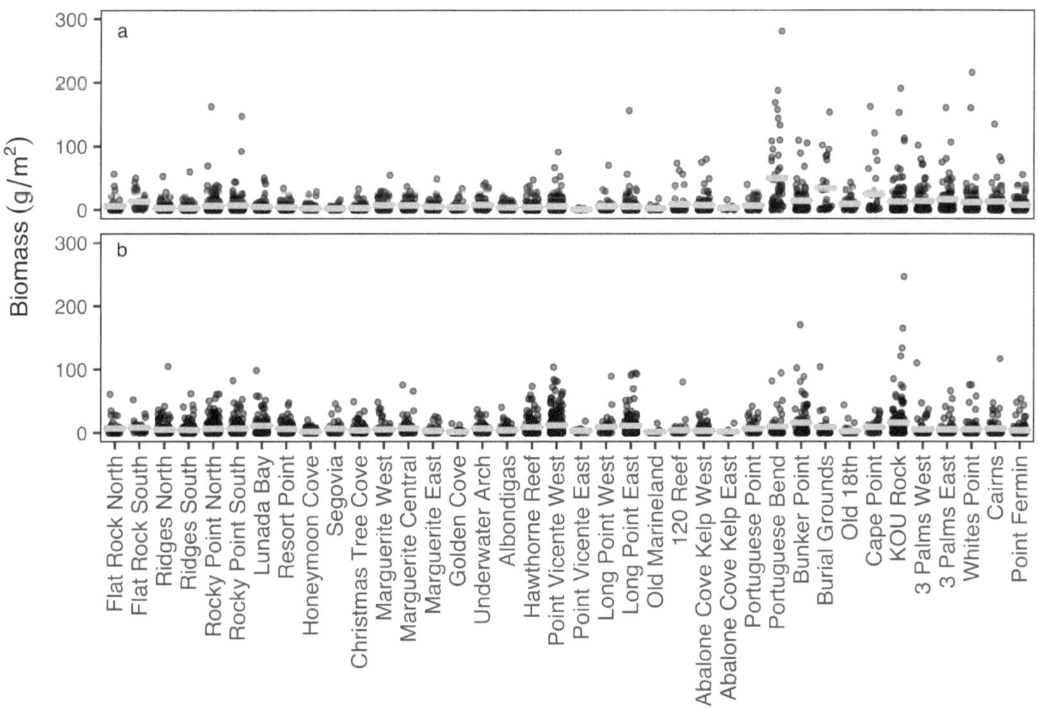

Figure 6. Biomass density of **(a)** Kelp Bass and **(b)** California Sheephead at each site along Palos Verdes Peninsula. Dots represent values for every transect, gray crossbars represent mean values.

nation plot of fish communities at all sites by depth zone clearly exhibits this depth stratification (Figure 7). The outer depth zone community is characterized by having the highest mean biomass density for most game fish, including Kelp Bass and California Sheephead. Then, more generally, an increase in biomass density with depth is observed for key planktivores like Blacksmith *Chromis punctipinnis* and Señorita *Oxyjulis californica*, as well as for Barred Sand Bass *Paralabrax nebulifer* and multiple species of rockfish *Sebastes* spp. (Figure 8). The total fish biomass density also tends to be high in the middle (~10 m) and outer (~15 m) depth zones at sites along the Palos Verdes Peninsula (Figure 9). Notable among all survey locations is the outer depth zone at the site KOU Rock, which consistently has the highest fish biomass density among anywhere on the peninsula, averaging more than 280 g/m^2. This semi-isolated pinnacle reef is inside the restoration project study area and subject to the same turbidity and sedimentation pressures as other reefs in the area; however, its high (~5 m) total relief prevents sediment accumulation and subsequent reef burial. This reef was and continues to serve as an ideal example for the replicated reef blocks that together form the proposed restoration reef at the study site.

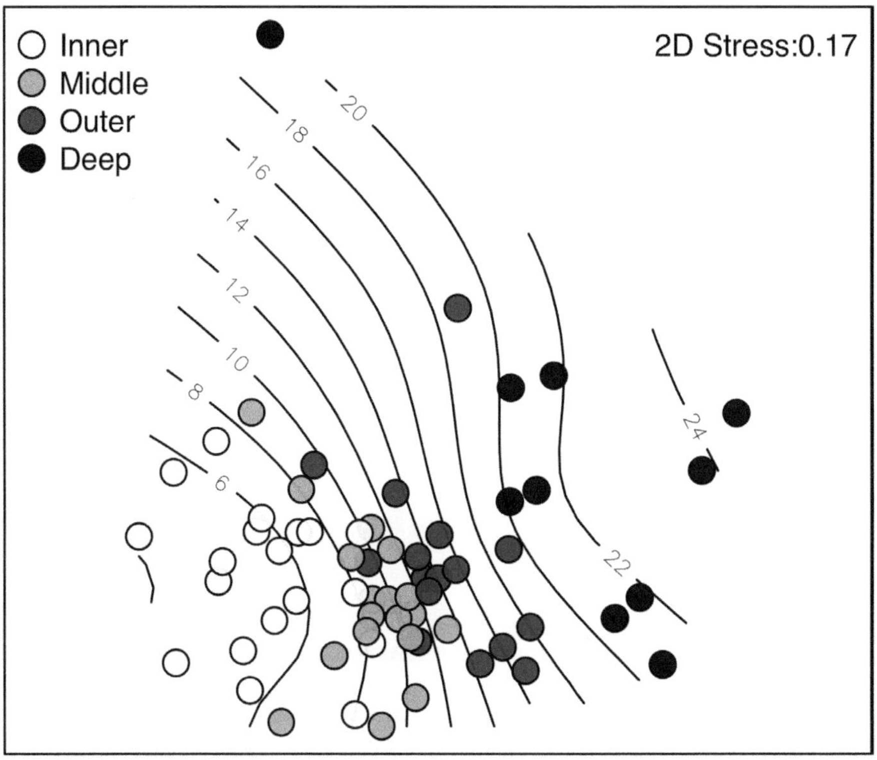

Figure 7. Nonmetric multidimensional scaling ordination plot of fish communities using the Bray–Curtis similarity index based on the square-root transformed mean taxa biomass density for each site and depth zone combination overlaid on a fitted mean depth (m) surface plot. Each site and depth zone combination is represented by a single point while depth zone designation is indicated by shade.

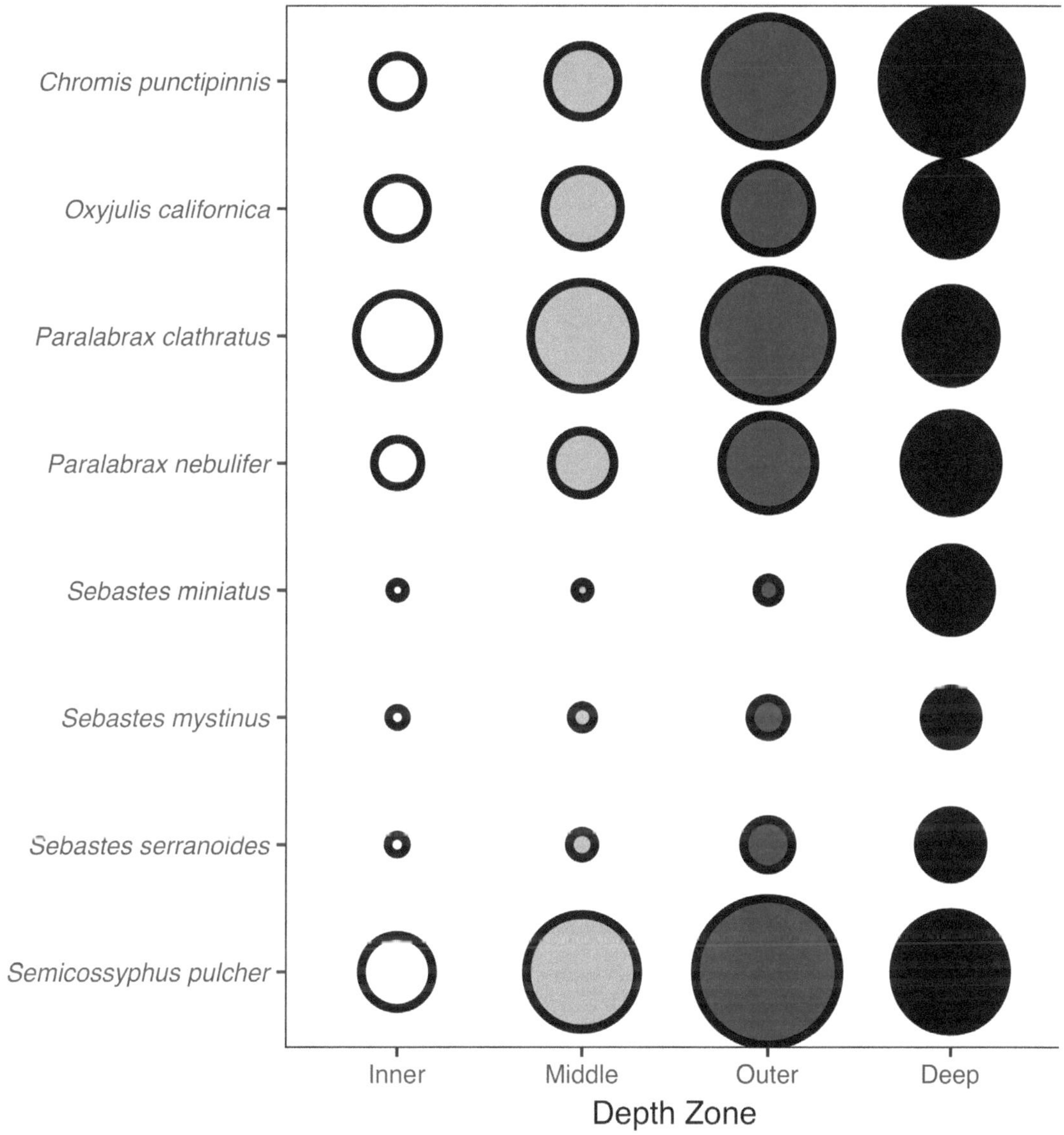

Figure 8. Relative biomass density of fishes scaled by circle size by depth zone for important game fishes and key planktivores at Palos Verdes Peninsula.

Reef Design

The policy of the California Department of Fish and Wildlife's Resources Agency historically has been to approve only quarry rock and concrete for artificial reef construction. In paired comparisons of these substrates, there were no significant difference in biological performance (Pondella et al. 2006; Reed et al. 2006); however, since the implementation of this policy, the cost and utility of concrete has changed because concrete is now recycled and is no longer cost-effective. Based on the current market, the budget allows for including 63,500 mt of quarry rock in this project. Our primary objective is to utilize this limited resource to create the most productive habitat by restoring the

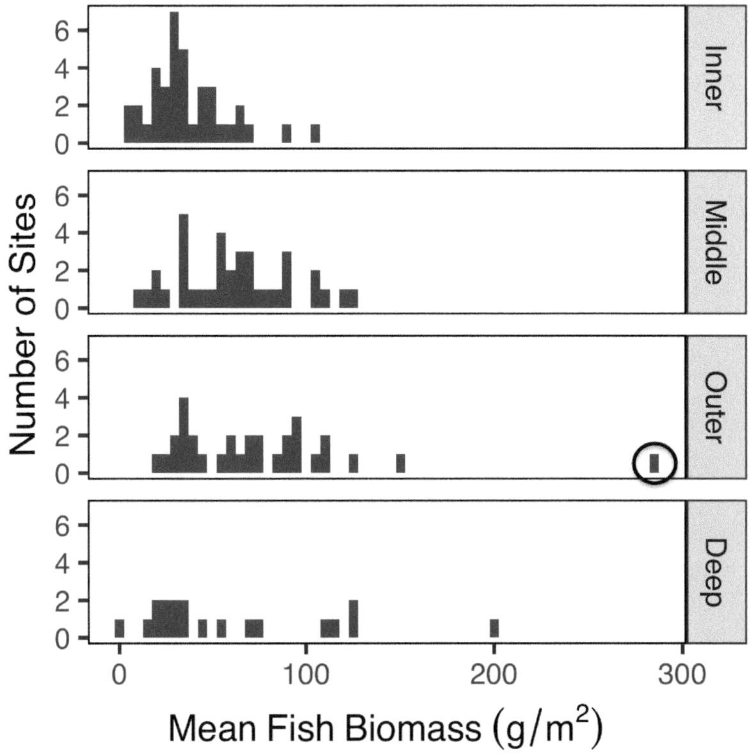

Figure 9. Distribution of total fish biomass at sites within specific depth zones. Note: age-0 and pelagic species are excluded from biomass estimates. The outer depth zone at the site KOU Rock is where the large semi-isolated pinnacle reef is located in the restoration project study area (circled). This high-relief reef in the outer depth zone (surveyed eight times from 2009 to 2017) is highlighted here as it served as the general model for the restoration reef design.

natural reef environment. To meet this objective, we considered the following criteria that balance engineering specifications with biological performance.

The first criterion to consider is the size of the rock produced by the quarry and the corresponding weight and void space. The quarry can produce rock sizes to meet an average size requirement with a specified allowable level of variation. Producing rocks within an extremely narrow size range can increase costs considerably. Using heterogeneously sized rocks was determined to be optimal because (1) this project's goal is to mimic natural reefs and natural reefs in the study area are not composed of single-sized rocks and (2) they are a less-expensive reef-building material. Designating an average size (weight) within the constraints of the quarry results in a distribution of rock sizes that vary from approximately 20% to 160% of the targeted weight (Figure 10). Previous studies compared elements of fish biomass on natural and artificial rocky reefs in southern California (Granneman 2011; Granneman and Steele 2014). They found that gonadal and somatic production (g/year) was positively correlated with the abundance of large boulders, large boulders defined as those at least 75 cm across. Production was lower on reefs with smaller boulders, most of these being natural reefs and the Wheeler

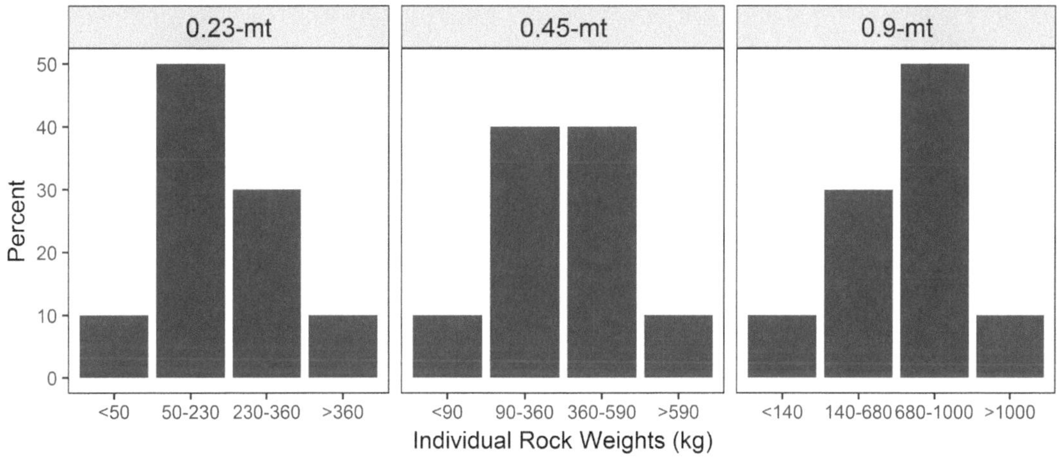

Figure 10. Percent size of quarry rock by weight based upon 0.23-, 0.45-, and 0.9-metric ton criteria.

North artificial reef. The authors explained that the Wheeler North artificial reef was designed with relatively low relief and rugosity. It was not designed to maximize fish production, but to mimic natural reefs in the southern Orange County region and to grow kelp. A higher proportion of larger boulders should also increase the likelihood of larger interstitial spaces between rocks in piles creating a variety (i.e., increase heterogeneity) of hole sizes for fishes and invertebrates that shelter within them (Friedlander and Parrish 1998). According to the quarry engineers, approximately 30% of the volume of rock is void space, independent of rock size. However, small rocks generally settle tightly, have smaller individual void spaces, and do not have as much biomass associated with them as larger rocks that have proportionally larger void spaces. The estimated average void space increases from 0.30 to 0.45 m as rock size transitions from 0.23 to 0.45 mt, and then from 0.45 to 0.61 m as rock size increases from 0.45–0.68 to 0.9 mt (Table 2).

Table 2. Quarry rock weight, dimensions, and estimated average void size.

Rock weight (kg)	Rock dimension (m)	Estimated average void size (m)
90	0.45 × 0.30 × 0.30	0.30
136	0.45 × 0.40 × 0.30	
181	0.49 × 0.49 × 0.30	
226	0.61 × 0.45 × 0.30	
453	0.91 × 0.76 × 0.61	0.45
680	1.07 × 0.91 × 0.61	
907	1.22 × 0.91 × 0.76	0.61
1,133	1.22 × 1.07 × 0.76	
1,360	1.37 × 1.07 × 0.76	
1,587	1.52 × 1.22 × 0.76	

Interstitial void space was also considered in our construction criteria due to the role interstitial void space plays in supporting the density and diversity of fishes in general, and because large void spaces are a critical habitat characteristic for California spiny lobster *Panulirus interruptus*. Void spaces are also critical for the shelter of Blacksmith. This nesting pomacentrid is a schooling, diurnal planktivore that sleeps in the reef and is the key fish taxon for incorporating the flux of zooplankton through this system into higher trophic levels (Bray et al. 1981, 1988). Building the reef of larger rocks will also minimize the chance that rocks located near the reef edge will become covered by sediment, thus creating more complex ecotone habitats at the sand/rock interface. A scalloped reef edge increases the perimeter of the reef and increases the contiguous ecotone. A trade-off to consider is that rocks can be simply modeled as a sphere, and as such, increasing their weight only incrementally increases their diameter, surface area, and void spaces (Figure 11). Note, 1.8- or 2.7-mt rocks are not substantially larger than a 0.9-mt rock, but they are two to three times the cost, respectively. Based on the above criteria, we chose 0.9-mt rock, which balances void space size against cost and maximizes fish biomass. In addition to optimizing rock sizes, the design of the components of the restoration reef, including sets of reef blocks and the "modules" that blocks are composed of (see Figure 12), are intended to maximize ecosystem benefits based upon a variety of physical and biological criteria. First, maximizing the amount of exposed surface area and reef perimeter increases the available habitat versus cost constraints. Engineering constraints dictate that reefs are constructed in a linear fashion as the 6-point barge anchoring sys-

Figure 11. 2.7-, 1.8-, and 0.9-metric ton quarry rocks arranged left to right.

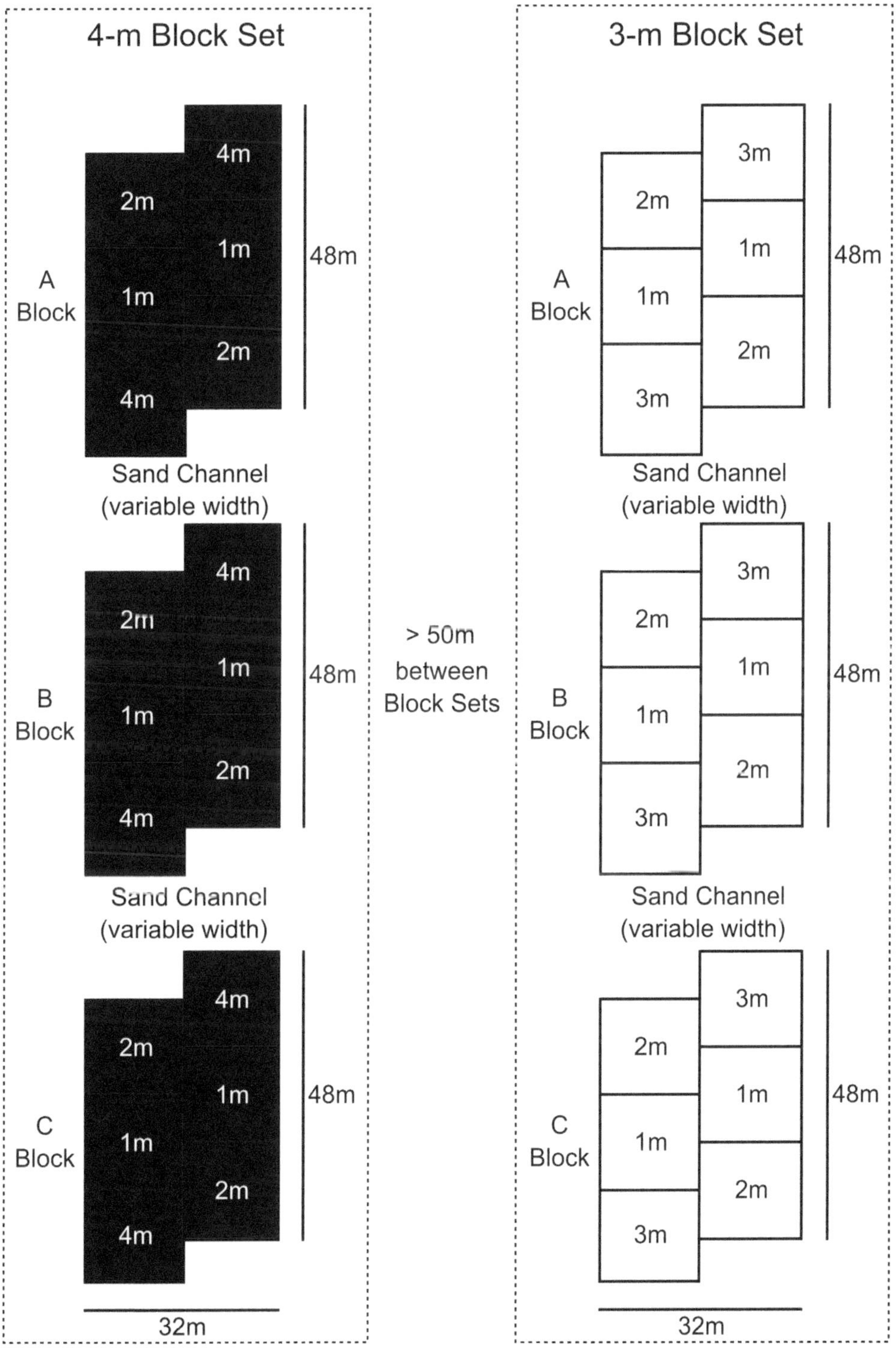

Figure 12. Design of 3-m and 4-m block sets. Each block set contains three blocks (A, B, C). Each block consists of a 3 x 2 set of modules, offset by ½ module length. Each module is a 16 x 16 m (footprint) pile of quarry rock with the overall height listed. There is a 10- to 20-m-wide sand channel between blocks and at least 50 m of space between block sets.

tems are used to construct the habitat. Our goal was to design a reef that maximized high-relief components, surface area, perimeter, flux, and ecotones and are consistent with the size and variability of natural reefs along the Palos Verdes Peninsula. High-relief reef components have a cost trade-off as they are more expensive (more weight per unit area) than low-relief reefs. And a critical consideration is how much rock is buried (and generally unavailable biologically) to create the high-relief components. Based on the construction constraints, the smallest unit of consideration are 16 × 16 m pyramid piles of quarry rock referred to here as reef modules (Figure 12). Considering our baseline assessment of the ecosystem, it was determined that a minimum vertical relief for a module of 1 m was necessary to minimize deleterious sedimentation impacts (e.g., burial, scouring).

Maximizing the surface area to volume ratio has two benefits: it is cost-effective as more rock is exposed and biologically available, and the greater surface area a reef has the more habitat and correlated production it will theoretically create. Thus, higher-relief modules were staggered within a reef block, placing higher relief modules proximate to lower-relief modules. Modules on each side of a block are also offset by one-half of the module length (Figure 12). In addition to creating more surface area, edge, and corresponding ecotone habitat, these strategies should increase diversification of water flow by limiting overlap of high relief modules, reducing the occurrence of one high-relief module being in the shadow of another high-relief module, and maximize the utilization of flux of nutrients through the system. Interspersing high-relief modules to the ends of each block should also buffer against sediment transport from potentially burying lower-relief components. These design features also increase heterogeneity, which may facilitate the creation of a mosaic of small-scale flow features, effectively facilitating microhabitat creation and diversification throughout the restoration area. Natural reefs on the Palos Verdes Peninsula exhibit this type of habitat heterogeneity.

With these design elements in mind, additional placement criteria were followed for positioning restoration reef blocks across the Bunker Point restoration area in the restoration reef design (Figure 13). First, blocks did not overlap with persistent kelp canopy. Persistent kelp canopy is an indication of stable rocky reef below that has not been covered by sediment. Blocks were placed at an optimal depth of 15–20 m as this is the most productive habitat in the region (see Field Surveys and Analysis above). At this depth, these reefs would also not impact surfing opportunities. Second, the orientation of each block set, and each individual block, was varied. This should further increase heterogeneity in reef characteristics, with respect to their relative orientation to the shoreline and to prevailing currents and wave action. Currents (e.g., longshore, tidal) in this nearshore arena are dynamic, seasonal, and variable. Thus, multiple orientations should optimize the interaction of these features under multiple current scenarios, again creating a mosaic of small-scale flow features. This variable orientation mimics the variable orientation of natural exposed and buried reef features across the Polos Verdes Peninsula. Third, to maintain connectivity with existing natural reefs, the ends of at least one block within a set were placed less than 30 m from existing nearshore natural exposed reef (kelp line) or existing (nonburied) rocky reefs so the blocks are not isolated islands in the soft substrate. Blocks were also positioned above a maximum of 1 m sediment depth to

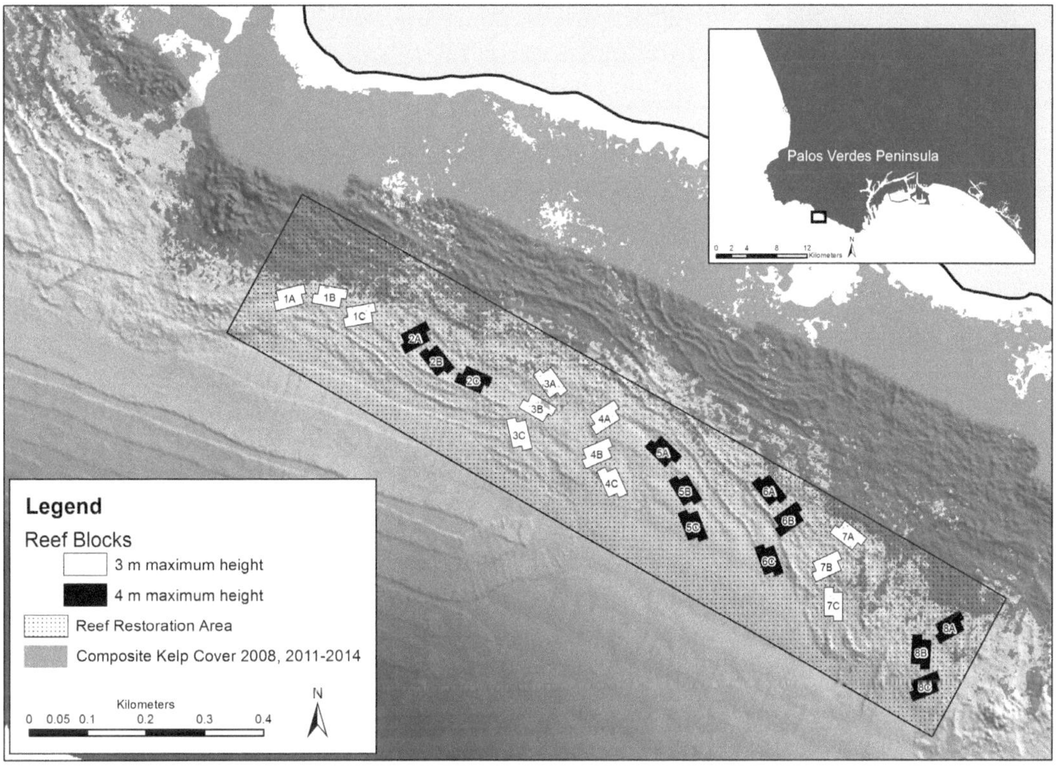

Figure 13. Proposed locations for the restoration reef blocks (1–8) at the Bunker Point restoration site study area with kelp canopy, side scan imagery. Each block set consists of three blocks (A–C). Block sets have a maximum reef height of either 3 m (white) or 4 m (black). Blocks were positioned so as to not overlap persistent kelp and orientated to follow contours created by existing natural (often buried/scoured) reef components.

limit long-term burial/sinking. Ten- to 20-m-wide sand channels between blocks within the eight sets were incorporated into the design to mimic natural reef characteristics and to promote sediment transport. Further, at least 50 m of space between block sets was incorporated. These distances were chosen due to the previously described halo effect around reef of ~30 m (Johnson et al. 1994). Reef modules separated by less than 30 m are more likely to function as a single reef for many species, while blocks separated by more than 30 m function more independently (Pondella et al. 2006). By separating the blocks and modules by the appropriate distances, a greater amount of reef perimeter and sand-rock ecotone habitat is created. Using a criterion of a minimum of 50 m between the eight replicate reef block sets can also increase their independence from one another.

A secondary focus of our reef design was to include replicated elements that could be studied to inform future restoration programs. This was balanced, however, with the primary goal of maximizing the potential for an effective restoration effort. A main question we are interested in examining is the effects of reef relief. Blocks will be in two forms, either with a 3-m maximum module height (relief) or a 4-m (Figure 12) maximum module height, with three replicate blocks per set and four replicate sets of each height. This will permit a comparison of the two reef heights' impact on fish biomass and production. Additionally, with the high level of heterogeneity but many repeated elements (for example, 1-, 2-, 3-, and 4-m modules, or blocks and block sets oriented at various angles relative to shore, or blocks in various seafloor depths), various other studies will also be possible. These could include studies of fine-scale fish or invertebrate habitat utilization patterns across various components of the restoration reef and will provide opportunities to inform future reef restoration programs.

The overall approach is to balance scientific study design considerations with maximizing the potential for an effective restoration effort across the range of important species and overall kelp forest biodiversity. Major motivations included incorporating heterogeneity throughout the restoration reef design both within (e.g., varying module heights within blocks) and among (e.g., varying block orientation) the reef blocks. Based upon the previously discussed criteria and engineering constraints, including the amount of rock material available for reef construction, the restoration reef was designed as eight sets of reef blocks (Figures 12 and 13). Each block set contains three blocks (A, B, and C). Each block consists of a 3 × 2 set of six modules. The three modules on each side of the block are offset by one-half of the module width (8 m). Each module is a 16 × 16 m pyramid pile of quarry rock with an overall height of 1, 2, 3, or 4 m (Figure 12). The blocks will be in two replicated designs, either with a 3-m maximum module height or a 4-m maximum module height.

Future Monitoring Plan

An important step in evaluating the effects of restoration actions along the Palos Verdes Peninsula is to develop an appropriate temporal and spatial sampling design for future monitoring. Short- and long-term monitoring of the restoration reef block sets and natural rocky reef sites across the Palos Verdes Peninsula will be critical for evaluating the success of this restoration project and for evaluating the effect of various restoration reef design elements on the associated biological community. Over the first months to years after construction of the restoration reef, we will have the opportunity to measure the level of attraction of adult fishes relocating from nearby reefs to the new reef habitat (Figure 14). Over the medium to long term (3–5 years, 10 years would be optimal), monitoring will provide the opportunity to estimate the increase in biomass of important species associated with the restoration reef blocks, as well as for the larger reef complex consisting of the restoration reef and the adjacent natural reefs. A before–after control–impact paired series (BACIPS) sampling design (Osenberg et al. 2002) is likely the most appropriate, particularly with respect to also assessing potential changes in biomass due to fish movements (relocation from nearby reefs). This model will help to account for year-to-year environmental variability when assessing changes in biomass. The restoration reef as a whole (all eight sets of reef blocks) would likely best be considered an unreplicated treatment in this context. While there will be multiple natural reef sites and block sets sampled within each treatment (i.e., restoration reef, adjacent natural reefs, and reference natural reefs), these mostly adjacent sites would not be independent (Table 3; Figure 14). A key to a BACIPS design is having multiple "before" sampling events across sites. Reef construction is currently planned for the fall of 2018. The proposed monitoring design would include four complete rounds of sampling

Table 3. Historical monitoring (with S indicating years sampled) and proposed monitoring (with X indicating sites to be sampled). Sites include Kelp Restoration (those involved in the kelp restoration project, which may contain urchin barrens, be active kelp restoration sites, or sites where the kelp has been restored), marine protected areas (MPAs; those within the MPAs that were implemented in 2012), Within Reef (dark-gray shaded; those located among the proposed sets of restoration reef blocks), Adjacent Reef (light-gray shaded; those located just north or south of the proposed restoration reef), or Reference (reef sites that do not currently involve any of the previously mentioned activities or designations). The monitoring plan includes sampling all sites in 2017 prior to reef construction, then once before and once after reef construction in 2018, then annually for at least 5 years after construction. Note that Cooperative Research and Assessment of Nearshore Ecosystems (CRANE) protocols require more than 50% coverage of rocky reef and the restoration area has not supported kelp or significant percentages of rocky substrate, thus precluding it from previous CRANE surveys.

		Historical monitoring												Proposed		
Site	Designation	2004	2007	2008	2009	2010	2011	2012	2013	2014	2015	2016	2017	2018 (pre)	2018 (post)	2019–2023
Flat Rock North	Reference		S	S							S	S	S	X		X
Flat Rock South	Reference		S								S	S	S	X		X
Ridges North	Reference	S	S		S	S	S	S	S	S	S	S	S	X		X
Ridges South	Reference	S	S	S				S	S	S	S	S	S	X		X
Rocky Point North	Reference	S	S	S	S		S	S	S	S	S	S	S	X		X
Rocky Point South	Reference		S			S	S	S	S	S	S	S	S	X		X
Lunada Bay	Reference						S	S	S		S	S	S	X	X	X
Resort Point	Kelp Restoration			S				S	S	S	S	S	S	X		X
Honeymoon Cove	Kelp Restoration						S	S	S	S	S	S	S	X		X
Segovia	Kelp Restoration							S	S		S		S	X		X
Christmas Tree Cove	Kelp Restoration					S	S	S	S	S	S	S	S	X	X	X
Marguerite West	Kelp Restoration						S	S	S	S	S	S	S	X		X
Marguerite Central	Kelp Restoration						S	S	S	S	S	S	S	X		X
Marguerite East	Kelp Restoration						S	S	S	S	S	S	S	X		X
Golden Cove	Kelp Restoration						S	S	S	S	S	S	S	X		X
Underwater Arch	Kelp Restoration					S		S	S		S	S	S	X		X
Albondigas	Kelp Restoration						S	S	S	S	S	S	S	X	X	X
Hawthorne Reef	Kelp Restoration		S		S	S	S	S	S	S	S	S	S	X	X	X
Point Vicente West	MPA	S	S	S	S	S	S	S	S	S	S	S	S	X		X
Point Vicente East	MPA		S													

Table 3. Continued.

Site	Designation	Historical monitoring												Proposed		
		2004	2007	2008	2009	2010	2011	2012	2013	2014	2015	2016	2017	2018 (pre)	2018 (post)	2019–2023
Long Point West	MPA		S					S	S		S	S	S	X		X
Long Point East	MPA		S		S	S	S	S	S	S	S	S	S	X	X	X
Old Marineland	MPA						S	S	S		S	S	S	X		X
120 Reef	MPA					S	S	S	S		S	S	S	X		X
Abalone Cove Kelp West	MPA					S	S	S	S	S	S	S	S	X		X
Abalone Cove Kelp East	MPA					S		S								
Portuguese Point	MPA					S		S	S		S	S	S	X		X
Portuguese Bend	Adjacent Reef								S		S	S	S	X	X	X
Bunker Point	Adjacent Reef		S				S	S	S	S	S	S	S	X	X	X
Burial Grounds	Within Reef										S	S	S	X	X	X
Old 18th	Within Reef								S		S	S	S	X	X	X
Cape Point	Within Reef										S	S	S	X	X	X
KOU Rock	Adjacent Reef				S		S	S	S	S	S	S	S	X	X	X
3 Palms West	Adjacent Reef		S						S	S	S	S	S	X	X	X
3 Palms East	Adjacent Reef		S	S	S					S	S	S	S	X	X	X
Whites Point	Reference		S	S		S	S	S	S	S	S	S	S	X	X	X
Cairns	Reference						S	S	S	S	S	S	S	X	X	X
Point Fermin	Reference		S	S		S	S	S	S	S	S	S	S	X	X	X

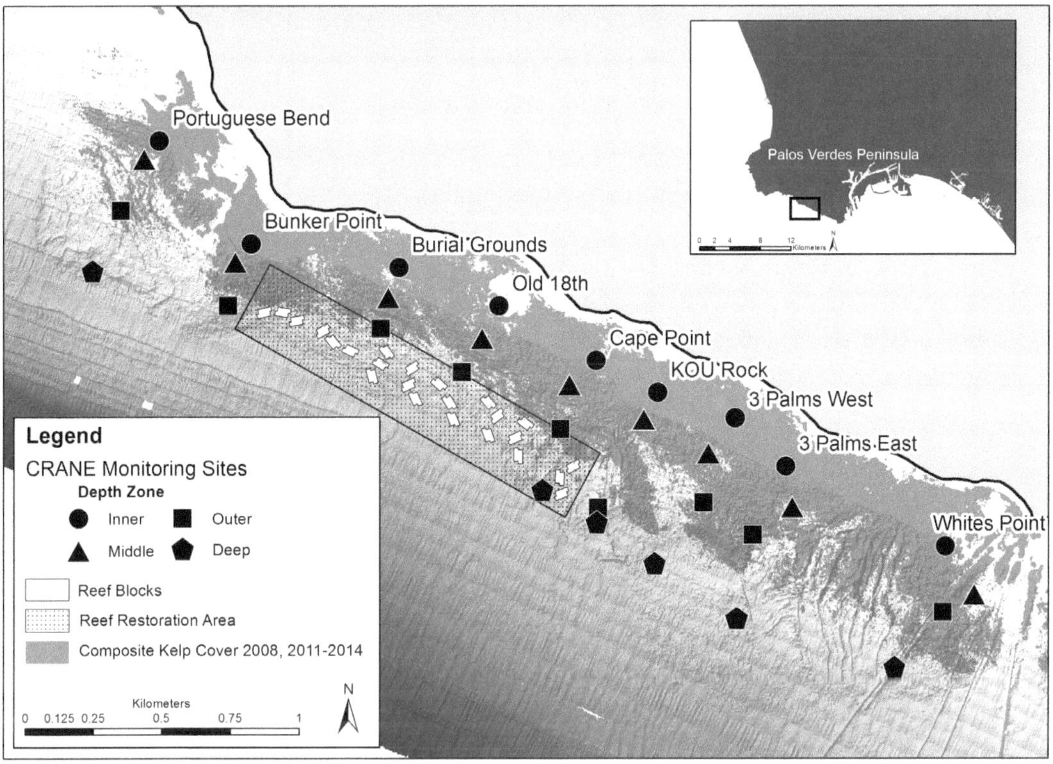

Figure 14. Restoration reef blocks and Cooperative Research and Assessment of Nearshore Ecosystems (CRANE) monitoring sites (individual depth zones indicated) at the Bunker Point restoration site study area with kelp canopy and side scan imagery.

before reef construction (2015–2018 [pre]; Table 3). The first round of "after" sampling would begin shortly after the completion of reef construction at the end of 2018 and would be completed in early (likely February) 2019. Subsequently, "after" sampling would be conducted annually for at least 5 years (Table 3). It will likely take at least this length of time for overall changes in biomass due to additional production to be observed (multiple years of larval/juvenile recruitment followed by a few years for those fishes and invertebrates to mature). The restoration reef modules will be sampled in a similar effort as is used to sample each depth zone at a natural reef site (Figure 14). At each module (A, B, and C) within each reef block (1–8), we will perform four fish transects (bottom/midwater/canopy portions per transect), two benthic UPC transects, and two benthic swath transects. The quantification of habitat characteristics performed in the CRANE protocol will also permit us to incorporate appropriate methods in the analyses to account for differences in habitat characteristics among sites and treatments (Miller and Russ 2014).

Another part of this assessment is estimating what proportion of biomass changes are due to fish movements to the restoration reef from the surrounding natural reefs compared with the increase in biomass from additional secondary production of fishes and invertebrates. The proposed sampling design will provide an opportunity to assess the degree to which increases in biomass on the new restoration reef are correlated with decreases in fish biomass on the adjacent natural reefs (Table 3; Figure 14; Osenberg et al. 2006; Osenberg et al. 2002), suggesting that some proportion of the fishes on the restoration reef relocated from nearby reefs. An increase in fish biomass on the restoration reefs and stable or increasing biomass on the adjacent reefs would suggest increased secondary production on entire reef complex. The monitoring data will also permit application of other novel analyses aimed at assessing the levels of local production and biomass flux within the restoration reef system (Smith et al. 2016). Performing additional studies would provide additional context from which to interpret the monitoring data and provide insight into the mechanisms behind changes in fish biomass in the system. These could include direct assessment of fish movements (e.g., traditional tagging, acoustic telemetry), which would be particularly informative if fishes on adjacent reefs could be tagged prior to reef construction. Other factors influencing fish production, such as increases in growth rates associated with higher-relief habitat (e.g., Granneman 2011; Granneman and Steele 2014), could be assessed directly (e.g., through otolith studies for fishes).

Finally, the proposed reef and monitoring designs, with multiple replicated elements, will also provide an opportunity for subsequent studies to examine the effects of restoration reef design features. A primary assessment would be the effect of block relief, 3-m versus 4-m maximum module heights, on the associated species biomass and habitat use patterns. Other features that can be assessed may include module orientation or position relative to the coast or dominant current pattern and block depth. Understanding how these factors impact fish and invertebrate habitat utilization patterns will provide an opportunity to inform future restoration programs in the state.

References

Bray, R. N., A. C. Miller, and G. G. Geesey. 1981. The fish connection: a trophic link between plankton and rocky reef communities? Science 214:204–205.

Bray, R. N., A. C. Miller, S. Johnson, P. R. Krause, D. L. Robertson, and A. M. Westcott. 1988. Ammonium excretion by macroinvertebrates and fishes on a subtidal rocky reef in Southern California. Marine Biology 100:21–30.

Claisse, J. T., D. J. Pondella, II, J. P. Williams, and J. Sadd. 2012. Using GIS mapping of the extent of nearshore rocky reefs to estimate the abundance and reproductive output of important fishery species. PLOS (Public Library of Science) ONE [online serial] 7(1):e30290.

Foster, M. S., and D. R. Schiel. 2010. Loss of predators and the collapse of Southern California kelp forests (?): alternatives, explanations and generalizations. Journal of Experimental Marine Biology and Ecology 393:59–70.

Friedlander, A. M., and J. D. Parrish. 1998. Habitat characteristics affecting fish assemblages on a Hawaiian coral reef. Journal of Experimental Marine Biology and Ecology 224:1–30.

Granneman, J. E. 2011. A comparison of fish production on artificial and natural reefs in Southern California. Master's thesis. California State University, Northridge.

Granneman, J. E., and M. A. Steele. 2014b. Fish growth, reproduction, and tissue production on artificial reefs relative to natural reefs. ICES Journal of Marine Science 71:2494–2504.

Johnson, T. D., A. M. Barnett, E. E. DeMartini, L. L. Craft, R. F. Ambrose, and L. J. Purcell. 1994. Fish production and habitat utilization on a Southern California artificial reef. Bulletin of Marine Science 55:709–723.

Kayen, R. E., H. J. Lee, and J. R. Hein. 2002. Influence of the Portuguese Bend landslide on the character of the effluent-affected sediment deposit, Palos Verdes margin, Southern California. Continental Shelf Research 22:911–922.

Kelner, J. J., R. Christensen, R. Clark, C. Caldow, and M. Coyne. 2005. Physical and oceanographic setting. Pages 89–134 *in* R. Clark, J. Christensen, C. Caldow, J. Allen, M. Murray, and S. MacWilliams, editors. A biogeographic assessment of the Channel Islands National Marine Sanctuary: a review of boundary expansion concepts for NOAA's National Marine Sanctuary Program. NOAA Technical Memorandum NOS NCCOS 21.

Miller, K. I., and G. R. Russ. 2014. Studies of no-take marine reserves: methods for differentiating reserve and habitat effects. Ocean and Coastal Management 96:51–60.

North, W. J. 1964. Ecology of the rocky nearshore environment in Southern California and possible influences of discharged wastes. Pages 247–262 *in* E. A. Pearson, editor. Advances in water pollution research, volume 3. Pergamon Press, Oxford, UK.

Osenberg, C. W., B. M. Bolker, J. White, C. M. S. Mary, and J. S. Shima. 2006. Statistical issues and study design in ecological restorations: lessons learned from marine reserves. Pages 280–302 *in* D. Falk, M. Palmer, and J. Zedler, editors. Foundations of restoration ecology. Island Press, Washington, D.C.

Osenberg, C. W., C. M. St. Mary, J. A. Wilson, and W. J. Lindberg. 2002. A quantitative framework to evaluate the attraction–production controversy. ICES Journal of Marine Science 59:S214–S221.

Pondella, D. J. 2009. Science based regulation: California's marine protected areas. Urban Coast 1:33–36.

Pondella, D. J., L. G. Allen, M. T. Craig, and B. Gintert. 2006. Evaluation of eelgrass mitigation and fishery enhancement structures in San Diego Bay, California. Bulletin of Marine Science 78:115–131.

Pondella, D. J., J. E. Casselle, J. T. Claisse, J. P. Williams, K. Davis, C. M. Williams, L. A. Zahn. 2015a. South Coast Baseline Program final report: kelp and shallow rock ecosystems. Available: https://caseagrant.ucsd.edu/sites/default/files/SCMPA-27-Final-Report_0.pdf. (April 2018).

Pondella, D. J., J. P. Williams, and J. T. Claisse. 2012. Biological and physical characteristics of the nearshore environment of the Bunker Point Restoration Area and the Palos Verdes Peninsula. Report by the Vantuna Research Group to the NOAA Restoration Center/Montrose Settlements Restoration Program, Long Beach, California.

Pondella, D. J., J. Williams, J. Claisse, B. Schaffner, K. Ritter, and K. Schiff. 2015b. The physical characteristics of nearshore rocky reefs in the Southern California Bight. Bulletin of the Southern California Academy of Sciences 114(3):105–122.

Reed, D. C., S. C. Schroeter, D. Huang, T. W. Anderson, and R. F. Ambrose. 2006. Quantitative assessment of different artificial reef designs in mitigating losses to kelp forest fishes. Bulletin of Marine Science 78:133–150.

Smith, J. A., M. B. Lowry, C. Champion, and I. M. Suthers. 2016. A designed artificial reef is among the most productive marine fish habitats: new metrics to address 'production versus attraction.' Marine Biology 163(9):1–8.

Stephens, J. S., Jr., D. J. Pondella, II, and P. Morris. 1996. Habitat value determination of the coastal zone off the city of Rancho Palos Verdes based on habitat-specific assemblage data. U.S. Corps of Engineers, Los Angeles District, Los Angeles, California.

Stull, J. K., K. A. Dryden, and P. A. Gregory. 1987. A historical review of fisheries statistics and environmental and societal influences off the Palos Verdes Peninsula, California. CalCOFI Reports 28:135–154.

Williams, C. M., J. P. Williams, J. T. Claisse, D. J. Pondella, II, and L. A. Zahn. 2013. Morphometric relationships of marine fishes common to central California and the Southern California bight. Bulletin of the Southern California Academy of Sciences 112:217–227.

Zahn, L. A., J. T. Claisse, J. P. Williams, C. M. Williams, and D. J. Pondella. 2016. The biogeography and community structure of kelp forest macroinvertebrates. Marine Ecology 37:770–785.

American Fisheries Society Symposium 86:187–196, 2018

The Charleston Deep Reef: Creating an Artificial Reef Marine Protected Area to Enhance Fisheries Resources

Robert M. Martore* and Melvin Bell
South Carolina Department of Natural Resources
217 Fort Johnson Road, Charleston, South Carolina 29412, USA

Abstract.—In support of the Magnuson–Stevens Conservation and Management Reauthorization Act of 2006, which tasked regional fisheries management councils with ending overfishing of numerous marine finfish species, the South Atlantic Fisheries Management Council established 8 deepwater (90–150 m [300–500 ft]) type II marine protected areas (MPAs) along the coastline of the southeastern United States. At the request of the South Carolina Department of Natural Resources (SCDNR), one of these MPAs was established on an undeveloped sand-bottom area previously permitted by SCDNR for artificial reef development. After monitoring the production potential of unfished artificial reefs for several years on shallower experimental reef sites, SCDNR staff proposed that a deeper location had the potential to become a highly productive spawning site, particularly for deepwater grouper species. Development of this permitted site began in 2014 when two 79-m (260 ft) barges with nearly 30 m (100 ft) of added profile were deployed. Subsequent monitoring of the site through remotely operated underwater vehicle video revealed colonization by several target species, including Warsaw Grouper *Hyporthodus nigritus*, Snowy Grouper *H. niveatus*, and Misty Grouper *H. mystacinus*. Due in part to the success of this deepwater MPA, the SCDNR was also granted spawning special management zone designation for its two previously established, undisclosed experimental artificial reef sites in federal waters off South Carolina in 2017.

Introduction

The Magnuson–Stevens Fishery Conservation and Management Act, first enacted in 1976, is the primary law governing marine fisheries management in U.S. federal waters. This act established eight regional fishery management councils that are responsible for developing fishery management plans to prevent overfishing and rebuild overfished stocks (U.S. Department of Commerce 2007). The act was strengthened with the passage of the Magnuson–Stevens Fishery Conservation and Management Reauthorization Act of 2006. This amendment added new requirements on regional councils to establish annual catch limits and accountability measures to achieve

* Corresponding author: martoreb@dnr.sc.gov

species-specific goals to eliminate overfishing. In 2007, to assist in meeting the management goals of these new requirements, the South Atlantic Fisheries Management Council (SAFMC) established eight deepwater type II marine protected areas (MPAs) in the Atlantic Ocean along the southeastern United States coast, specifically to protect populations and spawning sites of deepwater snapper and grouper species (Figure 1).

Designation as a type II MPA prohibits bottom fishing or the possession of snapper/grouper species while within the designated boundaries of the MPA; however, surface trolling for pelagic species such as dolphin or billfish is allowed. Most sites designated as MPAs encompassed areas of naturally occurring hard/live-bottom or rocky ledges inhabited by populations of snapper and grouper species of management interest. Consequently, the South Carolina Department of Natural Resources (SCDNR) requested that the SAFMC designate an additional site consisting primarily of flat, featureless sand bottom as a type II MPA. This site, called the Charleston Deep Reef, was previously permitted for deepwater artificial reef development by the state. After monitoring the production potential of unfished artificial reefs for several years on shallower experimental artificial reefs, SCDNR staff believed that a deeper location had the potential to become a highly productive spawning site, particularly for deepwater grouper species of management interest. This belief was also supported by the experiences of numerous commercial fishermen over the years who have encountered large populations of grouper on newly discovered shipwrecks.

Background

South Carolina Department of Natural Resources staff began researching the potential benefits of developing artificial reefs as

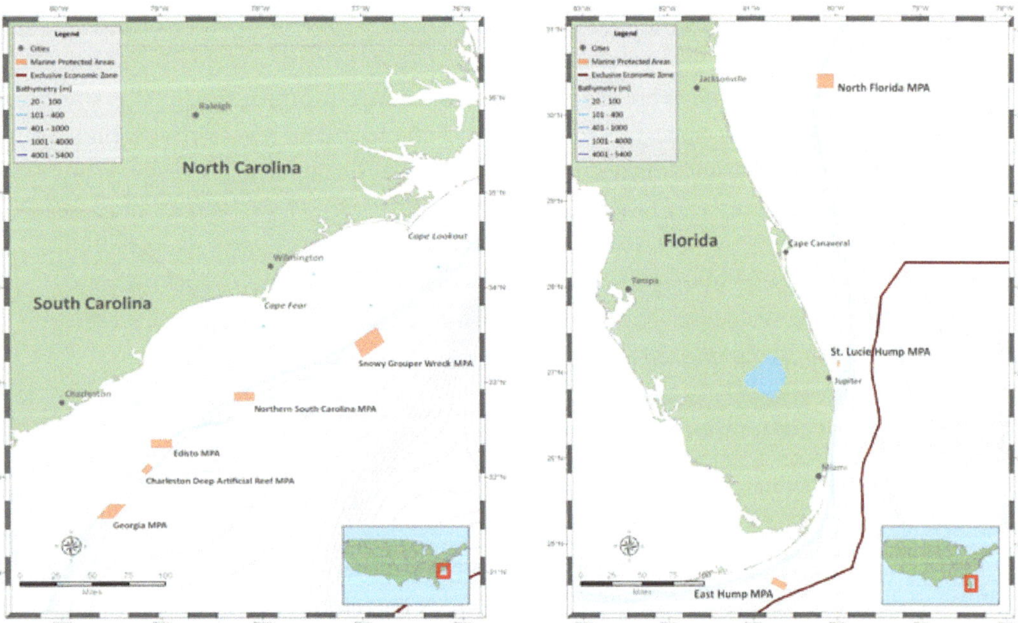

Figure 1. Deepwater marine protected areas (MPAs) established by the South Atlantic Fisheries Management Council in 2007.

marine reserves with the creation of an undisclosed research reef in 1998. The experimental artificial reef site, designated "Area 51," was initiated to investigate the feasibility of using artificial reefs as small-scale but potentially valuable MPAs. By monitoring and documenting the fish populations of an unfished artificial reef area over time and comparing their population parameters to those on regularly fished artificial reef areas of identical design, the potential value of nonexploited artificial reef MPAs as a supplement to traditional methods of utilizing artificial reefs could be evaluated. To more accurately measure the productivity potential of this newly created reef, it was necessary to eliminate unmeasured public fishing pressure on it by limiting public awareness and, consequently, public use of the site during the study period. The U.S. Army Corps of Engineers allowed SCDNR to utilize a special permitting process to bypass the standard public comment period normally required for artificial reef permitting. Several years of monitoring revealed that MPA reef sites indicated significantly higher abundances of commercially and recreationally important species (i.e., Black Sea Bass *Centropristis striata*, Gag *Mycteroperca microlepis*), recruitment of juveniles and subadults, spawning activity in several species, and minimal movement among nearby sites once populations became established (Gold 2001; Kauppert 2002).

Due in part to the results obtained from investigations at Area 51, the SAFMC provided funding in 2003 to replicate that study design in deeper water to specifically target other species of interest within the SAFMC's Snapper Grouper Management Unit. The permitting process and all reef parameters for the new site, designated "Area 53," were identical to those from Area 51 except that water depth for this site was 32 m (105 ft) compared to 21 m (70 ft) for Area 51. In addition to dart tags used in Area 51 to track fish movement and site fidelity, acoustic tags were also implanted in numerous fish of several species on Area 53. Receiver arrays were established on all four corners of the experimental site to better monitor site fidelity on the reef over time.

Observations from Area 53 were similar to those from Area 51, indicating the total abundances of Black Sea Bass, Gag, Scamp *Mycteroperca phenax*, and Gray Triggerfish *Balistes capriscus*, were significantly greater at unfished sites. Also, spawning activity within the reef complex was observed as well as an indication of high site fidelity for some species. In addition, protected reef sites had significantly larger-sized individuals and faster growth rates for some species (Kolmos 2007). Both sites clearly demonstrated that artificial reefs could be designed, permitted, and deployed to function specifically as MPAs and that this concept could be particularly useful in increasing the reproductive potential of designated species of management concern.

While neither experimental reef site had governmental regulatory protection from fishing during the study periods, the general public was unaware of the reef sites or the ongoing research activities. Although exploitation of the areas was certainly possible, there were no indications of public discovery of the sites from normal indicators of fishing activity, as evidenced by unexpected decreases in fish abundance, lost fishing tackle, or fouled boat anchors.

The Charleston Deep Reef

Observations from the Area 51 and Area 53 research sites led SCDNR staff to conclude that much deeper artificial reefs that excluded fishing might successfully function as habitat for large numbers of deepwater species and potentially serve as supplemental spawning locations to assist in the rebuilding of stocks deemed overfished. Additionally, there have been documented experiences of commercial fishermen harvesting large numbers of deepwater grouper species from previously undiscovered shipwrecks offshore. For example, a large shipwreck found 195 m (640 ft) deep off North Carolina was estimated to have an initial biomass of 31.3–32.9 metric tons (mt; 34.5–36.3 tons) whole weight of Snowy Grouper, with an estimated density of 11 kg/m^2 over the entire ship. Following its discovery by commercial fishermen, 17.2 mt (19 tons) of Snowy Grouper were harvested in approximately 3 months (Epperly and Dodrill 1995). This information suggested that there might be value in developing similar-scale artificial reefs specifically for the purpose of maintaining unexploited biomass of key species for spawning potential and stock enhancement.

From its inception, the objective of the Charleston Deep Reef was to be designated a special management zone (SMZ), and this objective was stated in the reef construction permit application to the U.S. Army Corps of Engineers. This objective was also made known to the SAFMC prior to the site selection process for the specific area to be permitted, and SAFMC staff assisted in selecting the actual deployment site. All previous bottom surveys of offshore South Carolina waters on file with SAFMC, along with those conducted by SCDNR, were reviewed to locate a suitable region devoid of hard- or live-bottom habitat. After selecting a site, side-scan sonar confirmed that substrate at the site was, in fact, flat sand bottom. The site selected was a 4 × 6 mi (6.4 × 9.7 km) region with water depths between 91 and 122 m (300–400 ft), located 83.7 km (52 mi) off the central South Carolina coast (Figure 2). The construction permit was approved in March 2005.

The original permit application for the Charleston Deep Reef indicated that the reef construction materials would be the steel trusses of two major bridges that spanned Charleston Harbor. The Grace Memorial and Silas Pearman bridges were scheduled for demolition at this time, and the steel spans, each nearly 2 mi (3.2 km) long and more than 30.5 m (100 ft) high, were specifically recommended for offshore reef deployment in the bridge demolition study (Dial Cordy and Associates 2001).

As the time for actual bridge demolition neared, the South Carolina Department of Transportation (SCDOT) recouped as much funding as possible by selling the steel trusses for scrap metal rather than allowing their use as reef material. This left the Charleston Deep Reef project with a permitted site but no significant material with which to construct the reef. Still intending to eventually deploy on the site, SCDNR proceeded with its request to the SAFMC for MPA designation and the site was included in the SAFMC's Amendment 14 to the Snapper/Grouper Fishery Management Plan submitted to NOAA Fisheries in 2007 and established as federal regulation in 2009. South Carolina now had a legally established artificial reef MPA but no actual reef material for the site.

Attempts were made throughout subsequent years to procure a steel-hulled vessel large enough to provide significant profile at the reef site's depths. Because there was no dedicated funding for the Charleston Deep Reef project through the SCDNR, there were no available vessels within the scope of the South Carolina Marine Artificial Reef Program's budget. Around the same time frame,

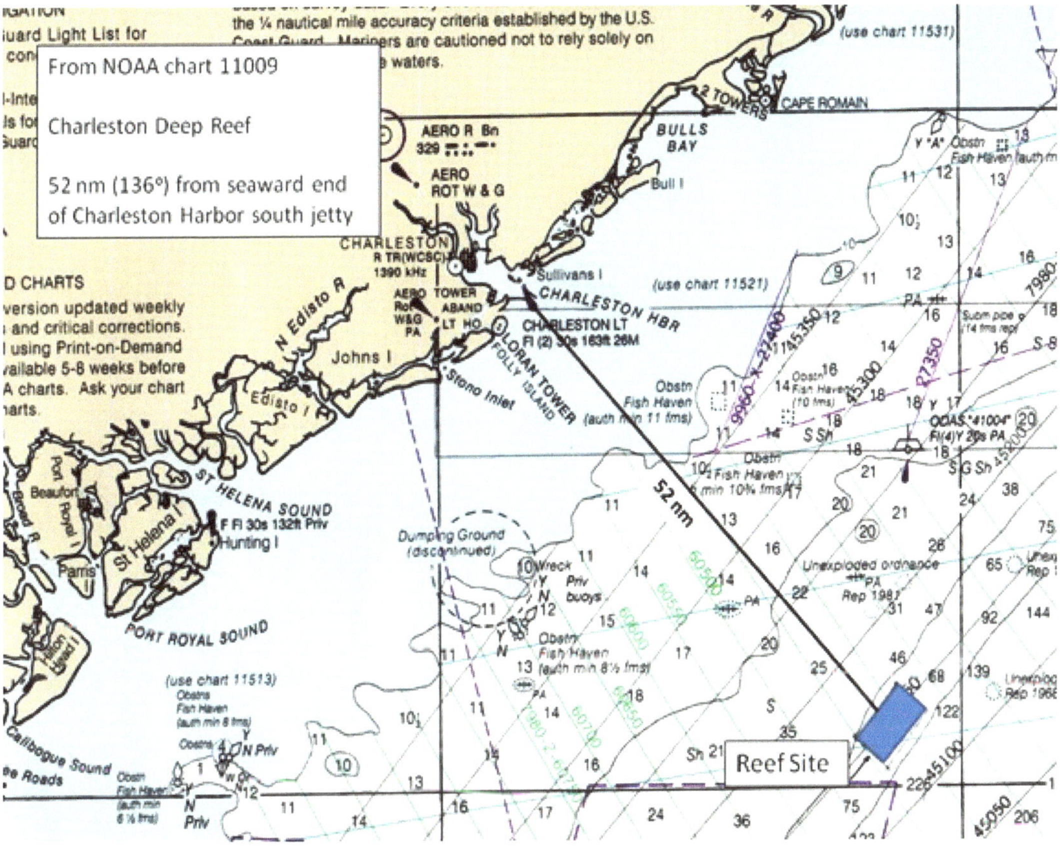

Figure 2. Location of the Charleston Deep Reef marine protected area.

a group of offshore fishermen approached SCDNR with the idea of raising funds to support a memorial reef to honor deceased offshore anglers. After numerous discussions, it was decided that the Charleston Deep Reef MPA would be an appropriate location for such a memorial. Consequently, an organization called the South Carolina Memorial Reef (www.scmemorialreef.com) was formed to raise funds to support the Charleston Deep Reef project. Donations of money and materials were received from individuals and businesses to support the project. Over several years, the South Carolina Memorial Reef conducted numerous fundraising activities that including fishing tournaments and auctions. Even with this newfound source of funding, most available larger vessels simply remained too expensive to procure, clean, and deploy.

Eventually, a local marine transportation company offered two large but relatively low profile steel-hulled deck barges, 79 × 16 m (260 × 52 ft). These vessels were purchased and their vertical profile increased by adding various steel structures donated to the project by local individuals and businesses. Among the contributed steel structures were a derelict derrick crane, steel I-beams, 20- and 40-ft steel shipping containers, truck chassis, and radio and cell towers. Plans for the design of the modified barges were approved as per the permit requirements and all structures were welded into place, creating nearly 30.5 m (100 ft) of profile on each of the barges (Figure 3).

Figure 3. The Charleston Deep Reef barges with added profile.

The first barge was deployed on May 4, 2014 and the second 3 weeks later on May 25. One month later, June 25, 2014, the NOAA vessel R/V *Pisces*, on its first research cruise to examine each of the designated MPA areas, used multibeam sonar and remotely operated vehicle (ROV) video to begin documentation of the newly created sites. Upon reviewing the data, it was noted that although both barges remained upright, the first barge had lost much of its added profile when it landed on the ocean bottom (Figure 4). The second barge remained intact. The only fish found on the

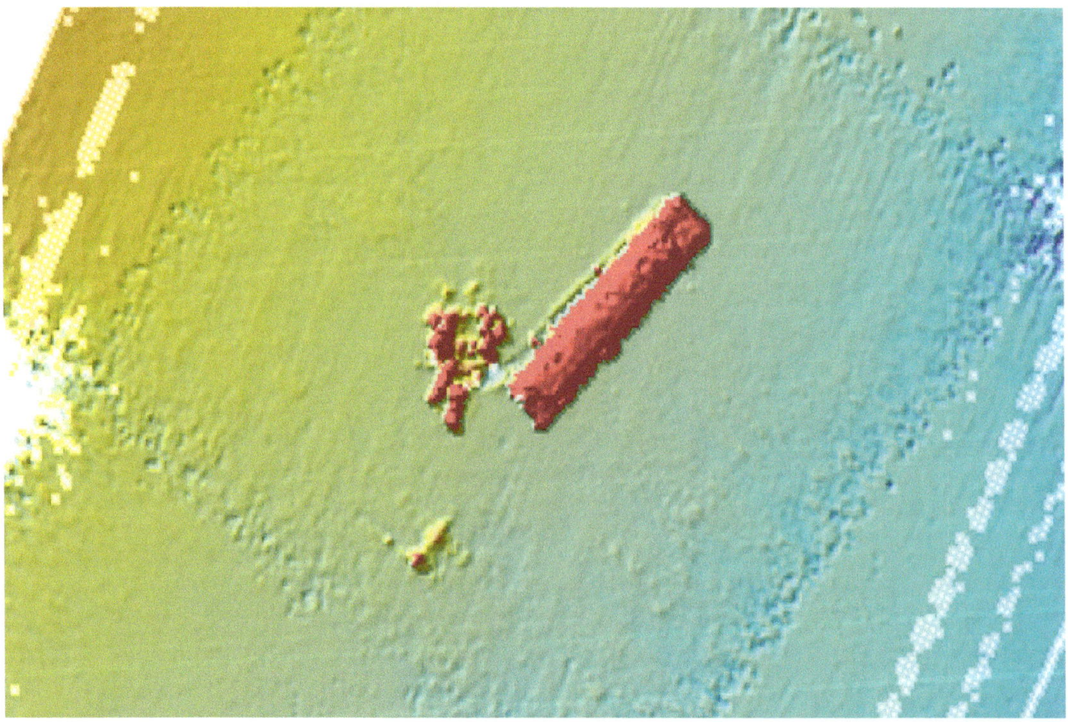

Figure 4. Multibeam image of barge 1 showing detached container boxes beside the upright barge.

barges at this time were amberjacks *Seriola* spp.

Two years later, June 13, 2016, a second survey of the Charleston Deep Reef was undertaken by the R/V *Pisces*. Video from one ROV dive on each of the barges revealed numerous species of interest on each, including Red Snapper *Lutjanus campechanus*, Scamp, and Yellowedge Grouper *Hyporthodus flavolimbatus* on barge 1 and Red Snapper, Scamp, Warsaw Grouper (Figure 5), Misty Grouper, Snowy Grouper, and Yellowedge Grouper on barge 2 (Table 1).

There was no indication of spawning activity during these surveys. However, colonization of the barges by multiple targeted species indicated that an artificial reef MPA could, in fact, become inhabited and potentially utilized for reproduction just as any other naturally occurring substrate. It is anticipated that future site monitoring will continue to show increases in fish biomass with concomitant spawning activity.

Colonization on the deployed reef materials at the Charleston Deep Reef by grouper species designated as overfished, coupled with the fishing success experienced by recreational anglers trolling for pelagic and highly migratory species in the waters overhead, led all involved with the reef's creation to continue its expansion through the addition of more reef materials. The South Carolina Memorial Reef Fund and the South Carolina Governors Cup Billfishing Series provided a

Figure 5. Warsaw Grouper on the Charleston Deep Reef.

Table 1. Fish abundances of all species observed on the two barges comprising the Charleston Deep Reef marine protected area. (From Harter et. al. 2016.)

Species	Barge 1 abundance	Species	Barge 2 abundance
Anthiinae	50	Bank Sea Bass *Centropristis ocyurus*	2
Calamus spp.	2	Jackknife-fish *Equetus lanceolatus*	2
Bank Sea Bass *Centropristis ocyurus*	3	Red Barbier *Baldwinella* (also *Hemanthias*) *vivanus*	350
Halichoeres spp.	7	Yellowedge Grouper *Hyporthodus flavolimbatus*	2
Graysby *Cephalopholis cruentata* (also *cruentatus*)	1	Misty Grouper *Hyporthodus mystacinus*	1
Reef Butterflyfish *Chaetodon sedentarius*	5	Warsaw Grouper *Hyporthodus nigritus*	5
Halichoeres spp.	10	Snowy Grouper *Hyporthodus niveatus*	26
Red Barbier *Baldwinella* (also *Hemanthias*) *vivianus*	150	Red Snapper *Lutjanus campechanus*	1
Blue Angelfish *Holacanthus bermudensis*	3	Scamp *Mycteroperca phenax*	15
Yellowedge Grouper *Hyporthodus flavolimbatus*	1	*Pareques* spp.	18
Wrasse Basslet *Liopropoma eukrines*	2	Roughtongue Bass *Pronotogrammus martinicensis*	2
Red Snapper *Lutjanus campechanus*	6	Red Lionfish *Pterois volitans*	22
Scamp *Mycteroperca phenax*	16	Vermilion Snapper *Rhomboplites aurorubens*	30
Mycteroperca spp.	4	*Seriola* spp.	35
Cubbyu *Pareques umbrosus*	7	Saddle Bass *Serranus notospilus*	3
Bank Butterflyfish *Prognathodes aya*	1		
Red Lionfish *Pterois volitans*	52		
Seriola spp.	15		
Saddle Bass *Serranus notospilus*	3		
Unknown	1		

mechanism to initiate additional fund-raising activities if appropriate materials could be identified for the site. In late 2016, a potential source of materials was located when the SCDOT announced plans to replace a steel swing bridge over the Wando River near Charleston, South Carolina with a fixed span bridge. Discussions with the general contractor for the demolition of the bridge produced a plan to remove the entire steel truss in one piece and place it on a barge at the bridge site. With the bridge welded into place on the barge, the entire structure, approximately 52 m (170 ft) long and 15 m (50 ft) tall, would be towed and sunk on the Charleston Deep Reef MPA site. The truss was removed in August 2017 and brought to the reef site in October (Figure 6). Underwater footage of this latest structure has not yet been obtained, but future ROV monitoring is planned.

Figure 6. A steel bridge truss welded to a deck barge deployed on the Charleston Deep Reef marine protected area.

Early indications were that the Charleston Deep Reef MPA would be successful in attracting deepwater species of management interest and that the site could be legally afforded protection through federal regulation. Consequently, SCDNR asked the SAFMC for similar protected status for its original two research reefs, Areas 51 and 53. This request was included in the council's Snapper Grouper Amendment 36, went through an extensive public review process, received final approval in March 2016, and was implemented as federal regulation in July 2017. These two research sites are now designated as spawning special management zones (SSMZs). Like the type II MPAs in deeper water, fishing for or possessing species from the Snapper Grouper Management Unit is prohibited within these areas while trolling for other pelagic species is allowed. South Carolina is now the first state in the USA with three artificial reefs deployed and maintained exclusively for the protection and enhancement of its reef fish fisheries resources.

With the successful precedent of artificial reef deployment specifically to enhance fisheries, it is hoped that additional support can be obtained for long-term monitoring, research, and evaluation activities on these and new MPA and SSMZ reef sites. These artificial-reef-enabled MPAs and SSMZs are necessary to further explore the potential benefits from developing artificial reefs specifically for the purposes of fisheries stock enhancement and to rebuild and maintain healthy populations of reef fishes along the coast. The concept of using artificial reefs primarily for stock enhancement and limiting or prohibiting fishing on them may seem somewhat alien to many anglers. Nevertheless, the very successful Charleston Deep Reef–South Carolina Memorial Reef partnership has resulted in a clearly beneficial situation for both fishery managers interested in conservation and rebuilding of deepwater grouper populations and blue water anglers interested in enhancing their pelagic and highly migratory fishing activities.

The ability to provide protection for artificial reef sites through the fishery management council process and federal regulation allows the biomass and spawning potential of deepwater grouper species to increase. These increases could potentially rise to the levels seen on newly discovered deep shipwrecks, such as the "snowy wreck" off North Carolina (Epperly and Dodrill 1995), and could allow a reef complex of strategically located and protected artificial reefs to serve in the rebuilding of overfished stocks of concern. To accomplish this, other states within the range of these species should also invest in and develop similar artificial reefs and seek appropriate regulatory protection for this purpose. Eventually, the development of a wider series of protected reef sites along the coast or around the nation may, in time, aid in the rebuilding of overfished stocks of many species.

References

Dial Cordy and Associates. 2001. Cooper River Bridges Demolition Section 206 study: environmental benefits analysis. Final Report prepared for U.S. Army Corps of Engineers, Charleston District, Charleston, South Carolina.

Epperly, S. P., and J. W. Dodrill. 1995. Catch rates of Snowy Grouper, *Epinephelus niveatus*, on the deep reefs of Onslow Bay, southeastern U.S.A. Bulletin of Marine Science 56:450–461.

Gold, H. 2001. Impact of the removal of Black Sea Bass (*Centropristis striata*) from artificial reef structures in the South Atlantic Bight. Master's thesis. College of Charleston, Charleston, South Carolina.

Harter S., J. Reed, S. Farrington, and A. David. 2016. NOAA CIOERT cruise report: South Atlantic MPAs and Oculina HAPC: characterization of benthic Habitat and biota, NOAA Ship Pisces Cruise 16–02, UNCW Mohawk ROV, June 7–22, 2016. National Oceanic and Atmospheric Administration (NOAA), Cooperative Institute for Ocean Exploration, Research & Technology (CIOERT), Florida Atlantic University, Boca Raton, Florida.

Kauppert, P. 2002. Feeding habits and trophic relationships of an assemblage of fishes associated with a newly established artificial reef off South Carolina. Master's thesis. College of Charleston, Charleston, South Carolina.

Kolmos, K. 2007. Succession and biodiversity of an artificial reef marine protected area: a comparison of fish assemblages on protected and unprotected habitats. Master's thesis. College of Charleston, Charleston, South Carolina.

U.S. Department of Commerce. 2007. Magnuson–Stevens Fishery Conservation and Management Act. Public Law 94–265.

American Fisheries Society Symposium 86:197–213, 2018

Artificial Reefs to Mitigate Human Impacts in the Marine Environment: The Wheeler North Reef as a Test Case

STEPHEN C. SCHROETER*
North San Diego County Laboratory, Marine Science Institute
University of California Santa Barbara
2270 Camino Vida Roble, Suite D, Carlsbad, California 92011, USA

DANIEL C. REED
Marine Science Institute, University of California Santa Barbara
Building 520, Room 4002, Floor 4L
Santa Barbara, California 93106, USA

PETER RAIMONDI
Center for Ocean Health, Long Marine Laboratory
University of California Santa Cruz
Santa Cruz, California 95060, USA

Abstract.—The Wheeler North Reef (WNR) is a large (70.4 ha [174 acre]) artificial reef in Southern California designed to mitigate the loss of kelp forest habitat and its associated community of algae, invertebrates, and fishes caused by the operation of the San Onofre Nuclear Generating Station (SONGS). Conditions of the SONGS' operating permit require that the success of WNR in compensating for the kelp forest resources destroyed by the 30 years of the power plant's operations be determined by long-term monitoring that is independent of the owners of SONGS. Performance standards pertaining to physical and ecological attributes of WNR are used as a basis for determining the success of WNR in meeting the mitigation objective to replace kelp forest resources in kind. We discuss details of the sampling design, evaluation criteria, and monitoring results and show how they are used to inform adaptive management that helps to ensure that the mitigation goals are met.

Introduction

One goal of fishery management is to account for the sources and relative magnitudes of natural and human-induced mortality of fish stocks of interest (Hilborn and Walters 1992; Quinn and Deriso 1999). One source of human-induced mortality, aside from fishing, is due to the degradation or loss of habitat caused by various human activities

* Corresponding author: stephen.schroeter@lifesci.ucsb.edu

(e.g., oil spills, coastal construction, or discharges of pollutants from ocean outfalls). While there are many examples of artificial reefs that have been used to enhance fishing (Carlisle et al. 1964; Bohnsack and Sutherland 1985; Bohnsack 1989; Wilson et al. 1990; Bortone 2006) there are relatively few artificial reef deployments with the specific goal of mitigating human-induced impacts to marine habitat (e.g., Ambrose 1994). This chapter focusses on the use of a large artificial reef to mitigate the loss of kelp forest habitat and its associated biota, including a reduction in the standing stock of reef fishes due to the operation of a coastal nuclear power plant in Southern California.

A critical aspect of any successful mitigation project is the ability to determine whether the project is meeting its stated goals and, if it is not, to determine the reasons why so that appropriate remedial actions can be implemented. Such determinations and actions require sufficient monitoring data. A recent review of compensatory mitigation in marine ecosystems concludes that most efforts to achieve the goal of no net loss of ecosystem services and ecological functions are questionable due to a combination of a brief period of assessment and data that were insufficient to assess the net effect of compensatory measures (Levrel et al. 2012). The reef mitigation project that we report on in this chapter addresses these issues (1) by having a clearly stated goal that requires the mitigation reef to replace the marine resources lost by the operation of the power plant, (2) through the use of well-defined performance standards linked to the impacts as a means of evaluating whether the reef is meeting its goal, and (3) with a rigorous long-term monitoring program with adequate statistical power to detect whether the performance standards are being met for the desired length of time. Monitoring data collected during this study are also used to determine the reasons why a performance standard is not being met and to provide a quantitative basis for implementing adaptive management and remedial actions in the event that the reef is not meeting its goals.

The marine habitat impacts for which the artificial reef was designed to mitigate here were caused by the single-pass seawater cooling system of the San Onofre Nuclear Generating Station (SONGS) Units 2 and 3, which began operation in 1983 and 1984, respectively. Both reactors were shut down in January 2012 due to excessive wear of the cooling system. In June 2013, both units were permanently retired.

Cooling the nuclear reactors raised the temperature of the intake water by an average of 10.5°C (19°F), discharging it through separate discharge lines that terminate between 1.87 and 2.59 km (6,150–8,500 ft) offshore. When the plant was fully operational, it collectively entrained and discharged about 9.1×10^9 L (2.4×10^9 gal) of seawater day (a volume equivalent to a square mile 12 ft deep). The discharge pipes were designed to rapidly mix the discharge water with about 10 times the volume of ambient water, reducing the initial discharge temperature from 19°F to 0°F above ambient beyond 300 ft (91.4 m) from the discharge pipes. The cooling efficiency of the diffusers (at the terminal end of the pipes) came with the unintended consequence of creating a turbid discharge plume that often extended over of the San Onofre kelp forest, which is located in the near vicinity of the two diffuser pipes (Figure 1). This plume reduced irradiance to the seafloor and thereby reduced recruitment of the microscopic stages of giant kelp *Macrocystis pyrifera* and growth and survival to the adult stages to a degree that resulted in the

Figure 1. Google Earth image showing the San Onofre Nuclear Generating Station (SONGS) Units 2 and 3 and its discharge plume, and the San Onofre kelp forest, outlined in yellow dashes, south of the discharge plume.

loss of a substantial amount of giant kelp forest habitat immediately downcoast of the discharge (Murdoch et al. 1989). The loss of giant kelp forest habitat in turn resulted in a 25.4 metric ton (mt; 28 U.S. tons) reduction in the standing stock of associated fish in the affected area (Murdoch et al. 1989).

After extensive review of studies of the power plant's marine impacts, the California Coastal Commission (CCC) amended the operating permit of the power plant in 1991 to require the operating utility company to construct an artificial reef large enough to sustain 60.7 ha (150 acres) of medium-to-high-density kelp bed that would support 25.4 mt (28 U.S. tons) of reef fish (which could result in a reef larger than 60.7 ha). Additional mitigation included funding for a mariculture/marine fish hatchery. Together, these measures were considered to compensate for the estimated loss of 72.4 ha (179 acres) of a medium-to-high-density kelp bed and associated community resulting from the discharge of SONGS' cooling water (California Coastal Commission 1997).

Construction of the artificial reef took place in two phases: an initial, small experimental reef (9 ha [22.4 acres]) and a subsequent, larger mitigation reef that was large enough to meet the 61-ha (150 acre) kelp and 25.4 mt (28 U.S. tons) fish standing stock requirements. In addition to the two phases of the artificial reef, the evaluation of many of the performance standards required comparisons to two nearby natural reefs (Figure 2). The purpose of the phase-1 experimental reef was to determine which combinations of substrate type and substrate coverage would most likely enable the mitigation reef to meet the performance standards specified in the SONGS coastal development permit.

The phase-1 reef was constructed in August and September 1999 on a chiefly sand bottom at depths ranging from 13 to 16 m, approximately 1 km offshore of the city of San Clemente, California, USA (Figure 2). It consisted of 56 modules clustered at seven locations (eight modules per location) spaced relatively evenly along 3.5 km of coastline (Figure 2). Each artificial reef module measured roughly 40 × 40 m, and the 56 modules collectively covered about 9 ha (22.4 acres) of the sea floor. The modules were comprised of quarry rock or concrete rubble and were constructed to form low-lying reefs (i.e., <1 m tall) that mimicked natural reefs in the region. These two types of materials were chosen because (1) they are the two materials most preferred by the California Department of Fish and Wildlife for building artificial reefs in California (Bedford 1997), and (2) little information existed on their relative effectiveness in supporting reef biota. Four modules at each location were constructed of quarry rock and four were constructed of concrete slabs and rubble for a total of eight modules at each location, as indicated above. These two construction materials differed with respect to their size and shape; the quarry rock was boulder-like in appearance while the concrete rubble consisted primarily of pieces of flat slabs that tended to be longer, wider, and slightly shorter than quarry rocks (Reed et al. 2006). The different sizes and shapes of the two materials resulted in rock and concrete modules that differed somewhat with respect to

Figure 2. Location of the artificial reef mitigation site (shown as the blue square) in relation to SONGS and the impacted San Onofre kelp forest and the naturally occurring kelp forests at San Mateo and Barn. Insert shows map of the Wheeler North Reef, showing phase-1 (blue squares) and phase-2 primary (blue) polygons. Black lines indicate transect locations.

small-scale topography (rugosity, interstitial void space, etc.). The slabs used to build concrete modules resulted in modules that had a greater proportion of horizontal substrate and a surface that was slightly more regular than modules constructed from quarry rock (Reed et al. 2006). By design, the amount of quarry rock and concrete slabs and rubble used to build the modules was systematically varied to produce three nominal bottom coverages of hard substrate (41%, 63%, and 81%) within the footprint of the modules of the two reef types within each location.

Results after 5 years of postconstruction monitoring were promising in that all six artificial reef designs (i.e., three nominal coverages × two reef types by material) and all seven locations (described below) showed a high tendency to meet many of the performance standards established for the mitigation reef (Reed et al. 2006). It was concluded from these findings that a low-relief slab and rubble concrete or quarry rock reef constructed off the coast of San Clemente, California that covered between 42% to 86% of the bottom could provide adequate mitigation for the loss of kelp forest biota caused by the operation of SONGS Units 2 and 3.

The design of the phase-2 mitigation reef followed from the analyses of data collected during the experimental phase-1 reef. Construction of the mitigation reef required about 5 months of effort and was completed in September 2008. Approximately 114,305 mt (126,000 U.S. tons) of boulder-sized quarry material was deposited in 18 polygons that collectively covered 61.5 ha (152 acres) of sea floor, as determined from bathymetry using multibeam sonar (Figure 2). The length, width, and height of the quarry rock boulders comprising the mitigation reef averaged 0.7 m (2.3 ft), 0.55 m (1.8 ft), and 0.43 m (1.4 ft), respectively. When added to the phase-1 experimental reef, a total of 70.4 ha (174 acres) of artificial reef (collectively named the Wheeler North Reef) were constructed with an average cover of quarry rock of 48%.

The mitigation requirement specifies that 1 year of mitigation credit would be given for each year that Wheeler North Reef met all of the performance standards. Because the Wheeler North Reef was required to provide compensation for damages to kelp forest resources for a period equal to the ~30-year lifetime operation of SONGS Units 2 and 3, fulfillment of the SONGS reef mitigation requirement would occur when the number of years of mitigation credit accrued by the Wheeler North Reef equaled the total years of operation of SONGS Units 2 and 3 during which the impacts occurred. Thus, despite the plant's shutdown in 2012, the 30-year period of required mitigation remains, which as of 2017 has yet to be achieved.

Methods

To achieve the mitigation goal of replacing the marine resources lost as a result of SONGS' operations, the CCC established a set of performance standards. These standards were designed to determine whether the mitigation goal has been met, including an extensive monitoring program capable of detecting (with a high degree of confidence) whether the Wheeler North Reef is meeting the performance standards. Importantly, the CCC required that a third party, independent of the utility company to avoid potential conflicts of interest, conduct the monitoring and performance evaluation.

Performance standards

There are two categories of performance standards: absolute standards, measured only at the Wheeler North Reef and requiring that the variable of interest attain or exceed a predetermined value (Table 1), and relative standards, requiring the value of the variable of interest on Wheeler North Reef to be similar to that measured on natural reference (control) reefs (Table 2). These performance standards require the Wheeler North Reef to support at least 61 ha (150 acres) of medium- to high-density kelp, 25.4 mt (28 U.S. tons) of fish, and assemblages of algae, invertebrates, and fishes that are similar to nearby natural reference reefs.

The evaluation of each absolute performance standard in any given year is based on the greater value obtained from either (1) data collected at the Wheeler North Reef that year or (2) a 4-year running average calculated from data collected at the Wheeler North Reef for that year and the preceding 3 years. A running average recognizes that short-term fluctuations in kelp forest biota are the norm, and the running average is used to allow mitigation credit for excess reef biota in good years to compensate for occasional years when values for the biota are slightly below those required by the absolute standards. All absolute standards must be met in a given year for that year to count towards compliance with the permit and to ensure replacement of lost resources.

In contrast to the absolute standards, the evaluation of each relative performance standard is based solely on a 4-year running average calculated from data collected at the Wheeler North Reef for that year and the previous 3 years. An either/or criterion (i.e., using data from either a single year or a 4-year running average) is not appropriate for the relative performance standard because the desired goal for the relative performance standards is not to achieve a specified value that is linked to estimated losses at the San Onofre kelp forest, but rather to evaluate whether the numbers of species and abundances kelp forest biota at the Wheeler North Reef are similar to those at the reference reefs. This is best accomplished using a short-term (4 years) running average that accounts for natural variation in time. Natural kelp forest communities vary greatly in their species composition and abundance, and it is likely that even the reference reefs will not consistently meet all the relative performance standards in a given year. Thus, to avoid requiring the Wheeler North Reef to perform better than the reference reefs, the Wheeler North Reef is required to meet at least as many of the relative standards as the lower-performing reference reef during the same year to count towards mitigation credit.

Table 1. List of absolute performance standards and their mitigation requirements.

Standard	Requirement
Hard substrate	At least 90% of the exposed hard substrate must remain available for attachment by reef biota.
Giant kelp *Macrocystis pyrifera*	The artificial reef(s) shall sustain 60.7 ha (150 acres) of medium-to-high-density giant kelp.
Kelp-bed fish standing stock	The standing stock of fish at the mitigation reef shall be at least 25.4 metric tons (28 U.S. tons).
Invasive and undesirable species	The important functions of the reef shall not be impaired by undesirable or invasive benthic species.

Table 2. List of relative performance standards. Compliance requires that each relative standard not be significantly less than the same standard at the lower performing of the two reference reefs.

Performance standard	Grouping
Algal percent cover	Benthos
Algal species richness	Benthos
Sessile invertebrate percent cover	Benthos
Mobile invertebrate abundance	Benthos
Invertebrate species richness	Benthos
Resident (=adult) fish density	Fish
Young-of-year fish density (fish recruitment)	Fish
Adult and young-of-year fish species richness	Fish
Fish reproductive rates	Fish
Fish production	Fish
Food chain support	Fish + benthos

Monitoring Design

Ninety-two monitoring stations (referred to as transects), each defined by a fixed 50 × 20 m area, were sampled at Wheeler North Reef (Figure 3). Eighty-two stations with the same transect dimensions were sampled at the San Mateo and Barn reefs. Transects on each reef are arranged in pairs with the two transects in each pair spaced 25 m apart. The exception to this are the single transects located on 12 of the phase-1 modules at Wheeler North Reef. Pairing of transects is done to increase sampling efficiency by reducing in half the number of anchoring events. Transects at Wheeler North Reef were allocated to the polygons and the existing experimental reef modules in proportion to their area. Maps of kelp persistence and the amount of hard substrate were used to strategically distribute the 41 transect pairs at the San Mateo and Barn reefs across areas known to support giant kelp.

Sampling at Wheeler North Reef, San Mateo, and Barn began in 2009 following the completion of construction of the phase-2 mitigation reef and is done annually and concurrently on the 70.4-ha (174 acre) Wheeler North Reef, San Mateo, and Barn from late spring to early autumn. Each sampling station is identified by unique differential GPS coordinates that marks the zero end of a 50-m transect and a compass heading along which divers lay out a measuring tape. A 20-m-wide swath centered along the 50-m transect defines the sample area at each sampling location. Different sized sampling units (e.g., 0.5, 1, 20, and 150 m^2) within this sampling area are used to evaluate different performance variables (Figure 3). Fish are sampled in 50 × 3 m band transects that extend 1.5 m off the bottom (outlined with a dotted line in Figure 3). Adult giant kelp more than 1 m tall, large understory algae, and large mobile invertebrates are counted in the five 10 × 2 m rectangular quadrats positioned perpendicular to the main transect at 10-m intervals (outlined with solid lines in Figure 3). The percent cover of invertebrates, algae, and bottom substrate is estimated using a grid of 20 points in the five 1 × 1 m quadrats (blue boxes in Figure 3). Smaller mobile invertebrates and small cryptic fish are counted either in 1 × 1 m or 1 × 0.5 m quadrats depending on their

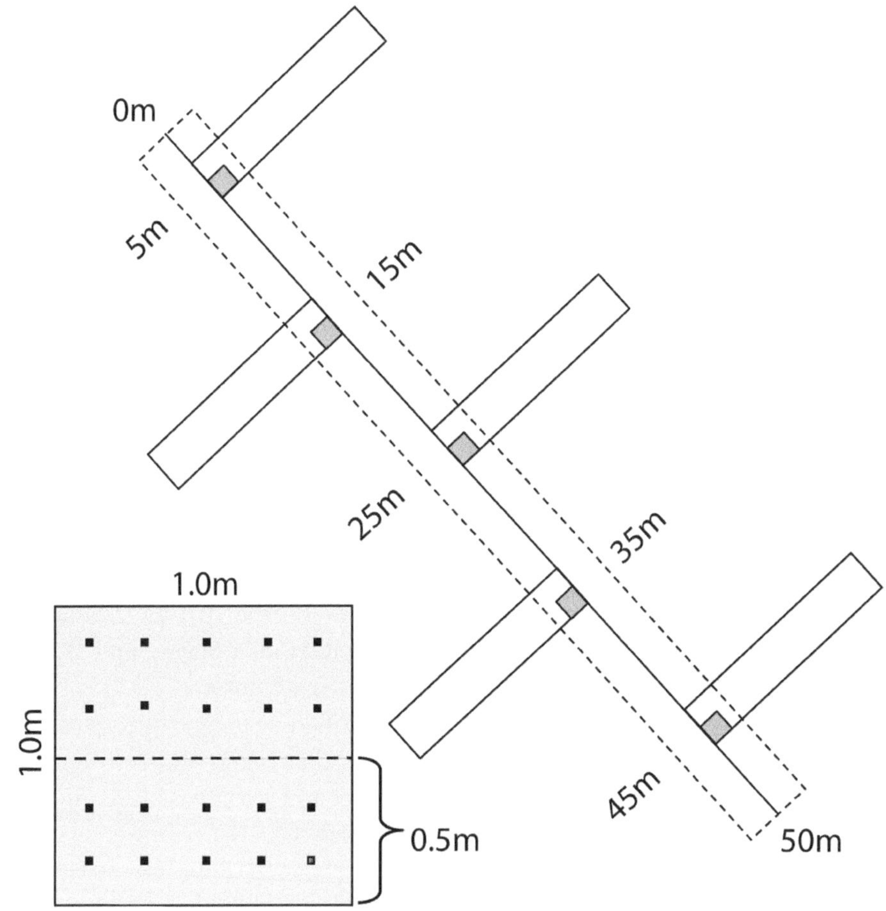

Figure 3. Schematic diagram of a sampling station. Fish are sampled in 50 × 3 m band transects that extend 1.5 m off the bottom (outlined with a dotted line). Adult giant kelp more than 1 m tall, large understory algae, and large mobile invertebrates are counted in the five 10 × 2 m rectangular quadrats positioned perpendicular to the main transect at 10-m intervals (outlined with solid lines). The percent cover of invertebrates, algae, and bottom substrate is estimated using a grid of 20 points in the five 1 × 1 m quadrats (shown in gray). Smaller mobile invertebrates and small cryptic fish are counted either in 1 × 1 m or 1 × 0.5 m quadrats, depending on their size and abundance.

size and abundance. Data used to evaluate the performance standards pertaining to fish reproductive rates (based on estimates of rates of egg and milt production), fish production (a rate which combines estimates of gonadal and somatic production), and food chain support, (a measurement based on the mass of the gut contents relative to the body mass of the fish) were collected for selected species of reef-associated fish that were collected on contemporaneously on Wheeler North Reef, San Mateo, and Barn.

Results of Reef Performance Monitoring

Results from monitoring show that the Wheeler North Reef met as many of the relative performance standards as the lower-performing reference reef (generally, the San Mateo kelp bed) since the beginning of compliance monitoring in 2009 (Table 3). Thus, the Wheeler North Reef has been successful in providing marine resources in a manner that is similar to natural kelp beds in the region. Importantly, it has consistently met all five of the relative standards pertaining to fish, including those that require the recruitment, production, and reproduction of fish on Wheeler North Reef to be similar to or greater than that of the two natural reference reefs (Figure 4a–c).

The Wheeler North Reef has generally performed well with regard to the absolute performance standards, meeting the hard substrate and invasive and harmful species standards during all 8 years of monitoring, and the standard for giant kelp in 6 of the 8 years (Table 3). In contrast, the artificial reef consistently failed to meet the performance standard, requiring it to support a fish standing stock of at least 25.4 mt (28 U.S. tons; Table 3; open symbols in Figure 5a, 5b). As a result, the Wheeler North Reef has yet to receive any mitigation credit.

Determining Reasons for Consistent Failure to Meet the Fish Standing Stock Standard

The standing stock of fish on a reef is influenced by the physical attributes of the reef, such as its footprint area, rock coverage, and topography. The Wheeler North Reef was designed to be low relief to mimic natural reefs in the region, including the reef at San Onofre that was damaged by SONGS's operations, and the reference reefs at San Mateo and Barn. Low-relief artificial reefs in Southern California are also more likely to support giant kelp, which was a major objective of the mitigation project. A critical issue in assessing the long-term performance of the Wheeler North Reef is whether its present configuration (70.4 ha [174 acres]) of low-relief rock that covers, on average, 48% of the bottom) is sufficient to sustain at least 25.4 mt (28 U.S. tons) of fish over the long term. Data collected from monitoring the quarry rock modules of phase-1 experimental reef and the refer-

Table 3. Summary of the Wheeler North Reef performance with respect to whether it met the four absolute performance standards and the collective group of relative performance standards for each year since 2009. To receive mitigation credit for a given year, the Wheeler North Reef must meet all four absolute performance standards and the collective group of relative performance standards.

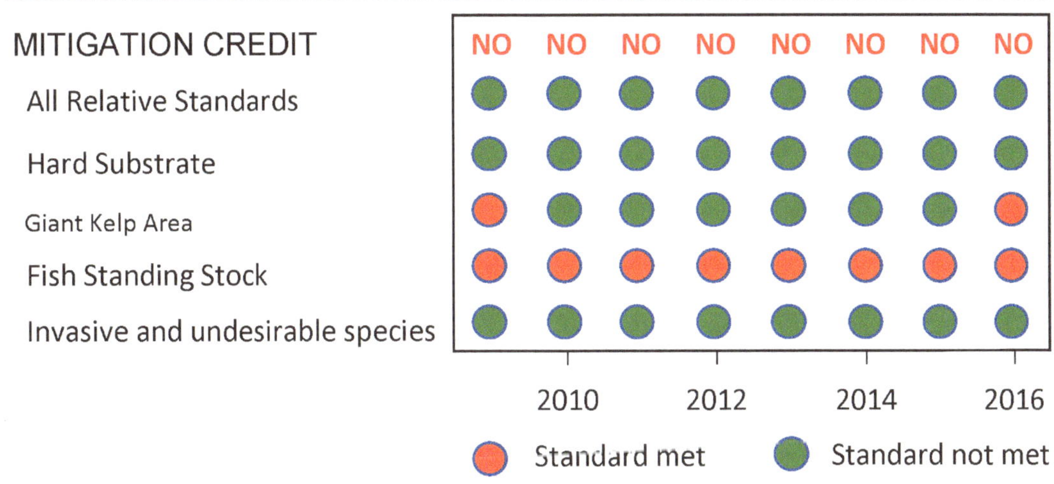

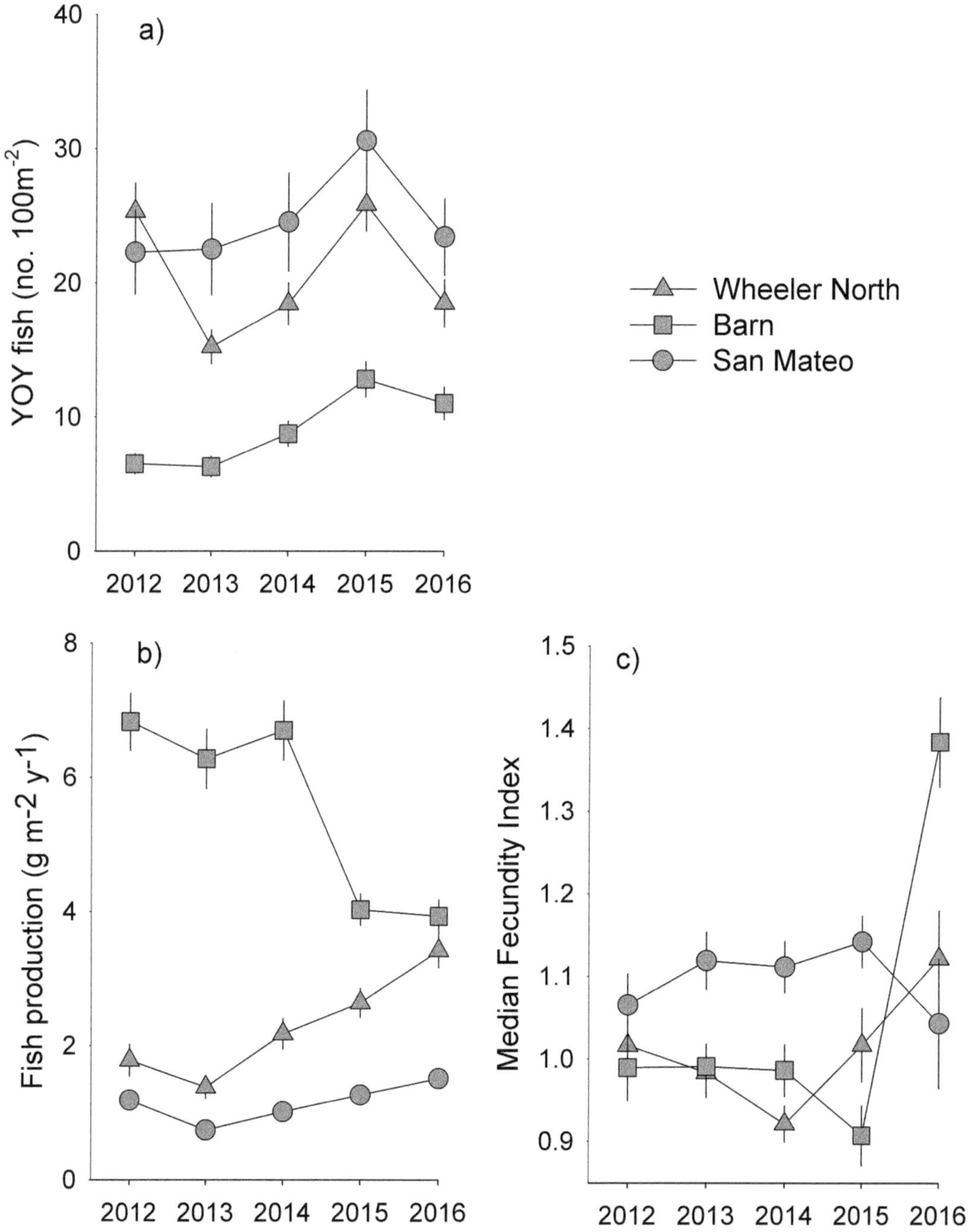

Figure 4. Mean **(a)** fish recruitment (young-of-year density—indicated as YOY), **(b)** production (the rate at which somatic and gonadal biomass is produced annually, expressed as g/m^2), and **(c)** reproduction (a fecundity index based on size-corrected annual egg production for three common indicator species) on the Wheeler North Reef and the two natural reference reefs, Barn and San Mateo. For details of methods to estimate production, reproduction, and recruitment, see Reed et al. 2017.

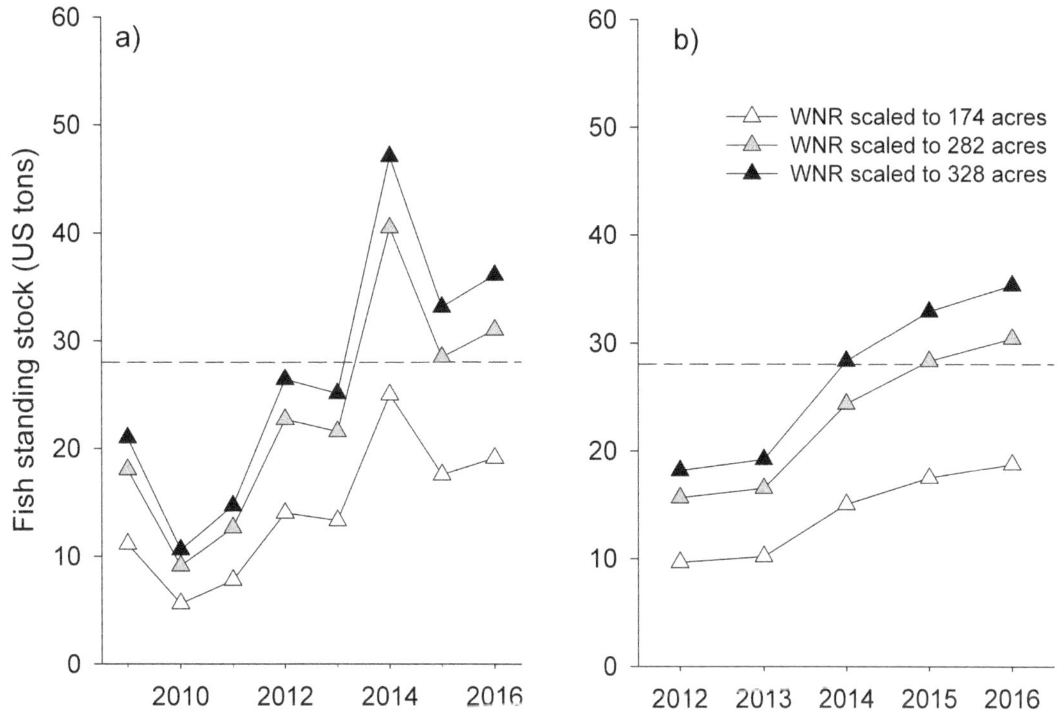

Figure 5. Time series of fish standing stock at the Wheeler North Reef (WNR); **(a)** annual values for 2009–2016, **(b)** 4-year running averages. The dashed horizontal line indicates 25.4 metric tons (28 U.S. tons). Open symbols indicate standing stock of Wheeler North Reef at current size (70.4 ha [174 acres]), gray symbols show standing stock scaled to size of San Mateo (114 ha [282 acres]), and black symbols show standing stock scaled to size of Barn (132.7 ha [328 acres]).

ence reefs since 2000 provides a reasonably long-term perspective for evaluating this issue.

To explore whether the footprint area of the Wheeler North Reef is sufficient to support a standing stock of 25.4 mt (28 U.S. tons), we used data on the biomass density of fish collected since 2000 to compare fish standing stock estimates on the two reference reefs based on their actual size (footprint) and on the size of the Wheeler North Reef. When standing stock was scaled to the actual size, the smaller of the two reference reefs (i.e., San Mateo) attained or exceeded 25.4 mt (28 U.S. tons) 5 out of the 14 years sampled, while Barn, the larger reference reef, supported 25.4 mt (28 U.S. tons) 8 out of 14 years (Figure 6a). When the standing stock of the two reference reefs was scaled to the size of the smaller, 70.4-ha (174 acre) Wheeler North Reef, San Mateo supported 25.4 mt (28 U.S. tons) of fish in only 3 of 14 years and Barn in only 2 of 14 years (Figure 6b).

We also explored whether the footprint area of the Wheeler North reef was sufficient to support a standing stock of 25.4 mt (28 U.S. tons) by scaling its standing stock the size to that of the two reference reefs using data collected from the entire 70.4-ha (174 acres) Wheeler North Reef since 2009. This analysis shows that the standing stock at Wheeler North Reef has been increasing and would have met the 25.4 mt (28 U.S. tons) fish stand-

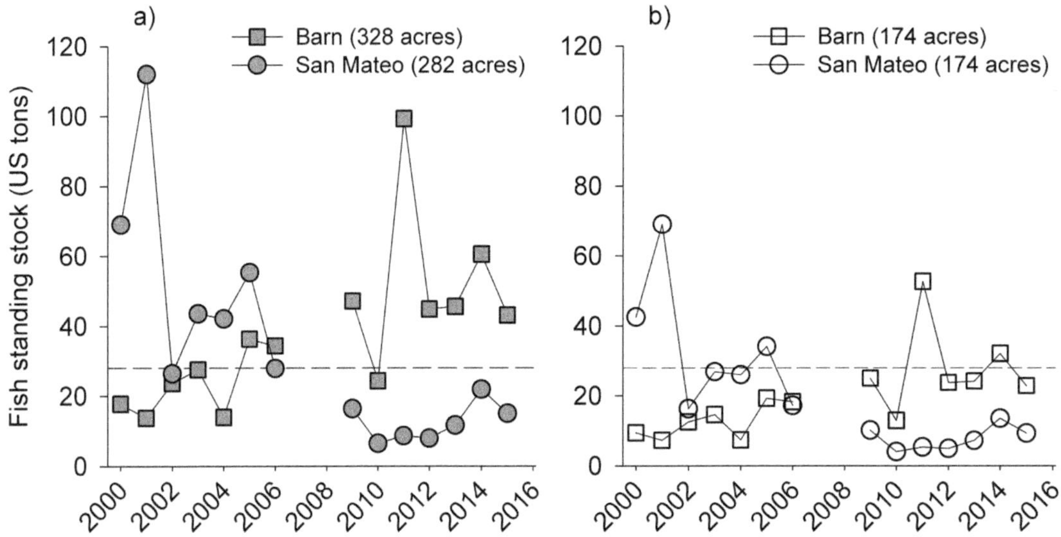

Figure 6. Time series of fish standing stock at the two reference reefs (San Mateo and Barn), **(a)** biomass scaled to the actual sizes of the reefs, and **(b)** biomass scaled to the size of the Wheeler North Reef. The dashed horizontal line indicates 25.4 metric tons (28 U.S. tons).

ing stock standard by 2014 had it been the size of the Barn reference reef or if in 2015 it had been the size of the San Mateo reference reef (Figure 5a, 5b). The rate of increase in the fish standing stock appears to be slowing, suggesting that the Wheeler North Reef is attaining its carrying capacity and, in its current configuration, is unlikely to consistently meet the 25.4-mt (28 U.S. tons) standard.

We also used the time series of rock coverage data from phase 1 to evaluate the extent to which the bottom cover of rock affects fish standing stock. We scaled data of fish biomass density collected on the experimental modules to the current size of the Wheeler North Reef and found that fish standing stock was positively related to rock coverage; 70.4 ha (174 acres) with 41% coverage of rock was projected to have met the 25.4-mt (28 U.S. tons) fish standing stock in only 1 of 13 years, 63% coverage of rock in 3 of 13 years, and 81% rock in 9 of 13 years (Figure 7).

Collectively, these results confirm that the reason for the consistent failure of the Wheeler North Reef to support a fish standing stock of 25.4 mt (28 U.S. tons) was because its footprint was not large enough given its average rock coverage of 48%. That fish recruitment, production (i.e., the rate of gonadal and somatic biomass produced per year). and reproduction on the Wheeler North Reef were similar to that of natural reefs (Figure 4a–c) further suggests that its failure to meet the 28-ton fish standing stock requirement is not because it provides poor quality habitat, but rather because it is not big enough.

Determining the Amount of Remediation Acreage

We used the monitoring data to gain insight into the number of additional acres needed for remediation to ensure that the Wheeler North Reef consistently supports 25.4 mt (28 U.S. tons) of reef fish. Monte Carlo simulation was done using the mean and standard deviation of standing stock collected from 2000 to 2016 on the phase-1 rock mod-

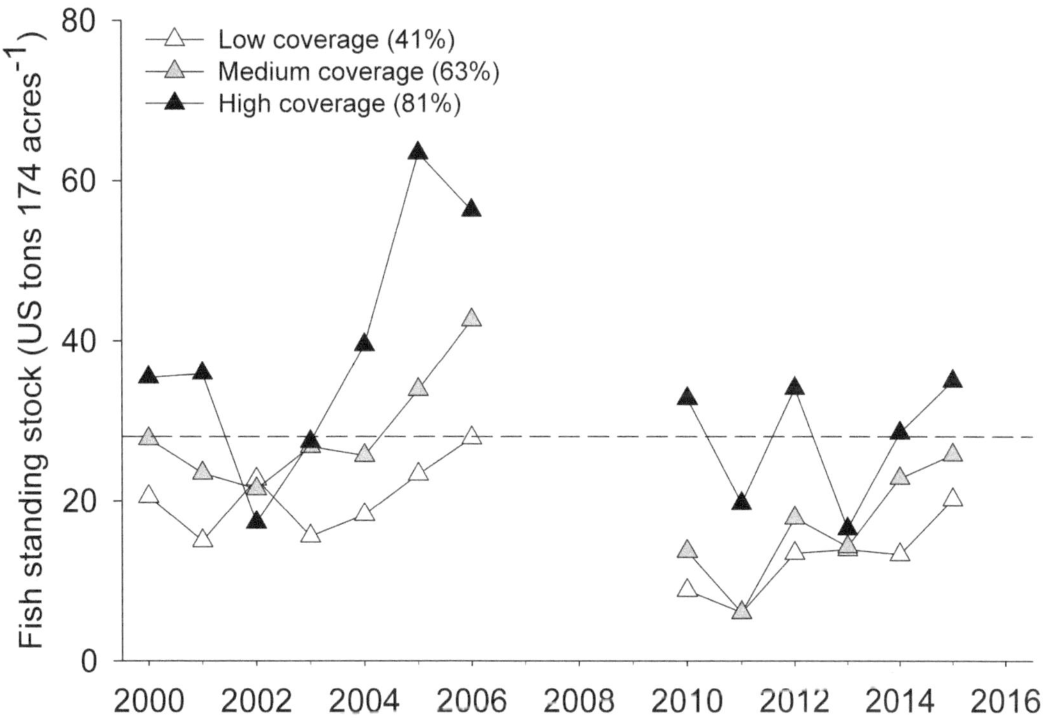

Figure 7. Fish standing stock biomass (scaled to 70.4 ha [174 acres]) on low-profile reefs with varying rock cover. The dashed horizontal line indicates 25.4 metric tons (28 U.S. tons).

ules. Prior to conducting the simulation, data collected during phase 1 were adjusted to reflect the difference between the 41% cover of rock on the phase-1 reef and the 48% cover on the 70.4-ha (174 acres) mitigation reef (which included both phase-1 modules and phase-2 polygons). Since evaluation of the fish standing stock standard is based on either the annual average or the 4-year running average, the simulations were done on both time series, and the larger of the two estimates was used as the basis for determining the remediation acreage. The analysis to estimate the acreage of the three low-relief designs was the same as the analysis for the 70.4-ha (174 acres) mitigation reef. However, the actual coverage of the low (41%), medium (63%), and high (81%) phase-1 reef designs was used without adjustment. Simulations were conducted to produce an estimate of the standing stock on the Wheeler North Reef with a 95% confidence.

Results of these analyses showed that the current 70.4-ha (174 acres) mitigation reef, with 48% rock cover, will support a standing stock of 12.9 mt (14.5 U.S. tons) of reef fish over the long term, 12 mt (13.5 U.S. tons) short of the required 25.4-mt (28 U.S. tons) target. Similar analyses estimating the additional acreage for low-relief reefs with low (41%), medium (63%), and high (81%) rock coverage would require 80.9, 50.6, and 42.5 ha (200, 125, and 105 acres), respectively (Table 4).

The tonnage of rock needed to construct the three reef designs in Table 4 is inversely related to the reef area required to meet the 25.4-mt (28 U.S. tons) fish standing stock standard (Figure 8). Cost scales to tonnage and material costs of a 80.9-ha (200 acres) low-cover reef would be about 60%

Table 4. The number of acres of new reef material needed for the Wheeler North Reef to have a 95% probability of meeting the 25.4-metric ton (mt; 28 U.S. tons) performance standard for fish standing stock in a given year for different configurations of new reef material using phase 1 as the data source. Phase-1 data were collected from 2000 to 2016 on 40 × 40 m low-relief rock modules that consisted of low, medium, and high rock cover. The added acreage is required to produce 12 mt (13.5 U.S. tons), which when added to the estimated 12.9 mt (14.5 U.S. tons) standing stock of the current 70.4-ha (174 acre) reef would equal 25.4 mt.

Configuration of new reef to be added for remediation	Additional hectares (acres) needed to support 25.4 mt of fish
Low relief (<1 m), low rock cover (41%)	80.9 (200)
Low relief (<1 m), medium rock cover (63%)	50.6 (125)
Low relief (<1 m), high rock cover (81%)	42.5 (105)

lower than a 42.5-ha (105 acres) high-cover reef, and it would provide about 90% more kelp forest habitat, which increases the probability of meeting the 60.7-ha (150 acres) performance standard for kelp area.

Discussion

Frequently, the primary goal of artificial reef programs is to enhance fisheries, presumably through an increase in standing stock

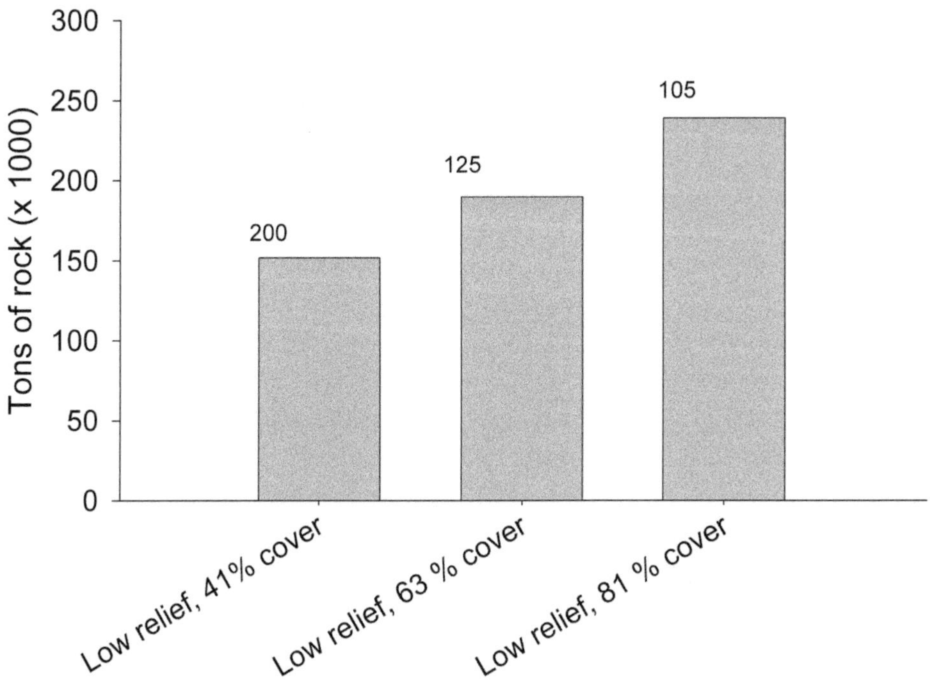

Figure 8. Tonnage of rock necessary to construct the three different reef designs, which (when added to the existing Wheeler North Reef) are expected to support a fish standing stock of 25 metric tons (28 U.S. tons) with a 95% level of confidence. Numbers above each bar are the number of acres of additional reef needed for each reef design, shown in Table 4.

(Bohnsack 1989; Wilson et al. 1990). Fishery-dependent data (e.g., catch or catch per unit effort) are most often used to determine whether this goal has been met, though such data may or may not be correlated with the fishery independent measures of stock size (Hilborn and Walters 1992). Several investigators have questioned whether artificial reefs actually increase regional biomass rather than acting as fish attractors (Bohnsack 1989; Polovina 1989).

Rather than focusing on fishery enhancement, one of the primary goals of the Wheeler North Reef was to mitigate for lost standing stock. Estimates of standing stock loss and replacement were based on fishery-independent measures of biomass (i.e., relative biomass density). Observations during and immediately following the construction of 61.5 ha (152 acres) of the phase-2 reef suggest that much, if not all, of the initial standing stock on the Wheeler North Reef was the result of attraction—the immigration of adult fish not originating from the newly constructed reef (Reed et al. 2010). Although our sampling regime cannot definitively separate effects of attraction from production, monitoring data show that fish recruitment, production (combined gonadal and somatic production), and reproduction are occurring on the Wheeler North Reef at rates similar to nearby natural reefs (Figure 5a–c). These results provide strong evidence that some proportion of the standing stock of the Wheeler North Reef is due to the addition to the standing biomass by resident fish and that this results in an increase in fish biomass (i.e., standing stock) over time. Although it is uncertain how long it will take, the monitoring data suggest that the fish standing stock at the existing 70.4-ha (174 acres) Wheeler North Reef is beginning to reach its carrying capacity after 8 years.

Distinguishing features of our study include (1) clear definitions of mitigation goals, (2) performance measures that adequately determine whether these mitigation goals have been met, (3) a sampling design with adequate statistical power to determine failure to meet the required performance standards, and (4) collection of data useful in designing any necessary remediation measures. These design features address some of the concerns regarding the evaluation of the effects of artificial reefs (Bohnsack 1989; Levrel et al. 2012) and allow us to assess mitigation success in two distinct categories. The first category, using the relative performance standards, addresses whether the artificial reef is performing similarly to natural reefs in the region. Our assessment concludes that Wheeler North Reef has met the target of similarity to natural reefs in all 8 years of postconstruction monitoring. The second category, using the absolute performance standards pertaining to giant kelp acreage and fish standing stock, addresses whether lost resources will be adequately replaced. Our assessment in this case shows that the Wheeler North Reef has consistently failed to meet the mitigation target for fish standing stock in all years since its construction. By contrast, the Wheeler North Reef has met the acreage requirement for giant kelp in 6 of the 8 years. In addition to assessing whether the Wheeler North Reef has met its mitigation targets, the monitoring data provided us with a mechanistic understanding for the cause of the Wheeler North Reef's failure to meet the fish standing stock standard and allowed us to and develop options for remedial measures that will with a high degree of confidence enable the Wheeler North Reef to meet its targeted goals in the future.

Acknowledgments

We gratefully acknowledge the following team of scientists from the University of California Santa Barbara (UCSB) Marine Science Institute that have worked on the San Onofre Nuclear Generation Station (SONGS) Mitigation Monitoring Project whose hard work and dedication in gathering and processing the data that made this publication possible: Lindsey Chamberlain, Dave Huang, Matt Silbert, Nate Spindel, and Denise Weisman. We thank Mia Adreani and Mark Steele for their work on fish production, reproduction, and food-chain support. We thank Rich Ambrose, Jonna Engel, Kate Huckelbridge, Mark Page, and Russ Schmitt for insightful discussions and advice and Susan Hansch, David Kay, and Patrick Tennant for their patience and support. The work contained herein was funded by Southern California Edison (SCE) as required by the California Coastal Commission under SCE's coastal development permit (No. 6-81-330-A, formerly 183-73) for Units 2 and 3 of SONGS.

References

Ambrose, R. F. 1994. Mitigating the effects of a coastal power plant on a kelp forest community: rationale and requirements for an artificial reef. Bulletin of Marine Science 55:694–708.

Bedford, W. D. 1997. Material specification guidelines and notification procedure for augmentation of artificial reef with surplus materials. California Department of Fish and Game, Marine Resources Region, Long Beach.

Bohnsack, J. A. 1989. Are high densities of fishes at artificial reefs the result of habitat limitation or behavioral preference? Bulletin of Marine Science 44:631–645.

Bohnsack, J. A., and D. L. Sutherland. 1985. Artificial reef research: a review and recommendations for future priorities. Bulletin of Marine Science 37:11–39.

Bortone, S. A. 2006. A perspective of artificial reef research, past, present, and future. Bulletin of Marine Science 78:1–8.

California Coastal Commission. 1997. SONGS permit 6-81-330-A (formerly 183-73). Mitigation Program adopted findings & conditions, May 14, 1997. California Coastal Commission, San Francisco.

Carlisle, J. G., Jr., C. H. Turner, and E. E. Ebert. 1964. Artificial habitat in the marine environment. California Department of Fish and Game Fish Bulletin 124.

Hilborn, R., and C. Walters, 1992. Quantitative fisheries stock assessment: choice, dynamics & uncertainty. Chapman and Hall, New York.

Levrel, H., S. Pioch, R. Spieler, 2012. Compensatory mitigation in marine ecosystems: which indicators for assessing the "no net loss" goal of ecosystem services and ecological functions? Marine Policy 36:1202–1210.

Murdoch, W. M., R. C. Fay, and B. J. Mechalas. 1989. Final report of the Marine Review Committee to the California Coastal Commission, August 1989. MRC Document No. 89-02. Available: http://marinemitigation.msi.ucsb.edu/documents/MRC_reports/final_report/mrc-final-rpt_to_ccc.pdf. (April 2018).

Polovina, J. J. 1989. Artificial reefs: nothing more than benthic fish aggregators. CalCOFI (California Cooperative Oceanic Fisheries Investigations) Reports 30:37–39.

Quinn, T. J., and R. B. Deriso. 1999. Quantitative fish dynamics. Oxford University Press. New York.

Reed, D. C., S. C. Schroeter, and D. Huang, 2006. An experimental investigation of the use of artificial reefs to mitigate the loss of giant kelp forest habitat. A case study of the San Onofre Nuclear Generating

Station artificial reef project. California Sea Grant College Program, University of California, San Diego.

Reed, D. C., S. C. Schroeter, and H. M. Page. 2010. Annual report of the status of condition C: kelp reef mitigation: San Onofre Nuclear Generating Station (Songs) Mitigation Program. Submitted to the California Coastal Commission, San Francisco.

Reed, D. C, S. C. Schroeter, H. M. Page, and M. Steele. 2017. Monitoring plan for the SONGS reef mitigation project. Prepared for the California Coastal Commission, San Francisco. Available: http://marinemitigation.msi.ucsb.edu/documents/artificial_reef/ucsb_%20mm_reports/mitigation_phase/monitoring_plan4reef-mitigation_project_rev_apr2017.pdf. (April 2018).

Wilson, K. C., R. D. Lewis, and H. A. Togstad, 1990. California Department of Fish and Game nearshore sport fish habitat enhancement program artificial reef plan for sportfish enhancement. California Department of Fish and Game, Administrative Report 90-15, Sacramento.

American Fisheries Society Symposium 86:215–234, 2018

An Integrative Artificial Reef Project for Conservation and Coastal Management in the Colombian Caribbean

Oscar Delgadillo-G*
MoAm S. A. S.
Calle 87 #2-53 Of, 101 Bocasalinas, Pozos Colorados, Santa Marta 470006, Colombia

Nirith N. Toro
Ecopetrol S. A.
Terminal Pozos Colorados, Santa Marta 470006, Colombia

Abstract.—Multiple coastal development activities coupled with unsustainable management have caused environmental degradation in the Santa Marta region of Colombia. To mitigate this impact, Ecopetrol entered into alliances with private and government institutions to initiate an integrative artificial reef project in Pozos Colorados Bay. To develop the project's framework, it was necessary to (1) establish context and objectives, (2) design plans and reef construction, (3) strengthen a target social population, and (4) conduct pre- and postdeployment ecological assessments. The achievement of each objective was met with delays and constraints, mainly due to administrative issues and legal requirements. Nevertheless, interventions and interactions among representatives of the 10 institutions involved in the project, as well as the strong commitment of fishers from three organizations in all stages of the process, were indicators of project's success. Together, these actions and contributions resulted in the deployment of the first six artificial reefs in a 137-ha area. Moreover, recorded changes in biological assemblages before and after reef deployment (richness: 3–37 species; abundance: 30.3–1,615.7 individuals), along with the presence of commercial, ecological, and endangered important species, support the concept of habitat enhancement procedures used here as a strategy for biodiversity conservation with potential for ecotourism activities. The utilization of this technology should be conducted in compliance with concerted schemes for coastal resource management and precautionary principles, directed towards the conformation of discrete marine reserves as future models of sustainable production in sensitive areas.

* Corresponding author: odelgadillo@moam.com.co

Introduction

Artificial reef technology around the globe has generally focused on the enhancement of artisanal and industrial fishery production as its main goal (Seaman and Jensen 2000; Bortone 2011). In recent decades, its objectives and applications have expanded to include ecological restoration, mitigation, mariculture, research, and biodiversity conservation (Lindberg and Relini 2000; Seaman 2002, 2007). Trends in world fisheries production have shown evidence of substantial fish stock depletion and species decline, especially in developing countries where appropriate management is lacking and data from small-scale fisheries have often been ignored (Costello et al. 2012; Pauly and Zeller 2016). Consequently, an alternative for the application of artificial reefs in a socioecological context, similar to that in Japan and France, discussed by Pioch et al. (2011), should be conducted. This approach should involve integrative management strategies to encourage local, vulnerable fishing communities to protect, restore, and make sustainable use of the marine natural resources.

Santa Marta, in the Caribbean Sea of Colombia, is an area of high diversity, both geographically and environmentally. As part of the largest coastal mountain system in the world, the Sierra Nevada of Santa Marta influences local weather and imparts structure to the coastline and continental shelf. These features, among others, provide a high variety to the coastal ecosystems and habitats, which include extensive protected areas and sanctuaries consisting of coral reefs, sea grass beds, and mangrove forests biotopes. Contributing to the heterogeneity of the environment, an upwelling event during the first months of each year, coupled with complex oceanographic and hydrological systems, increases the general productivity and fishery potential in this coastal region (Díaz et al. 2000, 2003; Idárraga-García et al. 2011; García et al. 2013).

The area has also experienced significant coastal development, including port facilities for hydrocarbons and coal transportation and an expansion of ecotourism. Pollution, owing to local development along with the unsustainable administration of natural resources and the effects of climate change, has led to environmental degradation, a reduction in fishery productivity, and the loss of traditional fishing grounds. All these factors have negatively affected fishing and fishing communities (Díaz et al. 2000, 2003; García et al. 2007; García 2010). Under these environmental and social conditions, Ecopetrol (i.e., the leading company on oil and gas business in Colombia, formerly known as the Colombian Petroleum Company), within its corporate social responsibility policy, locally contributes to the sustainable utilization of natural resources. Ecopetrol helped form alliances with a variety of stakeholders and cooperators to promote the governance of a plan to deploy artificial reefs. The goals of the strategy were met by strengthening the technical and organizational capacity of fishermen in the associated communities. Aiming to assess the feasibility and potential benefits of this integrative approach for management in the socioecological context, this plan emphasizes pathways for its implementation and preliminary results during the first stage.

Methods

Planning Context and Objectives

The first step in the process of achieving the plan's objectives was to determine the environmental, management, and socioeconomical characteristics of the potential site or location for artificial reef deployment. The second step was to identify particular problems in the target social population and to suggest remedies to resolve the problems. The third step was to establish potential interinstitutional agreements needed for participation, planning, and investment. Last, after the completion of the previous requirements, the management and research objectives of the project were defined (Lindberg and Relini 2000; Pioch et al. 2011).

Artificial Reef Building and Deployment Plans

To facilitate planning the materials, design, and sites for artificial reef deployment, the basic environmental characteristics of the area (i.e., physical, chemical, and biological) were evaluated through the review of already available information and field surveys. As is true for projects in any country when contemplating an intervention of this magnitude, licenses and permissions for artificial reef transport, deployment, research, and environmental viability were managed with different government institutions and stakeholders, including the establishment of industrial security protocols for these processes (Bortone et al. 2000; Sheng 2000; Wilding and Sayer 2002).

Strengthening of Target Fishing Communities

The various features, needs, and requirements of the target social group with regard to the fishery, education, and technical organization were identified to more easily achieve coastal stewardship and project appropriations. Accordingly, a variety of supporting activities, training, and technological transference were conducted.

Pre- and Postecological Assessments of Fishes and Mobile Macroinvertebrates

Besides conducting a review of already existing information, pre- and initial postdeployment assessments of fishes and mobile macroinvertebrates associated with the artificial reefs were performed using a random visual census (Bortone et al. 2000; Delgadillo-Garzón et al. 2004). These visual censuses consisted of two divers estimating faunal composition, species abundance, and size using scuba during five regular time intervals of 6 min each per station for a total of 30 min per diver. Species richness and relative abundance were calculated from the average of the 10 time intervals for each census day and station and were expressed as a count per unit of effort. The comparisons between pre- and postdeployment census were performed with a nonparametric Mann–Whitney U-test (Zar 2010).

Results

Environmental and Socioeconomical Context

Pozos Colorados Bay is located on the east side of the Gulf of Salamanca, between Punta Gloria and Punta La Loma, in the southwest sector of the city of Santa Marta (11°06'00"–11°09'00" N and 74°14'00"–74°16'00" W) in the Magdalena Department, Caribbean of Colombia (Figure 1). Pozos Colorados Bay has a continuous coastline length of 6.7 km, with depths between 0 and 30 m on the typically narrow continental shelf and a maximum width of 16 km. The area also receives the direct and indirect influence of Magdalena River and Ciénaga Grande of Santa Marta (coastal lagoon), with heavy runoff from these and other nearby continental sources. The bay has a limited amount of hard substrate (rock conglomerate) and sea grass beds, although there are some sites with rocky littoral zones. The bottom is predominantly mud (73%) and very fine sand (27%) with no readily apparent association of substrate type with depth.

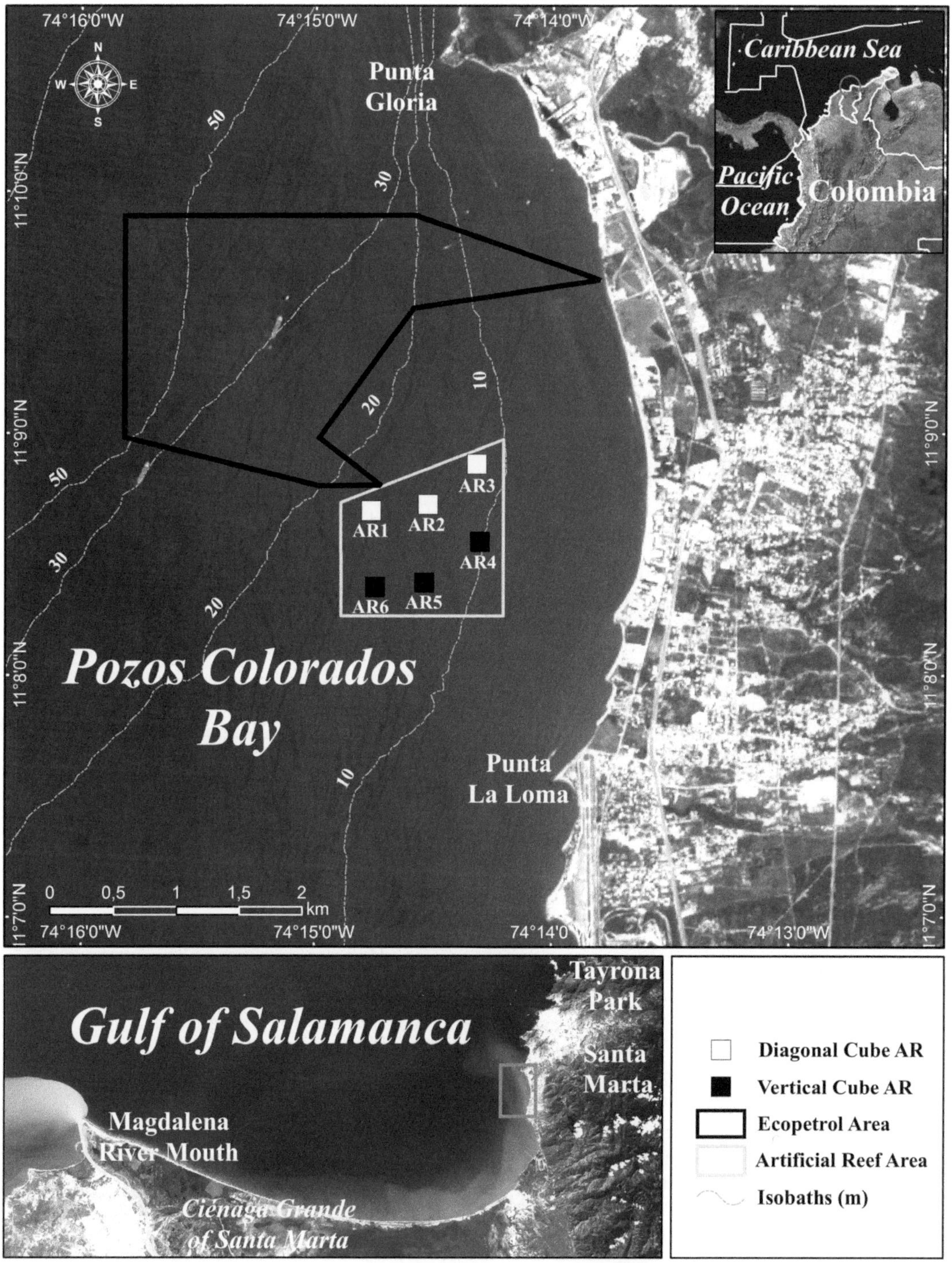

Figure 1. Location of Pozos Colorados Bay in the Gulf of Salamanca indicating the position of the artificial reefs.

The dry season extends from December to April with strong influence of Alisios winds from the north-northeast, which causes upwelling during these months. The wet season extends from April to November with winds from the south-southwest. Water temperature ranges from 21°C to 30°C and salinity from 27‰ to 38‰ (Guzmán-Alvis and Díaz 1996; Ramírez-Calle and Valencia-Vera 2005; García et al. 2013; Delgadillo-G and Flórez 2015).

Fishery resources within the continental shelf limits in the Gulf of Salamanca are highly diverse and productive, with at least 280 fish species of commercial and ecological importance (García and Armenteras 2015). Additionally, the artisanal fishing community is composed of no less than 1,800 fishers, excluding those fishers whose fishing activity is restricted to coastal lagoons and freshwater systems. The fishers use a wide variety of gears, such as hook and line, longline, gill nets, beach seines, shrimp beach seines, casting net, pots, speargun, and changa (small, modified artisanal trawls), conducting their fishing mainly from wood canoes with oak and sail or fiberglass boats with small, outboard engines (Rueda et al. 2010).

The environmental features of the region have facilitated port development, including the creation of restricted areas for industrial activities (e.g., pipes and monobuoys for oil transport, anchoring zones for vessels and coal transference, and industrial navigational routes). Additionally, in past years, Pozos Colorados Bay has experienced intense urban expansion along with an increase in the activities and infrastructure associated with tourism. In this context, the area of Ecopetrol marine operations within the bay (Figure 1) also includes the activities of artisanal fishing communities associated with several fishing organizations (e.g., Aspescoltur, Asopesmar, Copepazbe, Asocumar, Asopesbaga, Adimujer, and Asopozoscolorados). According to García (2010), these social groups are confronted with the effects of inshore and offshore industrial development, fishery overexploitation, pollution, climate change, and unsustainable management of natural resources. The effects of each of these is associated with a reduction in fishing productivity, the displacement of fishers from their traditional fishing grounds, environmental degradation, and a scarcity of government investment to mitigate these effects.

To initiate planning the artificial reef project, Ecopetrol entered into agreements and formed strategic alliances with 10 government and private stakeholders, including the University of Magdalena, Drummond Ltd., Foundation of Port Society of Santa Marta, INCODER (Colombian Institute of Rural Development), CORPAMAG (Regional Autonomous Corporation of Magdalena), AUNAP (National Authority of Aquaculture and Fishing), SENA (National Learning Service), Irotama Resort, Estelar Hotel, and MoAm, among others. Each agreement had partner-specific roles within the project. Consequently, one intended management objective was to strengthen the technical and organizational capacity of fishing communities while the research objectives for the artificial reefs were to determine their effectiveness in the conservation of biodiversity and their potential for promoting ecotourism.

Artificial Reef Design, Building, and Deployment

In accordance with the environmental characteristics of the study area and considering past successful performance with the first steel-pipe artificial reefs deployed in 2000 in the Gulf of Morrosquillo (Delgadillo et al. 2004; Delgadillo-Garzón and García 2009), refuse steel pipe (formerly used for oil transport) was used as a material of opportunity to create artificial reef modules. The design, configuration, and arrangement of the modules during the present project were designed to increase reef stability and complexity, which were identified as problematic for the recent (2007–2010) artificial reefs deployed in the Gulf of Morrosquillo (Delgadillo-Garzón et al. 2010).

Between May and October of 2011, 500 m of steel pipe (46 cm diameter, 1 cm thick) in 6-m-long sections were recycled to build the artificial reefs. Residual hydrocarbons (e.g., tar) were cleaned from each pipe by striking and scraping both internal and external surfaces. Subsequently, pipes were cut to length using oxyacetylene torch and high-pressure welding to assemble the pipes into a finished structure. Six artificial reefs designed by MoAm were built in a cubic-frame configuration of 4.5 × 4.5 × 4.5 m, three of them with crossing diagonal pipes between the main frame (diagonal cube model) and the other three with a row of eight vertical, 2-m pipes in the middle of the box (vertical cube model). To increase their complexity and for ecological research purposes, various pipes were drilled with holes of 5 and 20 cm diameter. The weight of the artificial reefs modules was between 9.9 and 11.2 metric tons (Table 1; Figure 2).

Licenses and permits were processed and obtained between 2011 and 2013 after the fulfillment of all the documentation required by different government entities (Table 2). A basic characterization of the bathymetric, oceanographic, and sediment features of the area determined that the minimum depth (i.e., clearance) of the artificial reefs (i.e., between the lowest water surface level and the highest part of the reef) did not represent any hazard to navigation. Additionally, the analysis of the prevailing currents and sediment dynamics relative to the arrangement and location of the structures did not indicate any appreciable negative effects on local oceanographic patterns and beach sediment deposition. Subsequent to these activities, the General Maritime Directorate of Colombia adopted Resolution No. 0018 in January 2013 with a maritime area of 137 ha for artificial reef development and requirements for their use. Regarding management, postplacement commitments for the utilization of the permitted area included posting signs on the polygon vortex, deploying buoys to mark the area for management, maintenance of the marking system, and the delivery of reports of the artificial reefs' performance.

After establishing security protocols for terrestrial and maritime transportation based on international procedures and local policy requirements, Drummond Ltd. conducted the deployment of the artificial reef modules using the tugboat *Michael T*, the crane vessel *Seaworthy*, and the participation of a maritime expert. The deployment operation at depths between 10 and 16 m in the permit area was successfully executed on the morning of June 7, 2013 (Figures 1 and 2; Table 1).

Strengthening Fishing Communities

The first stage of improving community organization was achieved by interacting with three target communities proximate to the area directly under the influence of Ecopetrol. The fishing associations Aspescoltur, Copepazbe, and Adimujer were comprised of a total of 75 fishermen. Different approaches were used to fulfill all their requirements and needs, as follows.

Education and Related Activities

SENA educated the community on business plans and cooperative management. As a re-

Table 1. Features of the artificial reefs deployed in Pozos Colorados Bcy (*D*: depth; *V*: volume; TH: total of holes; DC: distance to the coast; DNA: distance to the nearest artificial reef; DRR: distance to the rocky reef; ø: number of holes of 46, 20, and 5 cm of diameter).

Artificial reef (AR)	Model	Coordinates datum WGS84	*D* (m)	*V* (m^3)	TH	DC (m)	DNA (m)	DRR (m)	ø46	ø20	ø5
AR1	Diagonal cube	11° 08' 40.9" N–74° 14' 45.6" W	15.9	91	153	2,130	254	1,840	8	20	125
AR2	Diagonal cube	11° 08' 42.4" N–74° 14' 31.4" W	14.1	91	153	1,700	280	1,430	8	20	125
AR3	Diagonal cube	11° 08' 52.4" N–74° 14' 19.1" W	12.9	91	153	1,320	486	1,260	8	20	125
AR4	Vertical cube	11° 08' 33.1" N–74° 14' 18.3" W	10.6	91	44	1,280	421	978	24	20	0
AR5	Vertical cube	11° 08' 23.0" N–74° 14' 32.3" W	12.1	91	44	1,630	326	1,350	24	20	0
AR6	Vertical cube	11° 08' 21.9" N–74° 14' 44.6" W	14.6	91	44	1,990	253	1,690	24	20	0

Figure 2. Artificial reefs during the deployment operation in Pozos Colorados Bay: **(a)** diagonal cube module, **(b)** vertical cube module.

sult, the associations were legally formed and certified as an organization by the Chamber of Commerce. This facilitated the community's ability to receive materials, make use of facilities, and benefit from opportunities provided by public and private institutions. The University of Magdalena instructed the community in the processing and manufacture of food, seeking to teach each community how to generate aggregate value in the fishery products for commercial purposes.

Environmental education activities (i.e., workshops) were conducted by the university within the community for a wide age range of individuals (between 8 and 60 years old). Workshops included topics and activities that addressed (1) the management of wild fauna and threatened species, (2) ecosystem conservation, (3) the identification of marine and coastal species, and (4) beach cleaning and residue management. Additionally, together, INCODER and AUNAP developed a workshop on status, laws, and responsible fishing practices regarding artisanal fishery activities in Colombia. Likewise, MoAm performed workshops regarding artificial reefs and habitats, environmental management, and biodiversity.

Training and Equipment Support

SENA instructed fishermen in the proper use of outboard engines, including mechanics, function, utilization, and repairing. The objective here was to increase the fishermen's capacity to overcome any problems associated with fishing activities at sea as well as to reduce operation costs. In cooperation with a local scuba business, two fishermen received training in open-water diving to promote active participation and technical support in the activities of monitoring and future diving ecotourism.

To support and reinforce artisanal fishery capacity, fishing organizations associated with the project received loans of two fiberglass boats 7.6 m (25 ft) long, two 40-hp and two 15-hp outboard engines, and fishing gear such as gill nets, ropes, nylon materials, and various-sized hooks. INCODER and AUNAP also supported this project by providing equipment to enhance safety and facilitate navigation, such as life vests, buoyancy aids, and radios for communication.

Pre- and Postdeployment Ecological Assessment

The preassessment visual census surveys conducted by scientific divers at the pro-

Table 2. Summary of the dependent group and requirement responsibility attributable to each regulatory institution for the development of an artificial reef project in Colombia.

Regulatory institution	Dependency	Requirements	Final result
1. Ministry of Commerce, Industry and Tourism	Group of Planning and Sustainable Development of Tourism	• Project profile • Flat and geographic coordinates of artificial reef positions and estimated area required for the project	Certified of existence or not of touristic projects in the area
2. Ministry of Transport	Group of Infrastructure for Port Development and Logistics	• Project profile • Drawing of the specific location with flat coordinates of artificial reef positions and area required for the project	Certified for the procedure of concession for public use area
3. Ministry of the Interior	Missional Area, Direction of Previous Consult	• Project profile • Project cartographic location in shape files (shp) and Magna Sirgas of artificial reef position and area of the project	Certified of presence or not of ethnic groups
4. Ministry of Environment and Sustainable Development	National Agency of Environmental Licenses	• Project profile • Environmental characterization (i.e., physical, chemical, biological, oceanographic, geological) • Socioeconomical description • Navigational, industrial and governance characterization • Artificial reef description and evaluation plan	Environmental license
5. Regional Autonomous Corporation of Magdalena (Corpamag)	Subdirection of Environmental Management	• Project profile • Environmental characterization (i.e., physical, chemical, biological, oceanographic, geological) • Artificial reef description and evaluation plan	Environmental feasibility and research permit
6. General Maritime Directorate (DIMAR)	Area of Littorals	• Project profile • Artificial reef description, deployment, and evaluation plan • Certified and licenses from the regulatory institutions • Flat and geographic coordinates of artificial reef positions and area required for the project • Publication of an edict with basic information of the artificial reef project	Resolution with the area in concession and committments for management

posed artificial reefs deployment sites indicated three predominant fish species with a total relative abundance of 30.3 individuals per survey. Higher richness was recorded on site AR5 with a total of three species and an average of two species per survey (SD ± 1) while site AR1 had a total relative abundance of 17.8 individuals and a mean of 5.9 individuals per census (SD ± 5.3). The most abundant species was Blue Runner *Caranx crysos* with a total of 28.5 individuals per survey (Table 3; Figure 3).

After the artificial reef deployment, initial surveys during the first 3 months indicated changes in the biological variables of the area, with a total of 28 fish species and 1,592.1 organisms. Higher biological parameters were recorded at site AR6 with 17 species and site AR3 with 371.7 individuals. The average richness and relative abundance per site were 8.5 (SD ± 1.61) and 132.6 (SD ±48.7), respectively. The abundant fishes in descending order were Atlantic Bumper *Chloroscombrus chrysurus* (1,299.4), Atlantic Spadefish *Chaetodipterus faber* (81.7), and Atlantic Thread Herring *Opisthonema oglinum* (61.1) (Table 3; Figure 3). Two endangered species were recorded, according with the *Red Book of Marine Fishes of Colombia* (Chasqui et al. 2017): the critically endangered Goliath Grouper *Epinephelus itajara* and the vulnerable Mutton Snapper *Lutjanus analis*. Additionally, the Lane Snapper *L. synagris* was in the category of least concern. The Mann–Whitney test to compare pre- and postdeployment surveys indicated significant differences in richness ($W = 0.5$; $p < 0.001$) and relative abundance ($W = 0$; $p < 0.001$) of fishes.

Regarding mobile macroinvertebrates, none was observed in the preassessment survey. During postdeployment assessment, nine species and 23.6 individuals were observed associated with the artificial reefs, with the highest parameters recorded on site AR6 (7 species richness and 9.5 individuals). The mean values per artificial reef were 1.75 species (SD ± 1.17) and 1.96 individuals (SD ± 1.62). Abundant species in descending order were redbanded coral shrimp *Stenopus hispidus* (18.9) and bareye hermit *Dardanus fucosus* (1.3) (Table 4; Figure 4). The Mann–Whitney test to compare pre- and postdeployment survey indicated significant differences in richness ($W = 27$; $p < 0.001$) and relative abundance ($W = 27$; $p < 0.001$) of macroinvertebrates.

Discussion

Pozos Colorados Bay is an ideal area for an artificial reef project owing to its environmental conditions and socioeconomic circumstance. For example, the bay is absent natural, structure-based ecosystems or natural hard substrate and this inhibits some species from becoming established in the bay. Also, there are fishing communities in the bay that constantly exploit the diversity of fishery resources in the general area alongside. Last, there are already established regulations in place to control human activity in the area relative to coastal development. Hence, this provides the perfect scenario to demonstrate the positive effects of an integrative approach for habitat enhancement as an alternative strategy for concerted environmental management.

The high level of interaction among stakeholders is evidence of the public interest in supporting policies for local inhabitants. This interaction was similar to that observed in other areas of Colombia, such as the Gulf of Morrosquillo (Delgadillo-Garzón et al. 2010), and in other countries, like Japan and France (Pioch et al. 2011). Often, stakeholders take part in environmental projects according to their ability to accomplish the expected results. As was pro-

Table 3. Composition and relative abundance of fishes recorded during the pre- and postdeployment assessment of the artificial reefs in Pozos Colorados Bay (* species on the Red List of Colombia; *n*: number of surveys in each artificial reef; AR = artificial reef).

	Predeployment assessment ($n = 3$)							Postdeployment assessment ($n = 2$)						
Species	AR1	AR2	AR3	AR4	AR5	AR6	Total	AR1	AR2	AR3	AR4	AR5	AR6	Total
Blue Tang *Acanthurus coeruleus*	–	–	–	–	–	–	–	0.6	0.5	–	–	0.1	–	1.2
Redspotted Hawkfish *Amblycirrhitus pinos*	–	–	–	–	–	–	–	–	0.1	–	–	–	0.2	0.3
Porkfish *Anisotremus virginicus*	–	–	–	–	–	–	–	1	–	0.3	–	0.4	0.6	2.3
Sharpnose Puffer *Canthigaster rostrata*	–	–	–	–	–	–	–	10.9	4	17	15.2	7	1.6	55.7
Blue Runner *Caranx crysos*	17.8	–	0.6	–	10.1	–	28.5	0.4	2.2	–	0.3	2.2	–	5.1
Horse-eye Jack *Caranx latus*	–	–	–	–	–	–	–	–	–	0.1	–	–	–	0.1
Atlantic Spadefish *Chaetodipterus faber*	–	–	–	–	–	–	–	12.8	13	29.5	8.8	10	7.6	81.7
Spotfin Butterflyfish *Chaetodon ocellatus*	–	–	–	–	–	–	–	–	–	0.5	0.5	–	–	1
Reef Butterflyfish *Chaetodon sedentarius*	–	–	–	–	–	–	–	–	–	0.6	–	–	–	0.6
Atlantic Bumper *Chloroscombrus chrysurus*	–	–	–	–	–	–	–	75	298.6	270.1	255	250.6	150.1	1299.4
Dwarf Sand Perch *Diplectrum bivittatum*	–	–	–	–	–	–	–	0.5	0.5	0.6	–	0.7	1.1	3.4
Atlantic Goliath Grouper *Epinephelus itajara**	–	–	–	–	–	–	–	–	–	0.2	–	–	–	0.2
Eucinostomus sp.	–	–	–	–	–	–	–	–	1.1	–	–	–	5.7	6.8
Goldspot Goby *Gnatholepis thompsoni*	–	–	–	–	–	–	–	–	–	–	–	–	0.1	0.1
Tomtate *Haemulon aurolineatum*	–	–	–	–	–	–	–	7.8	2.1	–	2.8	–	10.1	22.8
Latin Grunt *Haemulon steindachneri*	–	–	–	–	–	–	–	3.8	4.7	2	15.7	4.5	1.3	32

Table 3. Continued.

Species	Predeployment assessment (n = 3)							Postdeployment assessment (n = 2)						
	AR1	AR2	AR3	AR4	AR5	AR6	Total	AR1	AR2	AR3	AR4	AR5	AR6	Total
Mutton Snapper *Lutjanus analis**	–	–	–	–	–	–	–	–	0.1	–	–	–	–	0.1
Lane Snapper *Lutjanus synagris**	–	–	–	–	–	–	–	0.3	0.7	–	0.2	–	0.8	2
Leatherjack *Oligoplites saurus*	–	–	–	0.2	0.3	–	0.5	–	–	–	–	–	–	–
Atlantic Thread Herring *Opisthonema oglinum*	–	–	–	–	1.3	–	1.3	–	–	50	10	0.2	0.9	61.1
Paralichthys sp.	–	–	–	–	–	–	–	0.3	–	–	–	–	–	0.3
Postlarvae	–	–	–	–	–	–	–	0.6	–	–	–	0.7	–	1.3
Red Lionfish *Pterois volitans*	–	–	–	–	–	–	–	1.3	–	0.7	0.5	0.4	0.7	3.6
Cobia *Rachycentron canadum*	–	–	–	–	–	–	–	–	–	–	–	–	0.1	0.1
Twinspot Bass *Serranus flaviventris*	–	–	–	–	–	–	–	–	0.1	–	0.6	–	0.3	1
Pelican Barracuda *Sphyraena guachancho*	–	–	–	–	–	–	–	–	–	0.1	–	–	–	0.1
Bicolor Damselfish *Stegastes partitus*	–	–	–	–	–	–	–	–	–	–	–	–	0.4	0.4
Black Brotula *Stygnobrotula latebricola*	–	–	–	–	–	–	–	–	–	–	–	0.3	–	0.3
Bluehead *Thalassoma bifasciatum*	–	–	–	–	–	–	–	–	6.4	–	–	1.2	1.5	9.1
Total relative abundance	17.8	0	0.6	0.2	11.7	0	30.3	115.3	334.1	371.7	309.6	278.3	183.1	1592.1
Total richness	1	0	1	1	3	0	3	13	14	13	11	13	17	28

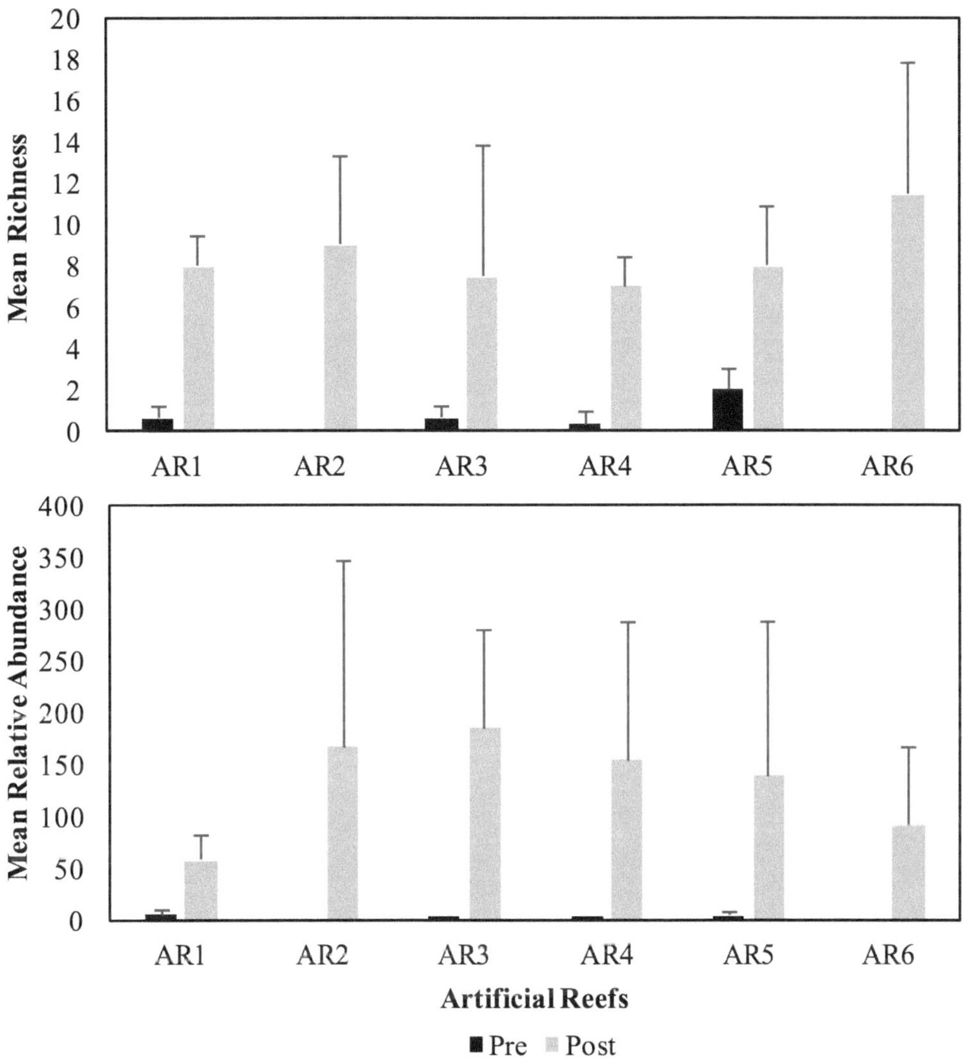

Figure 3. Mean richness and relative abundance of fishes recorded during pre- (Pre) and post- (Post) deployment assessments of the artificial reefs (AR) in Pozos Colorados Bay (bars represent standard deviation).

posed in the management objective here, the education, training, and economic support have facilitated the strong commitment of the target group to improve their technical-organizational capacity. This objective has been partially fulfilled due to the long-term intervention required to achieve most consistent results. The step-by-step integration between the institutions and stakeholders as well as the sense of ownership created by the involvement of fishers in the technical activities related with the processes were positive indicators of success.

Recycling steel pipe as a module of opportunity for building designed reefs is a promising procedure for the disposal and reutilization of waste material in an environmentally sound manner to enhance marine natural processes. The life span of steel in the seawater environment is up to 300 years,

Table 4. Composition and relative abundance of mobile macroinvertebrates recorded during pre- and postdeployment assessment of the artificial reefs in Pozos Colorados Bay (n: number of surveys in each artificial reef). AR = artificial reef.

	Predeployment assessment (n = 3)							Postdeployment assessment (n = 2)						
Species	AR1	AR2	AR3	AR4	AR5	AR6	Total	AR1	AR2	AR3	AR4	AR5	AR6	Total
Phylum Annelida														
Bearded fireworm *Hermodice carunculata*	–	–	–	–	–	–	–	–	–	–	–	0.1	0.3	0.4
Phylum Mollusca														
Western dondice *Dondice occidentalis*	–	–	–	–	–	–	–	–	–	–	–	–	0.4	0.4
Monoplex pilearis	–	–	–	–	–	–	–	–	–	0.1	–	–	0.2	0.3
Voluta virescens	–	–	–	–	–	–	–	–	0.4	–	0.4	–	–	0.8
Phylum Arthropoda														
Subphylum Crustacea														
Spiny hands *Charybdis hellerii*	–	–	–	–	–	–	–	–	–	–	0.5	–	–	0.5
Bareye hermit *Dardanus fucosus*	–	–	–	–	–	–	–	1.1	–	–	–	–	0.2	1.3
Mithrax sp.	–	–	–	–	–	–	–	–	–	–	–	–	0.2	0.2
Redbanded coral shrimp *Stenopus hispidus*	–	–	–	–	–	–	–	0.6	0.4	2	2.7	5.5	7.7	18.9
Yellowline arrow crab *Stenorhynchus seticornis*	–	–	–	–	–	–	–	–	–	–	–	0.3	0.5	0.8
Total relative abundance	–	–	–	–	–	–	–	1.7	0.8	2.1	3.6	5.9	9.5	23.6
Total richness	–	–	–	–	–	–	–	2	2	2	3	3	7	9

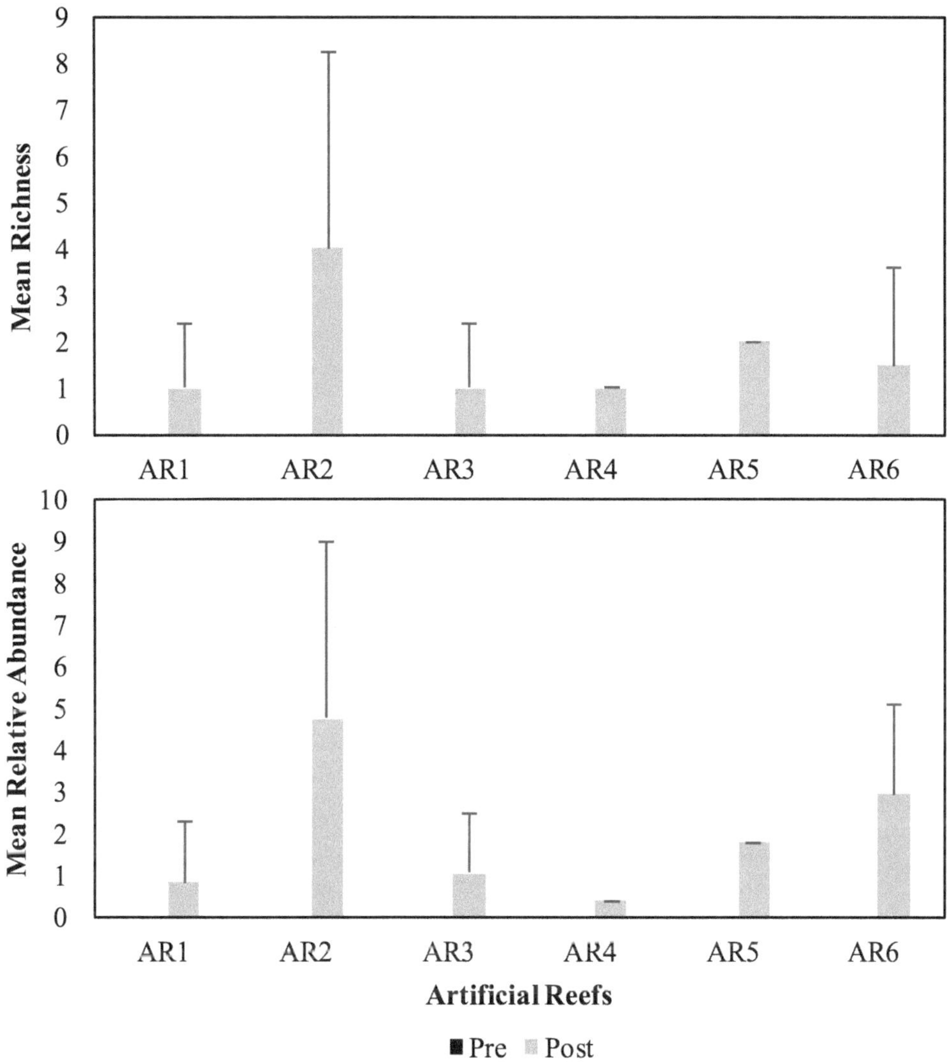

Figure 4. Mean richness and relative abundance of mobile macroinvertebrates recorded during pre- (Pre) and post- (Post) deployment assessment of the artificial reefs (AR) in Pozos Colorados Bay (bars represent standard deviation).

depending on the type and condition of the steel (Lukens and Selberg 2004). The availability of pipes with a variety of diameters from Ecopetrol offers the possibility to explore different applications of ecological design and engineering in restoration and mitigation, as stated by Grove et al. (1991) and Sheehy and Vik (1992). Pipe availability provides an opportunity to evaluate the implementation of specific designs to test ecological hypothesis and improve the requirements of target species (Lindberg et al. 2006; Caddy 2011; Ito 2011). This represents just part of an important potential utilization of steel-pipe artificial reefs for management and research purposes. Well-planned artificial reefs designs coupled with the modeling of worst-case scenarios on oceanographic dynamics must be prerequisites for the creation of durable and stable artificial reefs in order to realize the best return on investments for future projects.

The legal context for artificial reef project regulation in Colombia is still developing, as is the scenario in many other countries. Concomitantly, the fishing and environmental policies in Colombia, as well as in other nations, are still weak. Evidence indicates that efforts to manage this kind of large-scale environmental project are increasing, especially in recent years, as are regulatory rules for coastal development activities (e.g., creation of ports, dredging activities, submarine lines, installation of buoys and monobuoys, marinas, groins, and breakwaters). The procedures and requirements for licensing and permitting for artificial reefs deployment in Colombia are consistent with the models and guidelines of different countries across the world (Sayer and Wilding 2002; NOAA 2007; London Convention and Protocol/UNEP 2009; Lindberg and Seaman 2011; Pioch et al. 2011). Owing to growing interest in their application here, it is important that all the institutions involved in this process (Table 2) create a standard protocol or reference terms for artificial reefs deployments focused on Colombian environmental diversity. These protocols should be based on the procedures developed by other countries but modified to accommodate local community (i.e., environmental and civic) differences and requirements. Time demands, if not well managed to obtain all the permits and certifications, can delay the project timeline substantially. This can affect the accomplishment of the objectives and the credibility of participating institutions.

Independent of the intended goals of the specific artificial reef project, it is necessary to conduct research on artisanal fishing dynamics in the proximate area through time. This is because the expected direct or indirect effects of technology, in the long term, are likely to affect general productivity in the area. Fishing dynamics investigations are among the future objectives to improve fishing productivity, coupled with controlled harvesting by artisanal fishermen who potentially will use artificial reef technology.

Changes observed here in species richness and abundance between the pre- and postdeployment assessments attest to the idea that artificial reefs function as habitats that foster the settlement and recruitment of species. This is especially true for artificial reefs with complex vertical structures, relative to faunal diversity (Bohnsack et al. 1991; Rilov and Benayahu 2002; Delgadillo-Garzón et al. 2004). The number of species recorded here was relatively high in relation to the low sample effort and brief deployment. Many artificial reefs and habitats in the Colombian Caribbean were evaluated after longer deployment periods, but the richness and abundance of species here were comparable (Delgadillo-Garzón et al. 2004; Delgadillo-Garzón 2009). The preliminary outcome here indicates that the associated biotic assemblage responds positively to their interaction with the high structural complexity of the habitats in terms of volume, geometry, and holes, along with the area's high productivity and diversity, mentioned earlier. Therefore, these artificial reefs are a promising alternative for successful habitat enhancement in the environmental conditions described here. Long-term scientific monitoring will help to determine the fulfillment of the artificial reef objectives and cost-benefits of the project.

The presence of endangered and commercial species associated with the artificial reefs reinforces their significance in conservation. Moreover, the diversity of fishes and mobile macroinvertebrates (some of these common inhabitants of coral reefs) and the increasing complexity of the sessile biota suggest the potential these structures may have in the future development of ecotourism. Likewise, the application of specifically designed artificial reefs to provide protection, food, and survival for the recorded organisms and other target species will help to achieve the previously mentioned goals.

Recreational diving is increasing worldwide. Fortunately, it is an activity that appears to satisfy conservation targets and economic interests, as stated by Lindberg and Seaman (2011), Shani et al. (2011), and Kirkbride-Smith et al. (2013). Additionally, artificial reef deployments can add to the income of local community beyond that offered by commercial fishing (Brock 1994). This subject is very sensitive in the area of Santa Marta, which has experienced an expansion in recreational diving business and ecotourism, as these activities have been traditionally conducted inside a marine protected area, the Tayrona National Natural Park. Thus, if reef deployment is undertaken in other areas, the technology can be palliative when reducing the recreational diving pressure on natural coral reefs. The growing hotel infrastructure in Pozos Colorados Bay suggests the promising use of artificial reefs as an ecotourism alternative to the bay's actual social and environmental conditions. This technology is considered a good solution to support productive sustainable models for the diversification of fishermen's economy through an interaction with the tourism industry.

In short, there is evidence of the potential for habitat enhancement in an integrative concerted approach, aimed at marine and coastal management. The guidelines for reef development needs to incorporate the prolonged assessment of artificial reef performance (e.g., designs, fishery, ecology, biology, and oceanography) coupled with socioeconomic investment. Additionally, the stewardship of the artificial reefs by authorities and users should be based on precautionary principles. This may include the conformation of a network of discrete marine reserves as a commitment for future sustainability. Areas of critical coastal advancement and the presence of sensitive ecosystems should be the target of these interventions.

Acknowledgments

We thank the following fishing communities of Pozos Colorados Bay for their support and interest in developing this project: Aspescoltur, Copepazbe, and Adimujer, We also thank all the representatives of the institutions involved in the project. Ecopetrol, the University of Magdalena, Drummond Ltd., the Foundation of the Port Society of Santa Marta, CORPAMAG, SENA, INCODER, AUNAP, Irotama Resort, Estelar Hotel, and MoAm. We also thank Stephen A. Bortone and an anonymous researcher for a revision of the manuscript, which helped improve it.

References

Bohnsack, J. A., D. L. Johnson, and D. F. Ambrose. 1991. Ecology of artificial reef habitats and fishes. Pages 61–107 *in* W. Seaman, Jr. and L. M. Sprague, editors. Artificial habitats for marine and freshwater fisheries. Academic Press, San Diego, California.

Bortone, S. A. 2011. Introduction to the role of artificial reefs play in fisheries management. Pages 1–6 *in* S. A. Bortone, F. P. Brandini, G. Fabi, and S. Otake, editors. Artificial reefs in fisheries management. CRC Press, Boca Raton, Florida.

Bortone, S. A., M. A. Samoilys, and P. Francour. 2000. Fish and macroinvertebrate evaluation methods. Pages 127–164 *in* W.

Seaman, Jr., editor. Artificial reef evaluation with application to natural marine habitats. CRC Press, Boca Raton, Florida.

Brock, R. E. 1994. Beyond fisheries enhancement: artificial reefs and ecotourism. Bulletin of Marine Science 55:1181–1188.

Caddy, J. F. 2011. How artificial reefs could reduce the impacts of bottlenecks in reef fish productivity within natural fractal habitats. Pages 45–64 *in* S. A. Bortone, F. P. Brandini, G. Fabi, and S. Otake, editors. Artificial reefs in fisheries management. CRC Press, Boca Raton, Florida.

Chasqui, L., A. Polanco, P. Acero, P. A. Mejía-Falla, A. Navia, L. A. Zapata, and J. P. Caldas. 2017. Libro rojo de peces marinos de Colombia. [Red book of marine fishes of Colombia.] Instituto de Investigaciones Marinas y Costeras - Invemar, Ministerio de Ambiente y Desarrollo Sostenible, Serie de Publicaciones Generales de Invemar #93, Santa Marta, Colombia.

Costello, C., D. Ovando, R. Hilborn, S. D. Gaines, O. Deschenes, and S. E. Lester. 2012. Status and solutions for the world's unassessed fisheries. Science 338:517–520.

Delgadillo, O., C. B. García, and J. Correa. 2004. Dinámica temporal de la asociación de peces en dos arrecifes artificiales del Golfo de Morrosquillo, Caribe Colombiano. [Temporal dynamics of fishes associated with two artificial reefs in the Gulf of Morrosquillo, Caribbbean of Colombia.] Actualidades Biológicas 26(81):219–230.

Delgadillo-G, O., and P. Flórez. 2015. First records of phylum Bryozoa associated to artificial habitats in the Colombian Caribbean. Latin American Journal of Aquatic Research 43:33–45.

Delgadillo-Garzón, O., F. Delgado-Cuadros, R. M. Henríquez, L. M. García, T. Ruiz-Álvarez, A. Zamora-Guzmán, K. Martínez, F. Blanco, G. Coronado-Luna, M. J. Mogollón, F. Puerta-Padilla, J. C. Jiménez, L. Navarro, I. M. J. Vishnoff, and S. N. Perdomo-Ramos. 2010. Avances y desafíos del Programa Diáspora: arrecifes artificiales en el Golfo de Morrosquillo, Caribe Colombiano. [Advances and challenges of the Diaspora Program: artificial reefs in the Gulf of Morrosquillo, Colombian Caribbean.] Pages 561–656 *in* XIV seminario nacional de ciencias y tecnologías del mar. [XIV national seminar of sciences and technologies of the sea.] Instituto de Investigaciones Marinas y Costeras (INVEMAR), Serie de Publicaciones Especiales No. 21, Santa Marta, Colombia.

Delgadillo-Garzón, O. 2009. Fishes and mobile macroinvertebrates of artificial habitats in Taganga Bay, Colombian Caribbean. Boletín de Investigaciones Marinas y Costeras 38:197–204.

Delgadillo-Garzón, O., and C. B. García. 2009. Impact of two artificial reefs on diurnal artisanal fishing at Gulf of Morrosquillo, Colombian Caribbean. International Journal of Tropical Biology 57:993–1007.

Díaz, J. M., L. M. Barrios, and D. I. Gómez-López. 2003. Las praderas de pastos marinos en Colombia: estructura y distribución de un ecosistema estratégico. [The sea grass beds of Colombia: structure and distribution of a strategic ecosystem.] Instituto de Investigaciones Marinas y Costeras (INVEMAR), Serie de Publicaciones Especiales No. 10, Santa Marta, Colombia.

Díaz, J. M., L. Barrios, M. Cendales, J. Garzón-Ferreira, J. Geister, M. López-Victoria, G. Ospina, E. Parra-Velandia, J. Pinzón, B. Vargas-Angel, F. Zapata, and S. Zea. 2000. Áreas coralinas de Colombia. [Coralline areas of Colombia.] Instituto de Investigaciones Marinas y Costeras (INVEMAR), Serie de Publicaciones Especiales No. 5, Santa Marta, Colombia.

García, C. B., and D. Armenteras. 2015. Atlas de la ictiofauna demersal de fondos blandos del Caribe continental Colombiano:

aproximación a su diversidad. [Atlas of the soft-bottom demersal marine fish fauna from the continental Colombian Caribbean: an approximation of its diversity.] Universidad Nacional de Colombia, Bogotá.

García, C. B., L. O. Duarte, and G. Ramírez. 2013. Fisiografía y oceanografía del golfo de Salamanca (mar Caribe, Colombia). [Physiography and oceanography of the Gulf of Salmanca (Caribbean Sea, Colombia).] Pages 111–140 *in* L. M. Melgarejo and C. B. García, editors. Investigación en ciencias del mar: aportes de la Universidad Nacional de Colombia. [Research in marine sciences: contributions of the National University of Colombia.] Universidad Nacional de Colombia, Facultad de Ciencias. Red de Estudios del Mundo Marino (REMAR), Bogotá.

García, C. B. 2010. Conocimiento tradicional: lo que los pescadores artesanales del Caribe colombiano tienen para decirnos. [Traditional knowledge: what the Colombian Caribbean artisanal fishermen have to tell us.] Pan-American Journal of Aquatic Sciences 5:78–90.

García, C. B., L. O. Duarte, J. Altamar, and L. M. Manjarrés. 2007. Demersal fish density in the upwelling ecosystem off Colombia, Caribbean Sea: historic outlook. Fisheries Research 85:68–73.

Grove, R. S., S. J. Sonu, and M. Nakamura. 1991. Design and engineering of manufactured habitats for fisheries enhancement. Page 109–151 *in* W. Seaman, Jr. and L. M. Sprague, editors. Artificial habitats for marine and freshwater fisheries. Academic Press, San Diego, California.

Guzmán-Alvis, A. I., and J. M. Díaz. 1996. Soft bottom macrobenthic assemblages off Santa Marta, Caribbean coast of Colombia. Caribbean Journal of Science 32(2):176–186.

Idárraga-García, J., B. O. Posada, and G. Guzmán. 2011. Geomorfología de la zona costera adyacente al piedemonte occidental de la Sierra Nevada de Santa Marta entre los sectores de Pozos Colorados y el río Córdoba, Caribe Colombiano. [Geomorphology of the coastal zone adjacent to the western foothills of Sierra Nevada de Santa Maria between Pozos Colorados and the Cordóba River, Colombian Caribbean.] Boletín de Investigaciones Marinas y Costeras 40:41–58.

Ito, Y. 2011. Artificial reef function in fishing grounds off Japan. Pages 239–264 *in* S. A. Bortone, F. P. Brandini, G. Fabi, and S. Otake, editors. Artificial reefs in fisheries management. CRC Press, Boca Raton, Florida.

Kirkbride-Smith, A. E., P. M. Wheeler, and M. L. Johnson. 2013. The relationship between diver experience levels and perceptions of attractiveness of artificial reefs: examination of a potential management tool. PLOS (Public Library of Science) ONE [online serial] 8(7):e68899.

Lindberg, W. J., T. K. Frazer, K. M. Portier, F. Vose, J. Loftin, D. J. Murie, D. M. Mason, B. Nagy, and M. K. Hart. 2006. Density-dependent habitat selection and performance by a large mobile reef fish. Ecological Applications 16:731–746.

Lindberg, W. J., and G. Relini. 2000. Integrating evaluation into reef project planning. Pages 195–235 *in* W. Seaman, Jr., editor. Artificial reef evaluation with application to natural marine habitats. CRC Press, Boca Raton, Florida.

Lindberg, W. J., and W. Seaman, Jr. 2011. Guidelines and management practices for artificial reefs siting, use, construction and anchoring in southeast Florida. Florida Department of Environmental Protection, Miami.

London Convention and Protocol/UNEP (United Nations Environment Programme). 2009. Guidelines for the placement of artificial reefs. International Maritime Organization, London.

Lukens, R. R., and C. Selberg. 2004. Guidelines for marine artificial reef materials, 2nd edition. Gulf State Marine Fisheries Commission, Ocean Springs, Mississippi and Atlantic States Marine Fisheries Commission, Washington, D.C.

NOAA (National Oceanic and Atmospheric Administration. 2007. National artificial reef plan (as amended): guidelines for siting, construction, development, and assessment of artificial reefs. NOAA, Silver Spring, Maryland.

Pauly, D., and D. Zeller. 2016. Catch reconstructions reveal that global marine fisheries catches are higher than reported and declining. Nature Communications 7:10244.

Pioch, S., J. C. Raynal, and C. Aliaume. 2011. An integrated coastal area management strategy to deploy artificial reefs. Pages 65–75 *in* S. A. Bortone, F. P. Brandini, G. Fabi, and S. Otake, editors. Artificial reefs in fisheries management. CRC Press, Boca Raton, Florida.

Ramírez-Calle, L., and J. P. Valencia-Vera. 2005. Caracterización de los fondos marino-costeros, sector aeropuerto Simón Bolívar, Santa Marta, Caribe colombiano. [Characterization of the marine and coastal bottom, area of the Simón Bolívar Airport, Colombian Caribbean.] Undergraduate thesis. Universidad de Bogotá Jorge Tadeo Lozano, Santa Marta, Colombia.

Rilov, G., and Y. Benayahu. 2002. Rehabilitation of coral reef-fish communities: the importance of artificial reef relief to recruitment rates. Bulletin of Marine Science 70:185–197.

Rueda, M., D. Mármol, E. Viloria, O. Doncel, F. Rico-Mejía, L. García, and A. Girón. 2010. Identificación, ubicación y extensión de caladeros de pesca artesanal e industrial en el territorio marino-costero de Colombia. [Identification, location, and extension of the artisanal and industrial fishing grounds in the marine and coastal territory of Colombia.] Instituto de Investigaciones Marinas y Costeras (INVEMAR), Agencia Nacional de Hidrocarburos (ANH), Santa Marta, Colombia.

Sayer, M. D., and T. A. Wilding. 2002. Planning, licensing and stakeholder consultation in an artificial reef development: the Loch Linnhe reef, a case study. ICES Journal of Marine Science 59:178–185.

Seaman, W., Jr. 2002. Unifying trends and opportunities in global artificial reefs research, including evaluation. ICES Journal of Marine Science 59:514–516.

Seaman, W., Jr. 2007. Artificial habitats and the restoration of degraded marine ecosystems and fisheries. Hydrobiologia 580:143–155.

Seaman, W., Jr., and A. C. Jensen. 2000. Purposes and practices of artificial reef evaluation. Pages 1–19 *in* W. Seaman, Jr., editor. Artificial reef evaluation with application to natural marine habitats. CRC Press, Boca Raton, Florida.

Shani, A., O. Polak, and N. Shashar. 2011. Artificial reefs and mass marine ecotourism. Tourism Geographies: An International Journal of Tourism Space, Place and Environment 14:361–382.

Sheehy, D. J., and S. F. Vik. 1992. Developing prefabricated reefs: an ecological engineering approach. Pages 543–581 *in* G. W. Thayer, editor. Restoring the nation's marine environment. University of Maryland Sea Grant College, College Park.

Sheng, Y. P. 2000. Physical characteristics and engineering at reef sites. Pages 51–94 *in* W. Seaman, Jr., editor. Artificial reef evaluation with application to natural marine habitats. CRC Press, Boca Raton, Florida.

Wilding, T. A., and M. D. Sayer. 2002. Evaluating artificial reef performance: approaches to pre and post deployment research. ICES Journal of Marine Science 59:222–230.

Zar, J. H. 2010. Biostatistical analysis, 5th edition. Prentice Hall, Upper Saddle River, New Jersey.

American Fisheries Society Symposium 86:235–249, 2018

Re-Exploring the Application of Artificial Reefs for Community-Based Fishery Management in Malaysia

JARINA MOHD JANI*
Institute of Tropical Biodiversity and Sustainable Development, Universiti Malaysia Terengganu Kuala Nerus, Terengganu 21030, Malaysia

ELIZABETH OLSON
Department of Geography, University of North Carolina Carolina Hall, Campus Box 3220, Chapel Hill, North Carolina 27599, USA

GENEVIEVE PATENAUDE
School of Geosciences, University of Edinburgh Drummond Street, Edinburgh EH8 9XP, Scotland

Abstract.—Artificial reefs quickly became important in Malaysian fisheries management after their introduction in the 1970s. Their popularity lies in their potential for rehabilitating and enhancing coastal resources, particularly after the establishment of legal protection for coastal resources via regulations regarding fishing zonation in 1985. These human-made substrates were created from various materials, from tires to concrete structures. These structures were meant to create permanent sites conceptually similar to traditional artificial reefs deployed by local fishers that were more temporary in nature. Studies conducted up to the 1990s indicated that artificial reefs had the potential to bring resource users and managers together to establish a community-based management of small-scale coastal fisheries. However, the degree of progress made toward using artificial reefs to bring together users and managers remained obscure. Subsequently, a case study conducted between 2008 and 2010 used the sustainable livelihood approach to gain the perspectives of small-scale fishers on artificial reefs in Terengganu. This paper revisits the potential and challenges of using artificial reefs to implement a community-based fishery management approach. Focus is given to the institutional arrangements, both indigenous and official, that mediate the use of artificial reefs by small-scale fishers in Setiu. While many enabling factors exist, institutional weaknesses that impede a more inclusive management structure also persist. The importance of leveraging traditional fishing codes of conduct to manage access to artificial reefs is highlighted as the key to realizing the potential of deploying artificial reefs to enhance livelihoods based on sustainable fisheries resources.

* Corresponding author: jarina@umt.edu.my

Introduction

The idea of using artificial reefs to establish community-based fishery management in Malaysia was first suggested more than two decades ago. Effective natural resource management of aquatic resources is important in Malaysia as its coastal fisheries were, and continue to be, in crisis due to enforcement issues. Artificial reefs deployments would merge two fisheries management approaches that are intended to assist in improving natural fishery resources in Malaysia. The first approach is the popular use of artificial reefs that have been actively deployed to enhance fisheries resources since the 1970s. The second approach is community-based fisheries management (CBFM). Community-based fisheries management is a novel approach used as an alternative fisheries management model to solve problems related to the fisheries common-resource use that previously and unsuccessfully depended on centralized management models. The "glue" that is necessary to bind the two concepts together is the unique unjang practice that continues to be adhered to by local fishers in the country. This concept in the application of artificial reefs was found relevant more than 20 years ago (J. Yahaya, paper presented at the FAO/Japan expert consultation on the development of community-based coastal fishery management systems for Asia and the Pacific, 1993). This paper analyses the relevance of such a dualistic approach in the contemporary context of Malaysian coastal fisheries. These fisheries remain fraught with the same management issues that existed 20 years ago. Here, we examine a case study in Setiu, Terengganu's most important (i.e., largest) artificial reef site. Moreover, the fundamental CBFM elements and prerequisites are outlined as initially designed in the 1990s. This paper is based on findings made through sustainable, livelihood-based inquiries conducted between 2008 and 2010 on small-scale fishers in the area. The potential and challenges of using artificial reefs for CDFM implementation is also presented.

Artificial Reef Development in Malaysia: Of Old Notions and Modern Applications

In this paper, artificial reefs refers to human-made structures placed on the sea bottom to gather fish (FAO 2007). Under this broad definition, artificial reefs are also referred to as either fish aggregating devices (by the fisheries sector) or artificial habitats (by marine scientists). This latter appellation is preferred as it highlight the reefs' role in harboring marine life rather than exclusively aggregating fish (Seaman 2002; Relini et al. 2007). This variation in nomenclature illustrates an existing divide between those involved in artificial reef research (e.g., scientist, conservationists) and those involved in their development and deployment for the fishing industry, which predominates the situation in Malaysia (Ebber 2003). The Fisheries Development Authority of Malaysia (LKIM) implements artificial reef programs as part of the national fishing community development agenda. This group refers to the structure deployed to create alternative fishing grounds as a *unjam-unjam* or fish aggregating device (LKIM 2005). Unjam-unjam is also used by the Department of Fisheries Malaysia (DoFM) to identify the structures they create with the same objective of creating fishing grounds. However, the term "*tukun tiruan*," or artificial reefs, is used for those structures deployed to create habitats critical for providing protection and sites to facilitate feeding and breeding for fish species, as well as nursery grounds for invertebrates. This is done under the auspices of the Southeast Asian Fisheries Development

Center (SEAFDEC), the agency in charge of artificial reefs research and development for the DoFM (Wagiman et al. 1994). Generally, the structures deployed by the SEAFDEC are intended to serve as unfished or unexploited habitat.

Worldwide, artificial reefs increased in stature among fisheries managers when the structures and designs of reefs were improved by fisheries researchers, from the "backyard" (i.e., low technology) approach to a more sophisticated approach using modern technology (Jarina 2015). The same situation occurred in Malaysia with the small-scale experiments conducted by fisheries scientists in Pinang in the mid-1970s. The results of those studies indicated improvements in enhancing biomass through the deployment of artificial reefs. This led to an increase in funding to create more artificial reef sites along the coast of Malaysia. However, for future deployments to become part of the national agenda with annual budget allocations for artificial reef research and development, the efforts to enhance coastal resource areas must complement the country's efforts to alleviate poverty among coastal fisherfolk (Jarina 2015). Artificial reef deployment was such a popular fisheries enhancement activity in Malaysia that by the 1990s, it had become the most active country for artificial reefs programs in the region (Omar et al. 1994). Using waste materials such as tires and confiscated or derelict fishing vessels, the DoFM soon developed custom-made structures using new materials such as concrete and polyvinyl chloride pipes to create artificial reefs.

Local fishers from many parts of the world have long created artificial reefs in their effort to create new fish habitats or fish aggregating devices (Chou 1997; Ebber 2003; Relini et al. 2007). In Malaysia, fishers on the east coast of Peninsular Malaysia have been reportedly using artificial reefs since the 1900s by sinking derelict wooden boats, bundles of tree branches, twigs, and rocks (Hung 1988). Locally known as unjang, they are used to facilitate fishing operations, similar to payao, which are used by their fishing counterparts in the Philippines (Aprieto 1988; Delmendo 1991; Balgos 1995). These reefs are, however, mostly temporary due to the ephemeral nature of the materials used to make them. Hook-and-line and net fishers constructed small, moveable unjangs to attract specific types of fish, such as bream and scad. During British rule, Parry (1954) and Firth (1975) recorded how these structures are inspected daily by their owners. Meanwhile, bigger and more complex structures known as unjang besar (meaning big fish aggregating device) were constructed by larger traditional fishing operations using takur, or lift nets. These fishing operations often owned several unjangs that were left at sea until they rotted. Parry (1954) and Firth (1975) also remarked that although hook-and-line fishers usually fished independently, fishers also cooperated to build the unjang besar (some weighing as much as 5 metric tons), which was sunk to the sea bottom for use as a common fishing ground. The fish attracted to these structures became a common pool of fishery resources for those involved in deploying the device. According to the local code of conduct, nonowners could use unjangs with permission from the owner and for a share of the catch made at the unjangs.

Linking Artificial Reefs to Community-Based Fishery Management

The practice of deploying unjangs to enhance fisheries has survived to this day. Its conceptual link with artificial reefs was the reason Yahaya (unpublished) first proposed the incorporation of artificial reefs during

CBFM implementation in Malaysia. Community-based fishery management (i.e., a grassroots fisheries management approach) would create a new governance mechanism to replace the centralized form of management established since British colonial rule. Concomitantly, much of the local fishery, particularly along the east coast, has remained relatively traditional until the country's independence in 1957. However, the post-World War II modernization period resulted in an intensification of the local fishing industry through the introduction of a larger and more technology-based commercial fishery that incorporated purse seining and trawling (Shaari and Chnag 1993). These modern types of fishing, particularly trawling, are given the pejorative moniker "tiger net operators" by traditional fishers along the east coast of Peninsular Malaysia due to their perceived ruthless, predatory nature. Trawling had a significantly negative impact on the remarkably diverse traditional coastal fishery as it brought an end to the main types of large-scale traditional fishing operations, such as lift-net and Payang-net fishing, which were more sustainable but less efficient. Also, the trawl operations caused damage to the associated habitat of the coastal fisheries resources such as coral reefs (Jarina 2015). These factors angered small-scale, traditional coastal fishers that had been able to stay in operation but claimed to suffer irreparable losses of their livelihood due to the activities of these modern fisheries in the coastal zone (Yahaya 1988). As a response, fishing zonation under the Fisheries Act 1985 was established to demarcate the coastal fishing zone exclusively for traditional (i.e., small-scale) fisheries. The creation of fishing zones, however, only increased the anger among small-scale fishers due to lack of enforcement. Consequently, a solution for the continuous failure of conventional top-down governance to protect coastal fishery resources was needed. The CBFM model was developed as a bottom-up approach to fishery management that would allow shared authority in governing resources between users and the state fisheries agency. This model struck the right cord with many fisheries agencies in the Asia-Pacific region when introduced at "FAO/Japan Expert Consultation on the Development of Community-Based Coastal Fishery Management Systems for Asia and the Pacific," a 1992 fisheries symposium in Kobe, Japan. The model quickly gained popularity in Southeast Asia with varying degrees of success (Pomeroy 1995).

Based on her study of the institutional aspects of artificial reefs programs that existed in Malaysia, Yahaya (1994) suggested the use of artificial reefs to facilitate CBFM during the Indo-Pacific Fisheries Commission Symposium in 1993. A year earlier, at the "FAO/Japan Expert Consultation on the Development of Community-Based Coastal Fishery Management Systems for Asia and the Pacific" in Kobe, Japan from June 8 to 12, 1992 (one of the first workshops to discuss the implementation of CBFM in the region), Yahaya (unpublished) highlighted three categories of issues that could impede its successful implementation in Malaysia. The first of these were legal and institutional issues, such as the necessary changes to existing fisheries law and regulations. This included resolving the lack of enforcement authority and powers of the fishing community involved in CBFM, as well as the lack of experience and interest of the local fishermen's associations in managing fisheries resources. The second category involved socioeconomic issues, such as the difficulty in achieving equitable distribution of fishing rights due to cultural context. Here management costs may not be minimized but only transferred from government agencies to CBFM units. The third type were political and administrative issues due to the heavily centralized governance system in the country and the need for support and commitment from every level of the fisheries governance structure to embrace the grassroots-based management model.

Existing management practices made use of three types of local artificial reefs (i.e., privately owned unjang, LKIM-deployed unjam-unjam, and DoFM-deployed tukun tiruan. On review, Yahaya (1994) optimistically urged the Malaysian government to take advantage of the particularly unique governance aspect associated with the traditional use of unjang and practices related to its code of conduct and apply this to the private artificial reefs created by fishers. This was the key to a successful implementation of CBFM in Malaysia (Yahaya 1994). Yahaya's opinion established that the fundamental elements necessary for a successful CBFM framework, closely linked to the traditional unjang code of conduct in the area, were already in place. These were (1) the existence of a well-established local institution such as the Persatuan Nelayan (Fishermen's Association), (2) identifiable homogeneous fishing groups within the larger fishing communities, (3) self-imposed regulatory measures by users to achieve sustainable use of fisheries resources, and (4) a strong sense of sound management responsibilities and commitment among unjam-unjam users. She opined that "given these attributes coupled with strong commitment and support from fisheries managers, planners and politicians, the prospect of instituting community-based fisheries management through the deployment of unjam-unjam is indeed bright" (Yahaya 1994.323).

An Artificial Reef Twist in Malaysian Community-Based Fishery Management

The call for an artificial-reef-based CBFM model did not go unheeded. Noting the successful introduction of CBFM as an alternative management approach in Japan, Turkey, Canada, and some countries in the Pacific, the DoFM introduced a hybrid CBFM model. This was an adaptation of the original proposed CBFM framework (see Hotta's keynote speech in the Report on the Workshop on Smart Partnerships in the Fishing Industry in Chong and Madhu [1999]) to suit the local fisheries context (Salim 1999). In a presentation at a workshop, a DoFM representative (see Salim 1999) explicitly mentioned the deployment of community-shared artificial reefs as one of the collective actions that could be undertaken towards facilitating cooperation among fishers. This was an essential element for the successful implementation of CBFM. Required for implementation were four prerequisites: (1) a clear, defined fishery boundary; (2) an effective local institutional organization; (3) an effective information-gathering mechanism; and, finally, (4) some form of control mechanism to reward or punish fishers.

Now, two decades have passed and it is time to evaluate the progress of this management program. In fact, despite the expressed intent by the top management of the DoFM to ensure that CBFM becomes part of the local fishery governance framework, the implementation of CBFM has been weak (Nasuchon 2009). A literature review on CBFM implementation in Malaysia revealed a rather bleak picture with only two successful CBFM initiatives, and these were part of the artificial reef program (Buang et al. 2011).

Materials and Methods

The impetus for this paper came from a case study on the significance of artificial reef development on the livelihoods of small-scale fishers in Setiu, Terengganu. Located on the northwestern part of Peninsular Malaysia, Terengganu is a state with the most artificial reef sites in the country (Rahman 2009). Here, the use of the traditional unjang is prevalent (Jarina 2015). Practically, Setiu should be an ideal location to initiate the artificial-reef-based CBFM program (AR-CBFM). However, this management program has not yet been established in Setiu. To provide an updated evaluation of AR-CBFM potential in Setiu, all relevant information and findings from the case study are analyzed here to answer two questions: (1) Are the fundamental elements necessary for a successful implementation of CBFM present in Setiu? and (2) Does the Setiu case study have the four necessary CBFM prerequisites, as suggested by the DoFM in 1997.

Two villages were chosen as case study sites, namely Gong Batu and Rhu Sepuluh (Figure 1). These villages were selected because the coastal waters off Rhu Sepuluh have the most artificial reefs in the district of Setiu and the fewest artificial reefs are located off Gong Batu. This allows the evaluation to include extreme examples. Despite the omnipresence of artificial reefs in Setiu District, few reefs occurred in the waters of Gong Batu when the fieldwork was initiated in May 2008. The village of Gong Batu (5°4'0" N, 102°42'0" E) was created in 1964 as a new settlement for the villagers of Kuala Setiu. The villagers had to be relocated from their beachfront homes due to coastal erosion. Based on a document on the village's profile prepared by the Gong Batu Village Committee in 2007, 190 of 204 families in the village were involved in fisheries, either in capture fisheries, aquaculture, or both. The second research site, Rhu Sepuluh (5°34'51" N, 102°42'7" E), has a generic name that actually represents three kampongs (villages): Rhu Sepuluh, Bukit Chalok, and Pandan Jaya. According to the report by Rhu Sepuluh's Committee on Village Safety and Welfare in 2007, there were 131 households with a total population of 690, of which 30% of the inhabitants were involved in either fishing or fish farming or both. There was a fish-selling facility near the main boat landing area, just off the main road where the Setiu District's Malaysian Fisheries Development Authority (LKIM) and Fishermen's Association offices are located.

Framed by the sustainable livelihood approach (SLA; see Carney 1998), data were analyzed from in-depth interviews and ethnographic and participatory observations collected on five livelihood components (see Figure 2) from 40 small-scale fisher households in Setiu as well as representatives from related fisheries and artificial reefs stakeholders such as the Fishermen's Association, the DoFM, and fish buyers. The use of this SLA-based research framework allowed a broad understanding of the impact of artificial reefs on lives of their intended beneficiaries. The scope of this understanding is beyond the biological or socioeconomic impacts that often become the focus of artificial reef evaluations (Jarina 2015). Due to our interest in the link between the traditional unjang practice and artificial reefs for its potential application in CBFM, the focus here was on the second component (i.e., small-scale fishers' access to assets—particularly fishing grounds which are a form of natural capital). The second component also depends "on the social relations, institutions and organisations of which the households are part" (El-

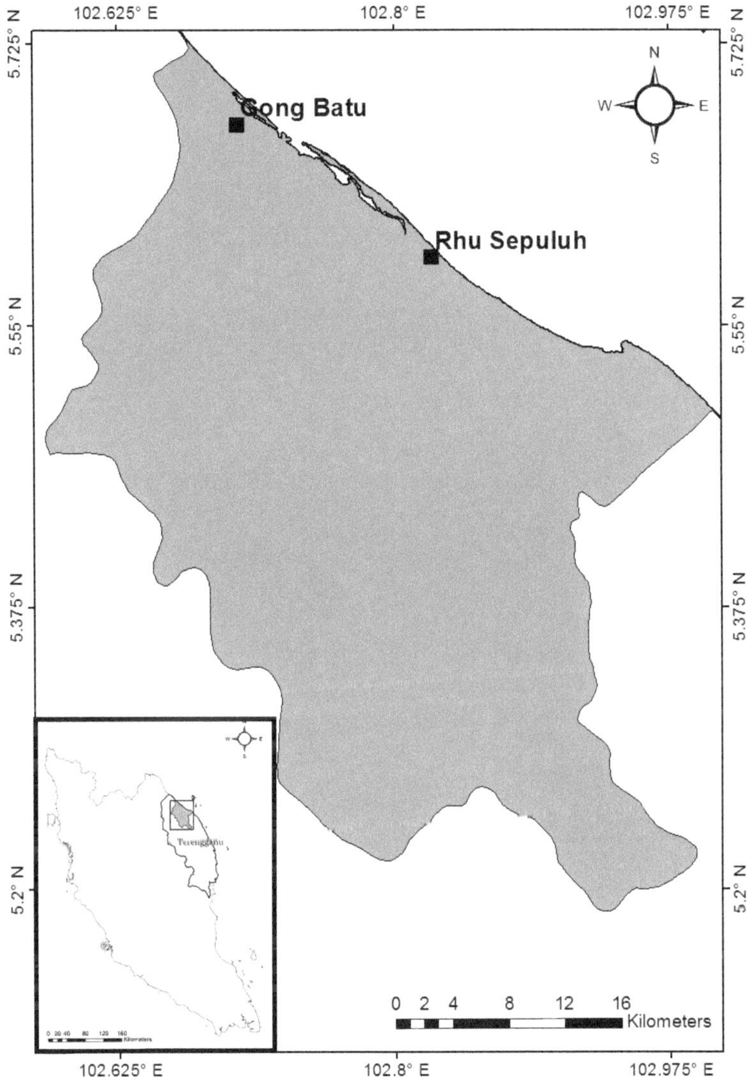

Figure 1. Case study sites in Setiu, Terengganu, Malaysia.

lis 2000:38). The later elements are also referred to as livelihood mediating factors and provide the link between assets (component I) and livelihood strategies (component IV) employed to attain the desired sustainable livelihood outcomes (component V). Component III (i.e., small-scale fishers' perceptions and expectations) was also analyzed as it contains important elements that inform the present and future aspects of an artificial reef-based CBFM.

Results and Discussion

The results of the analysis in the contemporary context of Setiu, based on the evaluation of Yahaya's fundamental elements and the CBFM prerequisites, are summarized in Tables 1 and 2, respectively. The analysis reveals that artificial reefs still have a high potential in contributing towards CBFM, although the same limitations (previously highlighted) continue to hinder its full realization. The factors that established the high

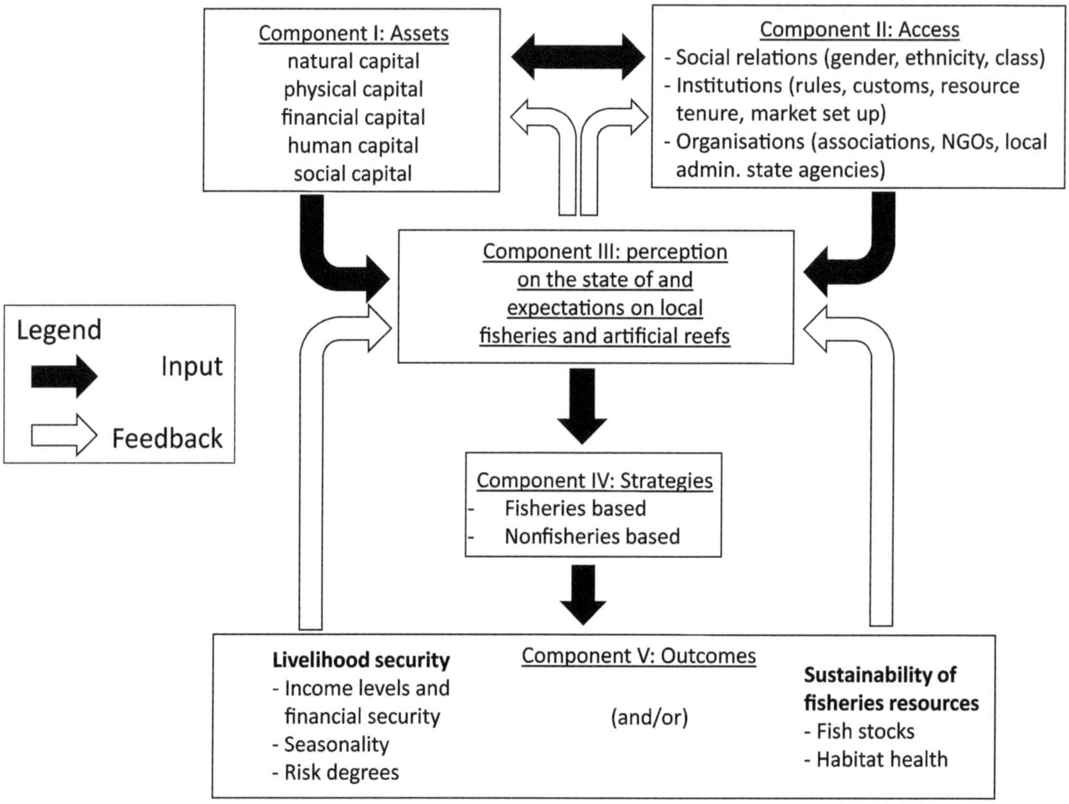

Figure 2. Sustainable-livelihood-approach-based research framework for artificial reefs evaluation in Setiu, Terengganu.

potential of artificial reefs in contributing towards CBFM in Setiu's small-scale fisheries (SSF) were as follows:

1. The Existence of Distinct "Communities of Fisher-Friends" with Strong Social Capital Bonds in Each Small-Scale Fishing Community

Although a local Fishermen's Association does exist, it does not play as effective a role as the Fishermen Associations in Japan, which are pivotal in managing Japan's local fisheries resources. This is because the Fishermen's Association in Malaysia is a top-down creation that is structurally and administratively patronized by the national fisheries development agency (i.e., LKIM). The LKIM office serves as the base for the Fishermen's Association, mainly to facilitate the distribution of various fisheries-related benefits such as subsidies and support from the government. To access these benefits, small-scale fishers in Setiu and many rural fisheries areas in the country have to become members of the Fishermen's Assocation. Indeed, the Fishermen's Association assisted in coordinating the designated group of fishers for artificial reef deployments in the 1990s, similar to the groups discussed by Yahaya (1994). Nevertheless, the role of the Fishermen's Association is now defunct due to the comprehensive involvement of artificial reef contractors. The association does not have any role in local fishery governance and, ironically, is sometimes headed by nonfishers. As explained previously, despite the centralized

Table 1. Results of the evaluation of successful artificial reef (AR) community-based fishery management (CBFM) in Setiu, based on Yahaya (1994).

Source	Fundamental elements for successful CBFM implementation found among AR users in Malaysia
Yahaya (1994)	Established local institution such as the Persatuan Nelayan (Fishermen's Association). Identifiable, homogenous group of fishers within the larger fishing community. Self-imposed regulatory measures towards sustainable resource use. Strong sense of sound management responsibility and commitment among AR users.
Setiu small-scale fishery (SSF)	Communities of fisher-friends among SSF who have strong social capital bonds with each other. They can be converted into local AR-CBFM. committees that can result in more inclusive decisions and cohesive actions The SSF consists of a community of fishers that share a common perspective on fishery and its related livelihoods despite their diverse fishing techniques. Having experienced the damages of their local coastal resources, the SFF denounces fishing activities that are detrimental to its conservation, such as using certain types of fishing that will damage the structures or hamper biomass productivity of AR. Based on their understanding of the unjang concept, and in response to the damages of their coastal fishery resources, SSFs acknowledge the need and are ready to actively participate in conserving their coastal resources via AR.

Table 2. Result of the evaluation of AR (artificial reef) community-based fishery management (CBFM) potential based on prerequisites of the Department of Fisheries Malaysia.

Source	Prerequisites for successful CBFM implementation in Malaysia
DoFM (1997)	Clear boundary. Effective local institution. Effective information gathering mechanism. Some form of control mechanism to reward or punish.
Setiu small-scale fishery (SSF)	The ARs have all been deployed in Zone A and are locally identified according to the villages in whose geographical waters they are placed. Currently, the Prsatuan Nelayan (Fisherman's Association) may not be effective enough. However, creating a AR-CBFM subcommittee is readily possible (see Table 1). Information is shared through traditional (pre- and postfishing exchanges at the local wakaf) or modern (cell phone) means The unjang code of conduct only has rewards for the structure owners and contraventions only render social ostracism, which may suffice in the past but not today

fishery management created by the British, most of the local fisheries participants comprise small- to mid-scale fishing operations. These fishers have remained traditional on the east coast of Peninsular Malaysia and are organized mainly according to social affiliations via family ties or friendship (Firth 1975). These fishers grouped into decentralized, flexible, and local institutions, termed "communities of fisher-friends," where the social bond between members is strong, creating a network of interrelationships among fishers from diverse types within the local fishery. Their community affiliation is linked to their sense of shared identity as fishers in their respective areas and the associated social reliance and cooperation that the bond provides. Although these small-scale fishers are technically individual fishers, they still cooperate both on land and at sea in their daily fishing activities. For example, the fishers assemble early in the morning on the beach to help each other launch their boats, often to a predetermined fishing ground, and return from the sea at the same time. While on land, they spend much bonding time exchanging information or cooperatively mending their fishing gear at the wakaf (shared beachfront shelters near where their boats are stationed). This interaction provides an opportunity for the younger generation of fishers to learn from their seniors (who often carry on fishing into old age) through listening and observation. When explaining the reason for this strong communal bond, Setiu small-scale fishers often say that: "fishing at sea is like looking for lost things—it not easy. The sea is a vast, dangerous place. So you need to have friends you can depend on." The conversion of this community of fisher-friends as a specific community-level committee (thus, less rigid) on AR-CBFM would be opportune as it would allow the establishment of a more authentic, grassroots institution that may be more functional towards deliberating socioculturally suitable and inclusive decisions and conducting more cohesive actions based on group solidarity.

2. The Setiu Small-Scale Fishery Is Already a Sustainable Model of Coastal Fisheries Resources Use, Where Each Operation Does Not Impede the Productivity of Natural and Artificial Reefs

The small-scale fishery in Setiu is diverse with fishing operations that use many different gears, from hook and line to various types of nets and fish cages. However, the fishers claim that none of their operations harms the sustainability of local coastal fisheries resources, unlike the large operators that use trawls and purse seines that the small-scale fishers strongly denounce as destructive to the both natural and artificial habitats through their indiscriminate approach to fishing. Among the small-scale fishers, only hook-and-line fishers and sometimes fish-cage fishers frequent artificial reef sites. Those who use nets target pelagic species and prefer deeper, open waters to avoid having their gear damaged on incompatible substrates. They use the legally permitted mesh size for their gill nets, construct fish cages that allow smaller fish to escape, and release smaller fish. Even their intensive, seasonal squid jigging is reportedly not harmful to the spawned squid eggs and rookeries below, ensuring the sustainability of the resource.

3. Readiness to Actively Participate in the Conservation of Their Coastal Fisheries Resources via Artificial Reefs

Based on their understanding of the unjang, which artificially created substrates to facilitate fishing and to promote regeneration of fishery resources (i.e., a natural capital that is valuable for their livelihood), Setiu small-scale fishers readily acknowledge the need to conserve local fishery resources though artificial reef programs, particularly to remedy the damage claimed to have been made by larger, commercial operations, such as trawl and purse-seine operations. Furthermore, due to the current prevalence of unjang for Setiu small-scale fishers, the already existing code of conduct is readily applicable to the management of fish associated with artificial reefs. This ensures the acceptance of the AR-CBFM-related responsibilities and commitments that are not dissimilar to those of unjang. This is also applicable for the communities of fisher-friends that ideally should lead its management, but also the larger fishing community.

4. Ideal Location of Setiu Artificial Reef within the Zone A (an Area Exclusively for Small-Scale Fishers) Boundary with Distinguishable Village Appropriation.

Via the unjang practice, a fisher is able to appropriate a portion of the sea as an exclusive fishing ground that is otherwise common to everyone. The site becomes the fisher's natural capital. As the same rules could be applied to an artificial reef, it is therefore important to clearly know which site belongs to whom because the structures would then represent an important livelihood asset to specific owner or owners, such as a CBFM unit. Artificial reefs in Setiu were deployed in coastal waters after these waters were designated as Zone A (i.e., areas for exclusive use by small-scale fishers). The creation of this zone was to enhance coastal fisheries resources that are protected from unsuitable operations such as trawl and purse-seine fisheries that often damage coastal habitats. Furthermore, these artificial reef sites are commonly identified with the geographically most proximate village. Thus, there already exists clear legal and locally accepted geographic limits that are necessary for the application of the AR-CBFM model in Setiu.

5. Effective Information Sharing among Fisher-Friends via Traditional and Modern Means

Fisher-friends are always seeking and exchanging information on fishing gear and locations. This exchange occurs while fishers are resting together at the end of the day at the wakaf and local coffee shops. Far from shunning modern technology, Setiu small-scale fishers embrace technology in their livelihood. The extent of their access to modern technology, however, depends on the financial and social capital that allow them to either buy or borrow this modern technology. One modern technology that is widely accessible and has proven its worth is the mobile (cell or satellite) phone. The mobile phone has made sharing of important information instantaneous, thus allowing other fishers access to good fishing spots or facilitating urgent rescue when necessary. Through these effective modes of sharing, important decisions on various aspects of their fishing operations are often made together. Instantaneous sharing of information could also facilitate activities related to AR-CBFM.

The factors presented above highlight the high potential of AR-CBFM in Setiu. The presence of these factors suggest that this important artificial reef site could also become an important site for successful CBFM implementation. The combined use of artificial reefs with CBFM shares many similar features with the two reported CBFM success stories in Langkawi (see Buang et al. 2011).

However, there is one very significant difference between the two reported CBFM successes and the Setiu case study that could threaten the fulfilment of its potential: the absence of control mechanisms other than those related to the traditional unjang practice. There is a control mechanism for reward and punishment among Setiu small-scale fishers, as required by the DoFM for the successful implementation of CBFM. The unjang code of conduct that could be applied to an artificial reef rewards those that create the sites by recognizing them as their exclusive natural capital to be shared with other fishers for a share of the catch—either in kind or in cash. Transgression of this simple rule (i.e., fishing without permission or without compensation at someone else's unjang) is strongly discouraged. This control mechanism could be implemented in coastal waters with an artificial reef where the reef becomes the natural capital of the AR-CBFM units that needs to be officially recognized by the fishery authority. However, while social ostracization may have been sufficient in the past when social acceptance was vital to community life in rural Malaysia, or in contemporary fishing communities in remote islands of the Pacific (Kitolelei and Sato 2016), it may no longer be sufficient to deter fishing transgressions or in discouraging abuses in Setiu. It is noted that Yahaya's (1994) seminal paper alluded to leaving enforcement to official authorities. This is similar to the rather low expectation of community control in the DoFM's prerequisites. However, enforcement was and continues to be the most prevalent management issue in Malaysian fishery management due to manpower and resource shortage at the DoFM. In Setiu, illegal fishing by larger operations (particularly purse-seine operators, who can effectively fish an artificial reef in coastal waters without incurring any damage) reportedly goes unpunished. Such contraventions are effectively avoided in the successful Langkawi CBFM units that consist of "Rakan FISH" (i.e., fishers who have the authority to enforce rules in their local waters; Buang et al. 2011). Indeed, without similar legal status, the CBFM units in Setiu will become no more than another grouping that cannot effectively control access to the coastal waters and artificial reef sites. Such a nonauthoritative entity will most likely fail to secure the full support of the Setiu fisher-friends communities that are eager to protect their natural capital.

The Fishers Are Ready—Can They Be Matched?

This chapter provides an evaluation on the potential of using artificial reefs for establishing CBFM in the contemporary context based on the findings of a case study in one the most important artificial reef sites in the country. In sum, a promising implementation of a CBFM for the future of Setiu small-scale fishers, as per the criteria mentioned by Yahaya (1994), is possible as all the prerequisites listed by the DoFM in 1997 have been met. At the core of this positive prospect are the structured communities of Setiu fisher-friends who are an invaluable asset in ensuring the successful implementation and sustainability of the AR-CBFM once it becomes established. These local, informal institutions conform to the Agrawal and Gibson (2001) definition of community because they (1) represent a small spatial dimension, (2) is a homogenous social structure, and (3) internally shares a global set of norms. Based on the findings presented here, it is suggested that a revival and full recognition of unjang-related codes of conduct in application to artificial reefs management would position small-scale fishers at center stage in coastal resource management. This code of conduct will also allow the use of their previously untapped traditional ecological knowledge in resource monitoring in the future development of artificial reef programs. More significantly, it will put in place a locally sensitive fisheries-based institution. This code also has the highest potential for acceptance and adherence by small-scale fishers as well as local commercial fishers due to the undisputed principle on which it is based: respect and recognition of effort made by fellow fisher-friends who oversee the AR-CBFM units. Once these fishers-friends are officially recognized and effectively assume their role as artificial reef managers, their exclusive right to harvest the associated fisheries will not be disputed. These fisher-friends will continue to contribute towards the maintenance, enhancement, and protection of their well-deserved ocean-born, natural capital at sea. However, these positive prospects could turn grim with a lack of recognition of their full legitimacy as the local management authority. Moreover, a lack of recognition will deter their participation in all aspects of management, from planning to enforcement. It is essential that the DoFM match the readiness of the small-scale fishers to jointly participate in the local management of the very resources on which their livelihood depends by delegating some aspects of resource management, such as enforcement, to them. This is particularly important when solving its most challenging issue (i.e., illegal fishing in coastal fishing areas). The potential of the unjang-inspired AR-CBFM is still as bright as it was more than two decades ago but will remain unrealized without the full legitimacy of the local management authority. Two decades ago, in explaining the hybrid CBFM model adopted for Malaysia, the DoFM explained that a full community participation model would be unsuitablefor Malaysia. This was due to unpreparedness at the grassroots level. So now that the grassroots level is ready, the question remains: is the government ready to revert to the ideal model where the community becomes the manager or, if not equal, co-manager in the governance of the local fisheries resources?

Acknowledgments

The main author gratefully acknowledges the Institut Pondok Perancis, Kuala Lumpur for the support given by hosting her at the institute during the preparation of the manuscript.

References

Agrawal, A., and C. C. Gibson. 2001. The role of community in natural resource conservation. Pages 1–31 *in* A. Agrawal and C. C. Gibson, editors. Communities and the environment: ethnicity, gender, and the state in community-based conservation. Rutgers University Press, New Brunswick, New Jersey

Aprieto, V. L. 1988. Aspects of management of artificial habitats for fisheries: the Philippines situation. Pages 102–110 *in* Report of the workshop on artificial reefs development and management. United Nations Development Programme, New York.

Balgos, M. C. 1995. Evaluation of artificial reef development in the Philippines. ICLARM (International Center for Living Aquatic Resources Management) Conference Proceedings 49, Manila, Philippines.

Buang, A., A. Habibah, J. Hamzah, and Y. S. Ratnawati. 2011. The promise of community-based fishery resource management. World Applied Sciences Journal 13:104–109.

Carney, D., editor. 1998. Sustainable rural livelihoods: what contribution can we make? Department for International Development, London.

Chong, K., and S. R. Madhu, 1999. Report of the workshop on smart partnerships for sustainability in the fishing industry. Bay of Bengal Programme (BOBP), BOBP/REP/91, Chennai, India.

Chou, L. M. 1997. Artificial reefs of Southeast Asia: do they enhance or degrade the environment? Environmental Monitoring and Assessment 44:45–52.

Delmendo, M. N. 1990. A review of artificial reefs development and the use of fish aggregating devices (FADs) in the ASEAN region. Pages 116–141 *in* V. L. C. Pietersz, editor. Symposium on artificial reefs and fish aggregating devices as tools for the management and enhancement of marine fishery resources. Food and Agriculture Organization of the United Nations, Regional Office for Asia and the Pacific, Bangkok.

DoFM (Department of Fisheries Malaysia). 1999. CBFM: an alternative approach to fisheries management in Malaysia. *In* K. Chong and S. R. Madhu, editors. Report of the workshop on smart partnerships for sustainability in the fishing industry. Bay of Bengal Programme, Chennai, India.

Ebbers, T. 2003. Reconciling fishing and environmental protection: resources enhancement strategies for the conservation and management of fisheries. Fish for the People 1(3):17–26.

Hung, E. W. F. 1988. Artificial reefs development and management in Malaysia. Pages 27–51 *in* Report of the workshop on artificial reefs development and management. United Nations Development Programme, New York.

Ellis, F. 2000. Rural livelihoods and diversity in developing countries. Oxford University Press, Oxford, UK.

Firth, R. 1975. Malay fishermen: their peasant economy. Norton Library, New York.

Jarina, M. J. 2015. Diverse fishing operations of Setiu small-scale fishery. Pages 177–198 in F. Mohamad, J. M. Salim, and J. M. Jani, editors. Setiu wetlands: species, ecosystems and livelihoods. UMT Publisher, Terengganu, Malaysia.

Kitolelei, J.V. and Sato, T., 2016. Analysis of perceptions and knowledge in managing coastal resources: a case study in Fiji. Frontiers in Marine Science [online serial] 3:189.

Nasuchon, N. 2009. Coastal management and community management in Malaysia, Vietnam, Cambodia and Thailand, with a case study of Thai fisheries management. United Nations, Office of Legal Affairs, Division for Ocean Affairs and the Law of the Sea, New York.

Omar, N. R., M. Raja, C. E. Kean, S. Wagiman, A. M. M. Hassan, M. Hussein, R. B. R. Hassan, and C. O. M.Hussin. 1994. Design and construction of artificial reefs in Malaysia. Bulletin of Marine Science 55:1050–1061.

Parry, M. L. 1954. The fishing methods of Kelantan and Terengganu. Journal of the Malayan British Royal Asiatic Society 25:77–144.

Pomeroy, R. S. 1995. Community-based and co-management institutions for sustainable coastal fisheries management in Southeast Asia. Ocean and Coastal Management 27(3):143–162.

Rahman, F.A. 2009. Status report of artificial reefs programs in Malaysia. Proceedings of workshops of artificial reefs for the enhancement fishery resources. Department of Fisheries Malaysia, Putrajaya.

Relini, G., M. Relini, G. Pandri, S. Merello, and E. Beccornia. 2007. History, ecology and trends for artificial reefs of the Liguarian Sea, Italy. Hydrobiologia 580:193–217.

Salim, K. H. 1999. CBFM: an alternative approach to fisheries management in Malaysia. Pages 77–84 *in* K. Chong and S. R. Madhu, editors. Report of the workshop on smart partnerships for sustainability in the fishing industry. Bay of Bengal Programme, BOBP/REP/81, Chennai, India.

Seaman, W., Jr. 2002. Unifying trends and opportunities in global artificial reef research, including evaluation. ICES Journal of Marine Science 59:S14–S16.

Shari, I., and Y. T. Chang. 1993. Technology and social change: the impact of technological development on fishing communities in east coast of Peninsular Malaysia. Penerbit UKN, Bangi, Malaysia.

Wagiman, S., R. Omar, C. Hussin, and R. Nor. 1994. Tukun tiruan Malaysia: habitat tiruan melestarikan perikanan laut. [Malaysian artificial reefs: sustaining marine fishery using artificial habitats.] Department of Fisheries Malaysia, Jabatan Perikanan.

Yahaya, J. 1988. Fishery management and regulation in Peninsular Malaysia: issues and constraints. Marine Resource Economics 5(2):83–98.

Yahaya, J. 1994. Fish-aggregating device (FAD) and community-based fisheries management in Malaysia. Pages 315–326 in Socio-economic issues in coastal fisheries management: proceedings of the IPFC symposium. Food and Agriculture Organization of the United Nations, Regional Office for Asia and the Pacific (RAPA), RAPA Publication 1994/8, Bangkok.

American Fisheries Society Symposium 86:251–264, 2018

Selection of Artificial Reef Deployment Sites by Using Evidential Reasoning

Nurul Haqimin Mohd Salleh*, Azifah Aqilah Aman, and Saharuddin Abdul Hamid
School of Maritime Business and Management, Universiti Malaysia Terengganu
Kuala Nerus, Terengganu 21030, Malaysia

Abstract.—Artificial reef programs can undeniably provide environmental benefits, especially when used to restore marine ecosystem and promote recreational scuba diving activities. The creation of artificial reefs is considered a key program for developing sustainable marine tourism and enhancing fish and invertebrate populations. Although there are many successful artificial reef programs that have been reported in the literature, some artificial reef programs are not effective due to a lack of decision-making protocols for selecting the best or optimal location for artificial reef deployment. This paper develops a model for selecting the best location for artificial reef deployment by using multiple criteria. Two different decision-making techniques (i.e., analytical hierarchy process and evidential reasoning) are employed in this model. To demonstrate the applicability of the proposed model, three potential sites off Perhentian Island (i.e., Pasir Tani, Tokong Laut, and Pulau Susu Dara Kecil), Malaysia were examined. Based on these test cases, the three most important criteria in selecting the best artificial reef site are sediment type, water depth, and presence of a local, natural reef. Based on the proposed model, Pulau Susu Dara Kecil was assessed as the best site for an artificial reef deployment compared to Pasir Tani and Tokong Laut. This proposed model is capable of assisting researchers and government agencies such as the Department of Fisheries Malaysia (DoFM) and the Department of Marine Park Malaysia (DMPM) in selecting the best sites for artificial reef deployment and can be modified based on decision makers' preferences and priorities.

Introduction

Malaysia is a country known to offer some of the best diving locations in the world along with modern and sophisticated diving facilities and infrastructure. The island destinations in Malaysia are also known for their calm waters and diverse marine life. The rapid growth of marine tourism has increased the number of scuba divers, and they may impact the environment and marine habitat (Stolk et al. 2005). According to Polak and Shashar (2012), the world's coral reefs are under threat and may be lost by the year 2050. One of the tourist activi-

* Corresponding author: haqimin@umt.edu.my

ties that have caused coral damage is scuba diving (Polak and Shashar 2012). The disturbance of coral by divers (e.g., touching coral, kneeling on benthic organisms, and dive fin motion) can result in coral breakage (Polak and Shashar 2012). One way to aid in the recovery of coral reefs is by deploying artificial reefs. An artificial reef can function as natural reef by protecting and facilitating coral regeneration and by enhancing populations of associated living marine resources (London Convention and Protocol/UNEP 2009). Artificial reefs can serve as substitutes for natural coral dive sites and, consequently, reduce recreational diver impacts on natural reefs (Stolk et al. 2005). However, questions have been raised about how to select the best sites to deploy artificial reefs. Optimal artificial reef deployment should include many criteria. This paper offers a model for selecting the best site for artificial reef deployment using multiple criteria.

Literature Review

The deployment of an artificial reef requires good planning to ensure success in achieving the objectives of the artificial reef program. Many countries have created and enforced guidelines for artificial reef deployment in order to sustain and protect the environment. As an example, the Department of Fisheries Malaysia referred to the *London Convention and Protocol/ UNEP Guidelines for the Placement of Artificial Reefs* (London Convention and Protocol/UNEP 2009). This document offers guidelines for Malaysia's artificial reef program. In another example, northeast Atlantic countries refer to the Convention for the Protection of the Marine Environment of the North-East Atlantic (OSPAR Commission 1991) for guidance in artificial reef development.

Several studies have been conducted on the subject of artificial reef deployment, such as those by Kennish et al. (2002), Shipley and Cowan (2011), Barber et al. (2009), and Mousavi et al. (2015). Kennish et al. (2002) used constraint mapping and prioritization methods to identify the incompatibility between an existing artificial reef site and the potential sites for future reef deployments. The criteria considered were cost efficiency and socioeconomics as well as broad environmental criteria. A systematic model developed by Barber et al. (2009) for artificial reef site selection consisted of seven steps (i.e., exclusion mapping, depth and slope verification, superficial substrate assessment, data weighting and the subsequent analysis, visual transect surveys, benthic air-lift sampling, and larval settlement collection). There has also been research conducted on artificial reef deployment for the enhancement of Red Snapper Lutjanus campechanus populations in the Gulf of Mexico (USA) where researchers used a fuzzy rough-set model that determined the optimal spacing of artificial reefs during their deployment (Shipley and Cowan 2011). Mousavi et al. (2015) investigated the suitability of the Kish Island's coastal area off Iran for artificial reef deployment by using the spatial multi-criteria decision making method. This new method was proposed by combining the multi-criteria decision making expert system and weighted liner combination methods. This study (Mousavi et al. 2015) also used the analytical hierarchy process (AHP) to prioritize the selected sites for reef deployment. The process of artificial reef deployment requires careful planning because without a proper plan, the existing biota may be harmed. Ultimately, this may nullify the effectiveness of the artificial reef program.

Local Environment

Planning for artificial reef deployment must consider local environmental criteria. Local environment can be defined as the environment surrounding a biotic community and consists of living and nonliving features. An artificial reef program that receives community support tends to be successful as a result of user group interaction with the program (Lindberg and Seamen 2011). Local environmental criteria include distance from the coast, shipping lanes and anchorage areas, and areas of historical importance. Artificial reef sites should be near major tourism centers and must allow easy access for diving activities. The proposed sites also should not interfere with the commercial marine traffic lanes to avoid danger to divers, not cause destruction of the artificial reef, and ensure safe navigation for vessels (London Convention and Protocol/UNEP 2009). Otherwise, there are no limitations when placing an artificial reef in areas of historical importance. There are places that have not been declared as historical areas, and consequently, there are no provisions to exclude artificial reefs from these areas.

Physical Environment

The physical environment also needs to be measured when selecting the best site to place an artificial reef. It is important to know the characteristics of the physical environment, such as water depth, sediment type, and tidal conditions, before the site can be selected. Water depth is important because by deploying the artificial reef at the appropriate water depth, there should be no damage to the reef and the reef's stability can be enhanced (Stephan et al. 1990). As an example, a Philippine artificial reef was submerged 5–30 m deep to attract fish by using fish aggregating devices (Delmendo 1991). Water depth should be measured by considering the tidal level, as an artificial reef should not be deployed where it could be exposed at the surface during low tide. In addition, the bottom sediment of the ocean plays an important role in enabling artificial stability. In coastal areas, there are many types of sediment, such as clay, gravel, silt, sand, and rock, but sand, notably, is the most effective sediment on which to place an artificial reef. Furthermore, the artificial reef should be placed at a suitable depth because if the water depth is too shallow or too deep the effectiveness of the artificial reef in enhancing coral growth and fish spawning will be reduced (O'Leary et al. 2001). According to O'Leary et al. (2001), suitable water depth may be necessary to allow for optimal growth of the artificial-reef-associated organisms because of light penetration.

Biological Environment

According to the London Convention and Protocol/UNEP (2009), a proposed artificial reef program should target specific marine species so that the restoration's effectiveness can be enhanced. This is because some marine species are specialized and have unique environmental requirements. The location of an artificial reef should be biologically suitable as spawning and nursery areas and not enable harm from illegal fishing activities (Ali et al. 2004). However, if the artificial reef is planned specifically for research purposes, the reef should be deployed in a site that is less populated by people (Fikes and Lehman 2010). In addition, deploying an artificial reef near a local, natural reef should be considered because the high density of fish on natural reefs can serve as a source of fish to colonize the artificial reef. For example, off Bluefields, Jamaica, the artificial reef site had a higher fish density (i.e., fish per unit area) in comparison to the nearby natural reef (Rudolph 2012).

In addition, the artificial reef had a higher species richness with 83% more species than the natural reef (Walker and Schlacher 2014).

Methodology

An AHP was employed to prioritize the criteria necessary to select the best or optimal site for an artificial reef. An AHP is a decision-making approach that evaluates the weight of each criterion and subcriterion that affects the final decision (Saaty 1977). Additionally, we also employed an evidential reasoning method to assess the selection criteria in the proposed model (Ruspini et al. 1992). As shown in Figure 1, seven steps have been developed to select the best placement site for artificial reefs.

Identification Criteria for the Artificial Reef Deployment (Step 1)

Step 1 is to identify the criteria for selecting the artificial reef deployment sites based on previous research. All criteria are identified from expert research papers, journals, and books, as mentioned in the literature review. These criteria are then categorized into three main environmental criteria (i.e., local, physical, and biological).

Generic Model Development (Hierarchal Structure) (Step 2)

In Step 2, a hierarchal structure model is developed by arranging the criteria that have been identified from literatures (Table 1) into a proper visual model. This hierarchal structure consists of main criteria and subcriteria. Initially, there were five main criteria and 25 subcriteria that were identified. However, these criteria were further revised and reduced by experts who were consulted in this research. The newly developed model is shown in Figure 2 below.

Weight Establishment for Selection Criteria (Step 3)

An AHP is used to assign a weight to all criteria identified in the model (Figure 2). Then,

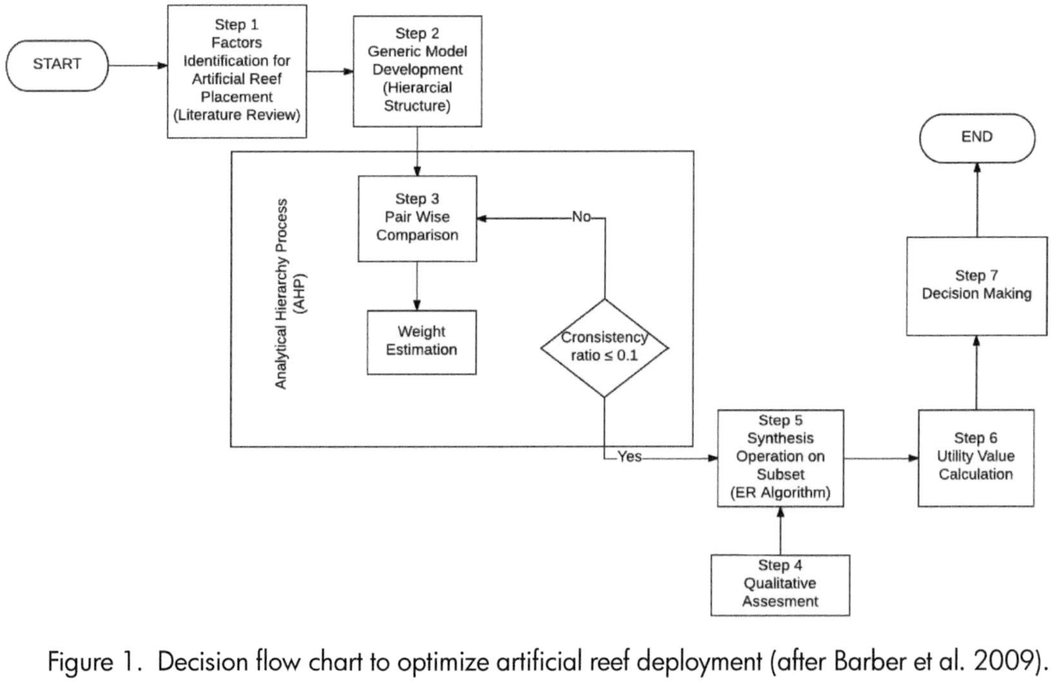

Figure 1. Decision flow chart to optimize artificial reef deployment (after Barber et al. 2009).

Table 1. Summary of criteria identified for selecting the best location for an artificial reef.

Main criteria	Subcriteria	References
Local environment	Distance to the nearest coast, shipping lanes and anchorage, and areas of historical importance	Edward (1988), IMO (2009), Lindberg and Seamen (2011)
Physical environment	Water depth, sediment type, and tide conditions	Stephen et al. (1990), Delmendo (1991), O'Learly et al. (2001).
Biological environment	Local natural reef, migration routes of fish and marine mammals, and protected spawning and nursery areas	Ali et al. (2004), Fikes and Lehman (2010), Lindberg and Seaman (2011), Rudolph (2012), Walker and Schlacher (2014)

these criteria are compared, based on the pairwise comparisons, using scales of the absolute number, as shown in Table 2. The data are obtained from questionnaires answered by the experts.

The judgments on pairs of attributes A_i and A_j are presented by an $n \times n$ matrix D. The entries a_{ij} are defined by entry rules, as follows (Saaty 1980; Mohd Salleh et al. 2014):

- Rule 1: If $a_{ij} = \alpha$, then $a_{ji} = \frac{1}{\alpha}$, $\alpha \neq 0$.
- Rule 2: If A_i is judged to be of equal relative importance as A_j, then $a_{ij} = a_{ji} = 1$.

According to the above rules, the matrix D is shown as follows:

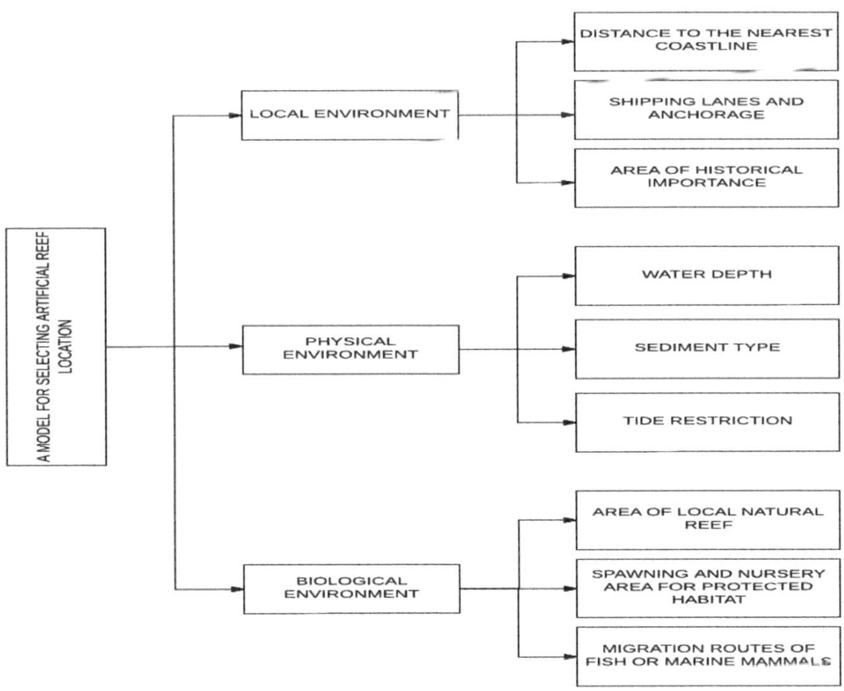

Figure 2. Model for selecting artificial reef sites.

Table 2. Scale of pairwise comparison.

Numerical assessment (scale)	Linguistic meaning
1	Equally important
3	Weakly important
5	Strongly important
7	Very strongly important
9	Extremely important
2, 4, 6, 8	Intermediate values between the two adjacent judgments

$$D = \begin{pmatrix} 1 & a_{12} & \ldots a_{2n} \\ \frac{1}{a_{12}} & 1 & \ldots a_{2n} \\ \frac{1}{a_{1n}} & \frac{1}{a_{2n}} & \ldots 1 \end{pmatrix}, \quad (1)$$

where $i, j = 1, 2, 3 \ldots n$ and each a_{ij} is the relative importance of attribute A_j. The quantified judgment of comparison on pair (A_i, A_j) is noted as a_{ij} in the matrix D; the next step is to allocate the weight vector for each criterion, as it indicates the prioritization of the criteria (Mohd Salleh et al. 2014).

A weight value w_k can be calculated by using equation (2) as follows:

$$w_k = \frac{1}{n}\sum_{j=1}^{n}\left[\frac{a_{kj}}{\sum_{i=1}^{n} a_{ij}}\right](k = 1,2,3,\ldots n), \quad (2)$$

where a_{ij} stands for the entry of row i and column j in a comparison matrix of order n.

Once weight for each criterion has been computed, a consistency ratio (CR) needs to be calculated. A reasonable value for the CR is 0.10 or less. If the ratio value is greater than 0.10, the ratio is considered unreasonable and there are inconsistencies during pairwise comparisons. The CR can be calculated by using equation (3), as follows:

$$\text{CR} = \frac{\text{CI}}{\text{RI}}, \quad (3)$$

where CI is the consistency index and RI is the random index, as shown in Table 3, while n is the number of items being compared and λ_{max} is the maximum weight value of the $n \times n$ comparison matrix D.

The CI value can be calculated by using equations (4)–(5) as follows:

$$\text{CI} = \frac{\lambda_{max} - n}{n - 1} \quad (4)$$

$$\lambda_{max} = \frac{\sum_{j=1}^{n}\left(\frac{\sum_{k=1}^{n} w_k a_{jk}}{w_j}\right)}{n} \quad (5)$$

Criteria Assessment (Step 4)

In Step 4, the assessment for selecting the best location of the artificial reef will be conducted by obtaining the answers from the domain experts. The assessment form (Table 4)

Table 3. Value of average random index (RI).

n	1	2	3	4	5	6	7	8	9	10
RI	0	0	0.58	0.9	1.12	1.24	1.32	1.41	1.45	1.49

Table 4. Criteria assessment form. AR = artificial reef; nm = nautical miles.

Criteria of AR site selection	AR site-selection assessment grades				
	Very low	Low	Medium	High	Very high
Distance to the nearest coastline	The location is within 0.1–0.2 nm from the nearest coastline.	The location is 0.3–0.4 nm from the nearest coastline.	The location is 0.5–0.6 nm from the nearest coastline	The location is 0.7–0.8 nm from the nearest coastline	The location is 0.9–1.0 nm from the nearest coastline.
Shipping lanes and anchorage	Artificial reef placement at this location is within the charted shipping lanes and anchorage areas for commercial vessels.	Artificial reef placement at this location is nearer within the charted shipping lanes and anchorage areas for commercial vessels.	Artificial reef placement at this location is near within the charted shipping lanes and anchorage areas for commercial vessels.	Artificial reef placement at this location is not within the charted shipping lanes and anchorage areas for commercial vessels.	Artificial reef placement at this location is far outside the charted shipping lanes and anchorage areas for commercial vessels.
Area of historical importance	The selected location contains plenty of archeological artifacts.	The selected location contains archeological artifacts.	The selected location contains fewer archeological artifacts.	The selected location hardly contains archeological artifacts.	The selected location does not contain archeological artifacts.
Water depth	The selected location has water depth within 0–4 m.	The selected location has water depth within 5–9 m.	The selected location has water depth within 10–14 m.	The selected location has water depth within 15–19 m.	The selected location has water depth within 20–24 m.
Sediment type	The selected location sediment is made up of soft and and muddy bottom.	The selected location sediment made up of loosely packed sand or silt.	The selected location sediment made up of gravel.	The selected location sediment is made up of hard clay sand.	The selected location sediment is made up of hard rock or hard sand.
Tidal restriction	The artificial reef is 100% exposed above the water	The artificial reef is 50% exposed to the water surface	The artificial reef is at the water surface during low tide and	The artificial reef is 50% not exposed at the water	The artificial reef is 100% far from the water surface

Table 4. Continued.

Criteria of AR site selection	AR site-selection assessment grades				
	Very low	Low	Medium	High	Very high
Tidal restriction (continued)	surface during low tide and is harmful for boat navigation.	during low tide and is harmful for boat navigation.	is harmful for boat navigation.	surface during low tide and is not harmful for boat navigation.	during water tide and is not hamful for boat navigation.
Local natural reef	There is plenty of natural reef at the selected location.	There is natural reef at the selected location.	There is less natural reef at the selected location.	There is hardly any natural reef at the selected location.	There is no natural reef at the selected location.
Area of spawning and nursery from protected habitat	The selected location contains plenty of sensitive or protected habitat.	The selected location contains sensitive or protected habitat.	The selected location contains less sensitive or protected habitat.	The selected location contains hardly any sensitive or protected habitat.	The selected location does not contain sensitive or protected habitat.
Migration routes of fish or marine mammals	The selected location does not consist of existing stocks of targeted species.	The selected location consists of fewer existing stocks of targeted species.	The selected location consists of existing stocks of targeted species.	The selected location consists of more existing stocks of targeted species.	The selected location consists of greater existing stocks of targeted species.

is used to obtain information on the site's environment. The experts give their evaluation according to the scale provided (i.e., very low, low, medium, high, and very high).

Synthesis Operation on Subset (Evidential Reasoning Algorithm) (Step 5)

An analysis of the evaluation scores can be conducted through manual calculation or use of computer software tool. Consequently, the Intelligent Decision System (IDS) software package is used for analyzing the belief degrees of linguistic variables for selecting the best site using the artificial reef criteria (Mohd Salleh et al. 2014).

Utility Value Calculation (Step 6)

The results of assessment of the artificial reef deployment sites is represented by five linguistic terms (i.e., very low, low, medium, high, and very high). From assessment, a single value can be used by professional decision makers for ranking the alternatives and for comparison purposes (Mohd Salleh et al. 2014). Consequently, a utility value concept developed by Xu and Yang (2001) was calculated in this study to obtain a single numerical score.

Decision Making (Step 7)

The information from the IDS determines which deployment site has a greater value when compared to other alternative sites. The site with the highest value will be chosen for artificial reef deployment.

Test Cases

Perhentian Island, Malaysia was selected as a case study area for selecting the best site for an artificial reef. Based on their experience in this field, 15 experts were chosen, most with 5–10 years of experience in artificial reef deployment. The experts were from the DoFM, the DMPM, the Institute of Oceanography and Environment at the University of Malaysia–Terengganu (Universiti Malaysia Terengganu [UMT]), the Fisheries Development Authority of Malaysia (Lembaga Kemajuan Ikan Malaysia [LKIM]), dive masters and instructors at Perhentian Island, and lecturers from the Marine Biology program, UMT. To obtain their evaluation, interviews and discussion sessions were conducted.

For the three main criteria, a 3 × 3 pairwise comparison matrix was developed for obtaining the weight for each criteria. As an example, M(LPB) is a matrix for comparing the relative priority of the local, physical, and biological environments. To obtain the aggregated comparison matrices, the geometric mean (GM) was used to aggregate the evaluation by experts within a group.

The GM of the importance criterion was calculated as follows:

$$\mathrm{GM}_{ij} = \left(e_{ij}^{1} \cdot e_{ij}^{2} \cdot e_{ij}^{3} \ldots e_{ij}^{k}\right)^{\frac{1}{k}}, \qquad (6)$$

where k is the number of experts, e_{ij}^{k} stand for the kth expert opinion for relative importance of the ith criterion to the jth criterion. The GM for the criteria local environment and physical environment ($\mathrm{GM}_{\mathrm{LP}}$) is

$$\mathrm{GM}_{\mathrm{LP}} = \begin{pmatrix} 7\times\frac{1}{4}\times\frac{1}{9}\times\frac{1}{9}\times\frac{1}{5}\times\frac{1}{9} \\ \times\frac{1}{6}\times\frac{1}{7}\times\frac{1}{5}\times\frac{1}{8}\times\frac{1}{5} \\ \times\frac{1}{9}\times 1\times\frac{1}{9}\times\frac{1}{7} \end{pmatrix}^{\frac{1}{15}} = 0.22.$$

The matrix D(LPB) for the main criteria (i.e., local, physical, and biological environments) for selecting the best site for an artificial reef is shown below:

$$\mathrm{LPB} = \begin{pmatrix} 1 & 0.22 & 0.33 \\ 4.64 & 1 & 1.39 \\ 2.99 & 0.72 & 1 \end{pmatrix}.$$

Weight calculation for each main criterion is demonstrated as follows:

$$W_L = \frac{1}{3}\left(\frac{1}{1+4.64+2.99}\right) + \left(\frac{0.22}{0.22+1+0.72}\right) + \left(\frac{0.33}{0.33+1.39+1}\right) = 0.1169$$

$$W_P = \frac{1}{3}\left(\frac{4.64}{1+4.63+2.99}\right) + \left(\frac{1}{0.22+1+0.72}\right) + \left(\frac{1.39}{0.33+1.39+1}\right) = 0.5213$$

$$W_B = \frac{1}{3}\left(\frac{2.99}{1+4.64+2.99}\right) + \left(\frac{0.72}{0.22+1+0.72}\right) + \left(\frac{1}{0.33+1.39+1}\right) = 0.3617$$

As a result, W_L, W_P, and W_B were 0.1169, 0.5213, and 0.3617, respectively. The next step is to calculate and check the consistency ratio of the pairwise comparison. First, λ_{max} was calculated to lead the RI and CR. Based on the equation below, the CR was calculated as follows:

$$L = (1 \times 0.1169) + (0.22 \times 0.5213) + (0.33 \times 0.3617) = 0.351$$

$$P = (4.64 \times 0.1169) + (1 \times 0.5213) + (1.39 \times 0.3617) = 1.5665$$

$$B = (2.99 \times 0.1169) + (0.72 \times 0.5213) + (1 \times 0.3617) = 1.0866$$

$$\lambda = \left[\frac{\left(\frac{0.351}{0.1169}\right) + \left(\frac{1.5665}{0.5213}\right) + \left(\frac{1.0866}{0.3617}\right)}{3} \right] = 3.0039$$

$$\text{CI} = \frac{\lambda_{max} - n}{n - 1} = \frac{3.0039 - 3}{3 - 1} = 0.00195$$

Based on Table 3, the RI for the three criteria was 0.58. As a result, the CR was calculated as follows:

$$\text{CR} = \frac{\text{CI}}{\text{RI}} = \frac{0.00195}{0.58} = 0.003$$

The CR value for the main criteria was 0.003. Saaty (1980) stated that a CR $\leq$ 0.1 indicates that the judgment is acceptable. The same calculation technique was applied to rank subcriteria in the model. The weights values and consistency ratio values for subcriteria are shown in Table 5.

The above calculation is for the local weights of the criteria, which were compared to each other within the same group (e.g., a group of main criteria consists of local, physical, and biological environments). To determine the most important criterion (i.e., lowest level criterion) in the model, a global weight is calculated by multiplying a local weight of these lowest level criteria and the local weights of an upper-level criterion or criteria. As an example, the global weight for distance to the nearest coast can be calculated as $GW_D = 0.4600 \times 0.1170 = 0.0538$. By using this same technique, all the calculated global weights were arranged according to their rank, as shown in Table 6.

Results and Discussion

Physical environment was the most important main criteria in selecting the artificial reef site (Table 5). Based on the AHP calculation, the weight for the physical environment was 0.5220, followed by the biological environment (0.3610) and the local environment (0.1170). The calculated value for the global weight is also shown in Table 6. The three most important criteria were sediment type (0.2573), water depth (0.2083), and local natural reef (0.1798). To ensure the judgments made by the experts are consistent, the CRs for main factors and its subfactors are calculated. The CR for the main factors are 0.003 where the CRs for subfactors of local environment, physical environment, and biological factor are 0.009, 0.001, and 0.001, respectively. Since all the CRs are below than 0.1, all the judgments made are consistent and reasonable. The ranking orders of other lowest-level criteria are shown in Table 6.

Table 5. Consistency ratios and weight values for all criteria and subcriteria in the artificial reef site-selection model.

Main criteria	Weight	Consistency ratio (≤ 0.1)	Subcriteria	Local weight	Global weight	Consistency ratio (≤ 0.1)
Local environment	0.1170		Distance to the nearest coast	0.4600	0.0538	0.009
			Shipping lanes and anchorage areas	0.4500	0.527	0.001
			Areas of historical importance			
Physical environment	0.5220	0.003	Water depth	0.3990	0.2083	0.001
			Sediment type	0.4930	0.2573	
			Tidal condition	0.1080	0.0564	
Biological criteria	0.3610		Local natural reef	0.4980	0.1798	0.001
			Spawning and nursery areas (protected habitat)	0.3640	0.1314	
			Migrataion routes of fish and marine mammals	0.1380	0.0498	

Based on Table 4, five experts assessed the three potential deployment sites, and these assessments were further analyzed using IDS software (i.e., evidential approach) to obtain the belief degrees. For example, Table 7 shows the degrees of belief from the assessment that was conducted at the Tokong Laut site.

As shown in Table 7, the belief degrees of the assessment criteria (nine lowest-level criteria in the model) were further analyzed to obtain the aggregated values for the main

Table 6. Ranking orders of the lowest level criteria.

Lowest-level criteria	Global weight	Rank
Sediment type	0.2573	1
Water depth	0.2083	2
Local natural reef presence	0.1798	3
Spawning and nursery areas (protected habitat)	0.1314	4
Tidal condition	0.0564	5
Distance to the nearest coast	0.0538	6
Shipping lanes and anchorage areas	0.0527	7
Migration routes of fish and marine mammals	0.0498	8
Areas of historical importance	0.0105	9

Table 7. Degrees of belief for the Tokong Laut site.

Criteria for site selection	Artificial reef site-selection assessment grades				
	Very low	Low	Medium	High	Very high
Distance to the nearest coast	0	1	0	0	0
Shipping lanes and anchorage area	0.0088	0.0088	0.2497	0.3702	0.3626
Areas of historical importance	0	0	0.2637	0.5604	0.1758
Water depth	0	0	0.0134	0.0534	0.9332
Sediment type	0.1016	0.4396	0.1529	0.1529	0.1529
Tide condition	0.1905	0	0.1905	0.1905	0.4286
Local natural reef presence	0.7585	0.1148	0.0792	0.0317	0.0158
Areas of spawning and nursery (protected habitat)	0	0.5022	0.0873	0.3755	0.0349
Migration routes of fish and marine mammals	0	0.0494	0.2924	0.6417	0.0165

criteria (i.e., local environment, physical environment, and biological environment). The belief degrees for these main criteria were then aggregated to obtain belief degrees for each deployment site. Based on Step 6, the utility values were calculated to obtain a single value for each deployment site for comparative purposes. The belief degrees and utility values for all three deployment sites are shown in Table 8 and illustrated in Figure 3.

To demonstrate the applicability of the proposed model, three deployment sites were assessed (i.e., Pasir Tani, Tokong Laut, and Pulau Susu Dara Kecil). The assessment results show that, among these three sites, Pulau Susu Dara Kecil is the most suitable for the artificial reef deployment compared

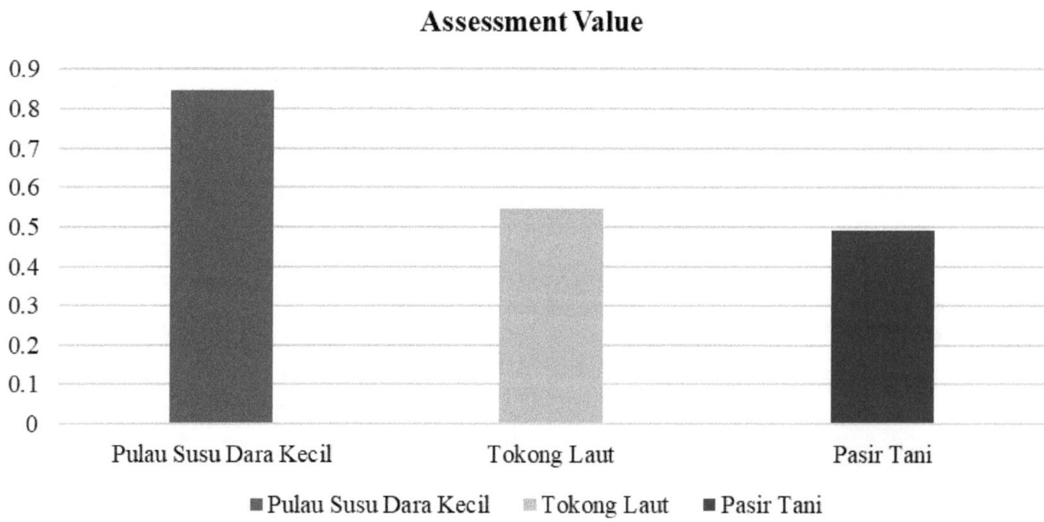

Figure 3. Assessment value for the three deployment sites.

with Tokong Laut and Pasir Tani. The hard sandy bottom of Pulau Susu Dara Kecil is potentially a key contributor to good stability for the artificial reef. In addition, water depth at the Pulau Susu Dara Kecil deployment site is 18 m, which is a reasonable depth for divers. In addition, there was less coral life at this site, and by placing the artificial reef at Palau Susu Dara Kecil, marine organisms should easily move to the artificial reef while not contributing to fishing pressure on the existing natural reef. The Tokong Laut deployment site is 25 m deep. Although the area is rich in marine life and has an abundance of hard and soft corals, this site is not suitable for artificial reef deployment because it slopes toward the naturally occurring black corals. In addition, the sediment is slightly muddy, and this can affect the stability of the artificial reef. Conditions at Tokong Laut were similar to those at the Pasir Tani site. At Pasir Tani, there is a Vietnamese refugee ship wreck. The area has become habitat for an abundance of marine life, such as barracudas and Bigeye Snapper *Lutjanus lutjanus*.

Conclusions

When selecting the best location for an artificial reef, it is important to facilitate the success and effectiveness of an artificial reef program. In this paper, a generic model for selecting an artificial reef site was developed. Three main criteria (i.e., local environment, physical environment, and biological environment were evaluated based on the AHP method to obtain weights for each criterion. Based on the AHP analysis, the physical environment was the most important criterion in placing the artificial reef, followed by the biological environment and the local environment. Based on the proposed model, Pulau Susu Dara Kecil was the best site for artificial reef deployment compared to Pasir Tani and Tokong Laut. This proposed model is capable of assisting researchers and government agencies such as the Department of Fisheries and the Department of Marine Park in selecting the best location for artificial reef deployment, and it can be modified based on decision makers' preferences.

References

Ali, A., M. M. Isa, S. A. Razak, and J. Ibrahim. 2004. A guide to make and set durable artificial reef fish aggregating devices (AR-FADs) for coastal areas. Marine Fishery Resources Development and Management Department, Southeast Asian Fisheries Development Center, Kuala Terengganu, Malaysia.

Barber, J. S., D. M. Chosid, R. P. Glenn, and K. A. Whitmore. 2009. A systematic model for artificial reef site selection. New Zealand Journal of Marine and Freshwater Research 43:283–297.

Delmendo, M. N. 1991. A review of artificial reefs development and use of fish aggregating devices (FADs) in the Asian region. Pages 116–141 *in* Food and Agriculture Organization of the United Nations (FAO). Papers presented at the symposium on artificial reefs and fish aggregating devices as tools for the management and enhancement of marine fishery resources. FAO, Regional Office for Asia and the Pacific (RAPA), RAPA Report 11, Bangkok.

Fikes, R. L., and R. L. Lehman. 2010. Recruitment and colonization of macro algae to a newly constructed rocky intertidal habitat in the northwest Gulf of Mexico. Gulf and Caribbean Research 22:9–20.

Kennish, R., K. D. Wilson, J. Lo, S. C. Clarke, and S. Laister, S. 2002. Selecting sites for large-scale deployment of artificial reefs in Hong Kong: constraint mapping and prioritization techniques. ICES Journal of Marine Science 59:S164–S170.

Lindberg, W. J., and W. Seaman. 2011. Guideline and management practices for artificial reef siting, use, construction, and anchoring in southeast Florida. Florida Department of Environmental Protection, Miami.

London Convention and Protocol/UNEP (United Nations Environment Programme). 2009. London Convention and Protokol/UNEP guidelines for the placement of artificial reef. International Maritime Organization, London.

Mohd Salleh, N. H., R. Riahi, Z. Yang, and J. Wang. 2014. Risk assessment of liner shipping from a business environment perspective. Pages 2320–2329 *in* M. Beer, S.-K. Au, and J. W. Hall, editors. Vulnerability, uncertainty, and risk: quantification, mitigation, and management. American Society of Civil Engineers, Reston, Virginia.

Mousavi, S. H., A. Danehkar, M. R. Shokri, H. Poorbagher, and D. Azhdari. 2015. Site selection for artificial reefs using a new combine multi-criteria decision-making (MCDM) tools for coral reefs in the Kish Island – Persian Gulf. Ocean and Coastal Management 111:92–102.

O'Leary, E., T. Hubbard, and D. O'Leary. 2001. Artificial reefs feasibility study. Marine Institute, Marine Resources Series No. 20, Galway, Ireland.

OSPAR Commission. 1991. OSPAR guidelines on artificial reef in relation to living marine resources. OSPAR Convention for the Protection of the Marine Environment of the North-East Atlantic, Annex 6. OSPAR, London

Polak, O., and N. Shashar. 2012. Can a small artificial reef reduce diving pressure from a natural coral reef? Lessons learned from Eilat, Red Sea. Ocean and Coastal Management 55:94–100.

Rudolph, J. H. 2012. Effects of artificial reef complementation on fish population in a marine protected area: Bluefields Bay, Jamaica. Doctoral dissertation. Missouri State University, Springfield.

Ruspini, E. H., J. D. Lowrance, and T. M. Strat. 1992. Understanding evidential reasoning 6:401–424.

Saaty, T. L. 1977. A scaling method for prioritise in hierarchical structures. Journal of Mathematical Psychology 15:234–281.

Saaty, T. L. 1980. The analytic hierarchy process. McGraw-Hill, New York.

Shipley, J. B., and J. H. Cowan. 2011. Artificial reef placement: a Red Snapper, *Lutjanus campechanus*, ecosystem and fuzzy rule-based model. Fisheries Management and Ecology 18:154–167.

Stephan, C .D., B. G. Dansby, H. R. Osburn, G. C. Matlock, R. K. Riechers, and R. Rayburn. 1990. Texas artificial reef fishery management plan. Texas Parks and Wildlife Department, Coastal Fisheries Branch, Fishery Management Plan Series Number 3, Austin.

Stolk, P., K. Markwell, and J. Jenkins. 2005. Perceptions of artificial reefs as scuba diving resources: A study of Australian recreational scuba divers. Annuals of Leisure Research 8(2–3):153–166.

Walker, S. J., and T. A. Schlacher. 2014. Limited habitat and conservation value of a young artificial reef. Biodiversity and Conservation 23:433–447.

Xu, L. and J.B. Yang. 2001. Introduction to multi-criteria decision making and the evidential reasoning approach. Manchester School of Management, Manchester, UK.

American Fisheries Society Symposium 86:265–277, 2018

Methods to Quantify Recreational Angling Effort on Artificial Reefs off Florida's Gulf of Mexico Coast

Tiffanie A. Cross*, Beverly Sauls, and Rachel Germeroth
Florida Fish and Wildlife Conservation Commission, Fish and Wildlife Research Institute
100 Eighth Avenue SE, St. Petersburg, Florida 33701, USA

Keith Mille
Florida Fish and Wildlife Conservation Commission, Division of Marine Fisheries Management
620 South Meridian Street, Box 4B2, Tallahassee, Florida 32399, USA

Abstract.—Florida boasts an abundance of natural and artificial reefs that support a large and diverse recreational fishery off the Gulf of Mexico coast. Recent efforts to mitigate the effects of the *Deepwater Horizon* oil spill have increased the number of artificial reefs deployed off Florida's Gulf Coast. Fisheries-dependent data are needed to assess whether artificial reef programs are meeting intended objectives and to understand changes in recreational angler behavior, which may influence catch per unit effort and biomass of landed fish. The objective of this study was to quantify the use of artificial reefs by recreational anglers targeting reef fishes in the Gulf of Mexico off the west coast of Florida. We utilized an existing survey designed to monitor recreational fishing effort by anglers that target reef fishes from private boats. Over a 20-month period, an estimated total of 776,026 (SD ±27,540) angler trips targeted reef fishes off the Gulf Coast of Florida, of which 46% utilized artificial reefs. Approximately two-thirds of all reef angling trips took place nearshore in state-managed waters, and 70% of trips that utilized artificial reefs occurred in this area. Regionally, the highest proportion of angler trips targeting reef-associated species on artificial reefs took place in the panhandle of Florida, where Red Snapper *Lutjanus campechanus* are most abundant. Seasonally, state and federal fishing regulations also had an apparent influence on fishing effort and artificial reef use by recreational anglers. The method of assessment we present here could be useful for monitoring future trends in recreational fishing effort with respect to artificial reefs.

* Corresponding author: tiffanie.cross@myfwc.com

Introduction

Florida's mix of natural and artificial reef fish habitats along the coastal Gulf of Mexico provides for a diverse recreational reef fish fishery. Off the northwest panhandle of Florida (i.e., northern Gulf of Mexico), recreational anglers have nearshore access to deepwater reef habitat where Red Snapper *Lutjanus campechanus*, Vermilion Snapper *Rhomboplites aurorubens*, and Gray Triggerfish *Balistes capriscus* are abundant (Strelcheck et al. 2007; Saul et al. 2013). East of the panhandle (i.e., the Big Bend region), Gag *Mycteroperca microlepis* and Red Grouper *Epinephelus morio* are more abundant; however, Red Snapper abundance is increasing as the overfished stock continues to rebuild in the eastern Gulf of Mexico (Saul et al. 2013). In the Big Bend region, the nearshore area is dominated by expansive sea grass beds while reef habitats are further offshore and less accessible to recreational anglers (Coleman et al. 1996; Koenig et al. 2000; Switzer et al. 2012). Further south, the broad slope of the west Florida shelf is characterized by low-relief, natural hard-bottom habitats that support abundant grouper populations. The nearshore recreational fishery frequently targets grouper species in shallow water, but anglers must travel farther distances to access depths where Greater Amberjack *Seriola dumerili* and Red Snapper may also be caught.

Artificial reefs have been used to enhance stocks of marine fisheries, thus providing increased opportunities for the recreational angler. Various reef-associated fishes are attracted to high-relief structure (e.g., artificial reefs) that may be easily located with electronic fish finders and GPS coordinates and repeatedly targeted by recreational anglers. Generally, increased abundance of reef fishes on artificial reefs improves catch per unit effort (CPUE) and thus enhances recreational fishing for reef fishes (Grossman et al. 1997; Karnauskas et al. 2017). However, habitat preferences and life history strategies differ among species. Red Snapper and Gray Triggerfish exhibit an affinity for artificial reef structure with high site fidelity (Strelcheck et al. 2007), whereas pelagic amberjacks *Seriola* spp. are more transient (Dance et al. 2011). In contrast, groupers associate more with natural hard bottom. Red grouper have a high affinity for low relief limestone karst, and Gag are associated with small patches of hard bottom and rocky ridges (Lindberg et al. 2006; Coleman et al. 2010; Wall et al. 2011).

The establishment of artificial reefs in Florida began in the mid-1930s. The first permit was issued in 1936 by the U.S. Army Corps of Engineers (Seaman 1982), and the Florida Fish and Wildlife Conservation Commission's (FWC) Artificial Reef Program was initiated in 1982. Currently, FWC reports more than 2,200 artificial reef deployments in Florida's Gulf waters, including 1,222 within the boundaries of the state's fishery management jurisdiction (up to 9 nautical miles [16.7 km] from shore off the Gulf of Mexico coast) and 980 reefs in the exclusive economic zone. Artificial reefs are concentrated in the northwest panhandle region (Figure 1), where mitigation from potential environmental damage related to the *Deepwater Horizon* oil spill has increased deployment efforts. Artificial reefs are less concentrated in the Big Bend region, especially in state waters, and are widely dispersed off the southwest peninsula (Figure 1).

Understanding the human use of artificial reefs is required to measure both the economic and ecological success of an artificial reef program. In the past, perceived benefits have been assumed. However, there is an emerging need for studies that quantify

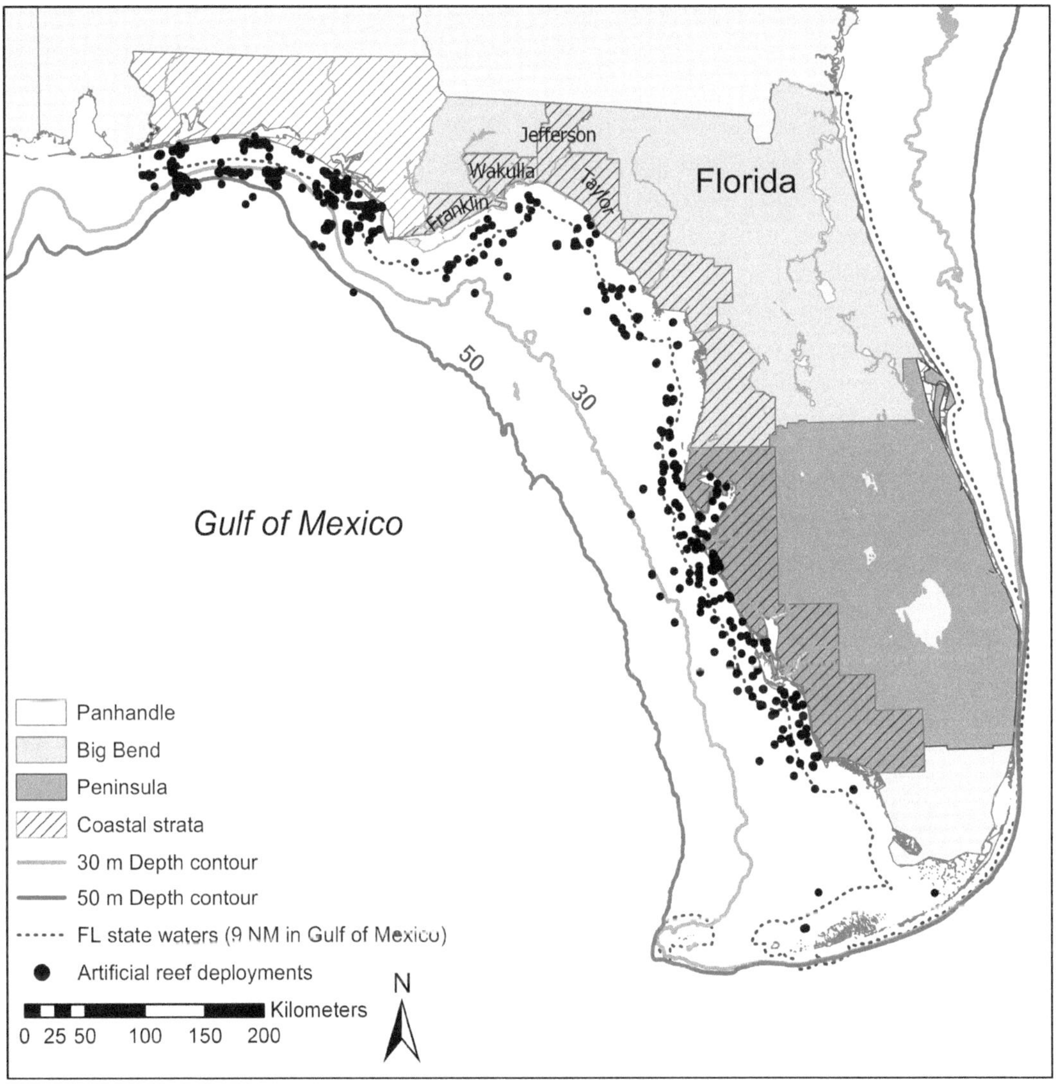

Figure 1. Filled circles represent artificial reef locations off Florida's Gulf Coast. State residents subscribed to the Gulf Reef Fish Survey were partitioned by region and subregion (coastal and noncoastal counties) prior to stratified random sampling for the mail survey. For each trip, respondents were asked to report whether fishing took place adjacent to the panhandle (white), Big Bend (medium gray), or peninsula (dark gray) regions. Hatched areas represent coastal counties adjacent to the Gulf of Mexico. Franklin, Wakulla, Jefferson, and Taylor counties in the Big Bend region have special regulations for Gag.

how well artificial reefs are meeting intended marine fisheries management objectives and evaluate whether the costs associated with implementing artificial reef programs are returned through economic benefits. Fisheries independent studies have evaluated fish distribution and production on artificial reefs (Grossman et al. 1997; Powers et al. 2003; Strelcheck et al. 2007; Dance et al. 2011; Addis et al. 2013; Karnauskas et al. 2017), and others still have assessed recreational vessel visitation rates at artificial reefs for use in economic and ecological valuation of artificial reefs (Adams et al. 2006; Tinsman and Whitmore 2006; Huth et al. 2014; Simard et al. 2016). Monitoring the recreational fishing component of an artificial reef program is also important in gaining an understanding of how large-scale projects influence fishing behavior, change CPUE, and potentially alter the average size of fish caught. Additionally, investigations of recreational fishing on artificial reefs may explain trends in overall landings of various reef fish species, both in number and in biomass. To date, we are unaware of a study that has reported recreational fishing effort on artificial reefs.

To assess the value and success of Florida's artificial reef program, it is necessary to quantify human use of artificial reefs deployed off Florida's Gulf Coast. The objective of this study was to quantify artificial reef use by recreational anglers targeting reef fishes in the Gulf of Mexico off the west coast of Florida. Here, we took advantage of an existing survey designed to monitor recreational fishing effort for reef fishes off the west coast of Florida. The Gulf Reef Fish Survey was initiated by the state of Florida in 2015 to improve the timeliness and precision of recreational catch-and-effort statistics for a suite of reef fish species commonly targeted by anglers. The survey methodology utilizes an established database of saltwater anglers that participate in the fishery. In 2016, new questions were added to the questionnaire for the effort portion of the survey to quantify the portion of recreational fishing trips that take advantage of artificial reefs.

Methods

Gulf Reef Fish Survey

The Gulf Reef Fish Survey provides a directory of private boat anglers who fish for reef fishes in the Gulf of Mexico. Anglers, 16 years of age and older, fishing from private recreational boats off the west coast of Florida (excluding Monroe County) are required to register for the Gulf Reef Fish Survey to legally harvest 10 species in the reef fish complex that includes Vermilion Snapper, Gray Triggerfish, Gag, Red Grouper, Black Grouper *Mycteroperca bonaci*, and amberjacks, including Greater Amberjack, Lesser Amberjack *Seriola fasciata*, Almaco Jack *S. rivoliana*, and Banded Rudderfish *S. zonata*. Applicable state saltwater fishing license requirements and state and/or federal size and bag limits and harvest seasons also apply. Anglers are subscribed to the Gulf Reef Fish Survey during the purchase or renewal of a state of Florida saltwater fishing license for no additional fee, or as a separate and free transaction if no license was purchased. During a transaction through the state fishing license system, anglers are required to provide a driver's license number or other acceptable form of identification, which is used to confirm residency status (in state or out of state) and automatically populate the database with the current mailing address for individuals.

Each month, anglers with a current subscription to the Gulf Reef Fish Survey were randomly selected to receive a mail survey. Prior to sample selection, anglers were partitioned into 14 separate survey groups based on their residence and whether the address could be matched with a recreational boat registered in the state of Florida (Table 1). A sample of 10% or 500, whichever was less, was selected from each survey group. This stratified random sample design was developed to account for potential differences in response rates among anglers with different fishing avidities. For example, anglers that live near the Gulf Coast and have direct access to a registered boat may be more likely to participate in the reef fish fishery in the

Table 1. Mail survey strata, defined by residence of Gulf Reef Fish Survey subscribers. Regions and subregions for Florida residents (strata 1–12) are depicted in Figure 1.

Stratum	Region	Subregion	Boat registration
1	Florida NW panhandle	Coastal	No
2	Florida NW panhandle	Coastal	Yes
3	Florida Big Bend	Noncoastal	No
4	Florida Big Bend	Noncoastal	Yes
5	Florida Big Bend	Coastal	No
6	Florida SW peninsula	Coastal	Yes
7	Florida SW peninsula	Noncoastal	No
8	Florida SW peninsula	Noncoastal	Yes
9	Florida SW peninsula	Coastal	No
10	Florida SW peninsula	Coastal	Yes
11	Keys and SE Florida	Noncoastal	No
12	Keys and SE Florida	Noncoastal	Yes
13	Alabama and Georgia	Noncoastal	–
14	Other states	Noncoastal	–

Gulf of Mexico and may also show a higher propensity to respond to the voluntary survey (avidity bias). Stratification also accounts for regional variability (Figure 1). For example, fishing effort may vary regionally in response to seasonal openings and closures for different species that are abundant, such as Red Snapper in northern regions and Gag in southern regions. By surveying groups of anglers with different participation rates separately, survey responses may be weighted to account for potential bias associated with varied response rates.

Included in the packet mailed to selected anglers was a cover letter explaining the purpose for the survey, the questionnaire with a map to assist in reporting the general region where each trip took place, a full-color guide with pictures for each Gulf reef fish species to aid in positive identification, and a postage-paid return envelope. One week after the survey packets were mailed, a reminder postcard was sent to all selected anglers to prompt a timely response, and 2 weeks later, a second questionnaire was mailed to anglers that had not yet returned the first survey. Response to the mail survey was voluntary.

The survey questionnaire prompted anglers to recall their fishing activity over the previous month and provide specific details for up to 12 trips. For each trip, the angler was asked to report the region fished (Figure 1), the percent of time (if any) spent fishing within the state of Florida's jurisdictional boundary (defined as ≤9 nautical miles off the Gulf Coast), and the types of fish that the angler kept, released, or tried to catch (selected from a list provided). A reported trip was counted as a Gulf reef fish trip if one or more of the required species was selected.

In May 2016, additional questions were added to the Gulf Reef Fish Survey questionnaire to quantify artificial reef use during recreational fishing. One question asked whether the angler ever fishes recreationally on artificial reefs, and for each fishing trip reported for the survey month, respondents were also asked to record whether they fished on an artificial reef.

Effort Estimation

Survey responses were used to generate monthly estimates of the total number of Gulf reef fish trips taken by registered participants during a given month, as well as the number of those fishing trips that used artificial reefs, using the methods for stratified random sampling described by Kish (1965). The sample weight for individuals selected to receive a survey from a given stratum (*h*) was calculated as

$$W_h = \frac{N_h}{n_h},$$

where N_h is the total number of registered Gulf reef fish anglers in the angler population stratum, and n_h is the number of anglers selected from the stratum to receive the mail survey. To account for anglers that did not return the completed questionnaire, the survey response rate in each stratum was calculated as

$$R_h = \frac{\sum_{i=1}^{n_h} W_h r_{h,i}}{\sum_{i=1}^{n_h} W_h},$$

where r_i is the binomial variable indicating whether selected individual *i* responded to the survey (1 = yes, 0 = no).

The mean number of Gulf reef fish trips per response in each stratum was calculated as

$$\bar{y}_h = \frac{\sum_{i=1}^{n} y_{h,i}}{n_h},$$

where $y_{h,j}$ is the number of Gulf reef fish trips reported in state or federal waters by the *i*th respondent in stratum *h* and n_h is the number of people in stratum *h* that responded to the survey. This value was also calculated separately, including only reported trips where the respondent indicated that an artificial reef was utilized. Variance was calculated as

$$\text{var}(\bar{y}_h) = \frac{\sum_{i=1}^{n} \left(y_{h,i} - \bar{y}_h\right)^2}{n_h - 1}.$$

The overall mean weighted number of trips per response across all 14 strata was calculated as

$$\bar{y} = \sum_{h=1}^{14} W_h \frac{1}{R_h} \bar{y}_h.$$

And overall variance was calculated as

$$\text{var}(\bar{y}) = \sum_{h=1}^{14} \left(W_h \frac{1}{R_h} \right)^2 \text{var}(\bar{y}_h).$$

The total number of trips taken by all anglers registered for the Gulf Reef Fish Survey during a given month was calculated as

$$\hat{Y} = N\bar{y},$$

where *N* is the total number of Gulf reef fish anglers that were registered during the survey month. Variance is calculated as

$$\text{var}(\hat{Y}) = N^2 \text{var}(\bar{y}).$$

Results

Each month, an average of 6,304 individuals were selected to receive an effort survey questionnaire in the mail, and the percentage that returned the survey ranged from 14.1% to 23.4% (Table 2). Between May 2016 and December 2017, anglers subscribed to the Gulf Reef Fish Survey took an estimated 776,026

Table 2. Total numbers of individuals registered for the Gulf Reef Fish Survey, sample sizes, and numbers of response rates by survey month and year.

Year	Survey month	Number registered	Sample Sample size	Number of responses	Raw response rate (%)	Weighted response rate (%)
2016	May	402,161	6,212	1,171	18.9	18.5
	June	412,084	6,184	1,120	18.1	19.2
	July	431,712	6,204	1,074	17.3	17.5
	August	410,066	6,151	1,088	17.7	16.9
	September	407,233	6,159	1,067	17.3	17.5
	October	405,002	6,159	1,114	18.1	17.7
	November	405,451	6,116	1,152	18.8	19.2
	December	401,418	5,835	1,031	21.7	23.4
2017	January	387,680	5,814	1,114	19.2	18.9
	February	413,807	5,856	1,132	19.3	19.0
	March	440,703	6,235	1,091	17.5	17.0
	April	445,885	6,249	1,093	17.5	18.1
	May	450,984	6,782	1,166	17.2	17.5
	June	458,700	6,434	1,122	17.4	17.4
	July	478,369	6,817	1,139	16.7	17.5
	August	451,708	5,797	889	14.4	14.1
	September	437,955	6,796	1,144	16.8	17.2
	October	431,341	6,778	1,139	16.8	16.5
	November	434,802	6,742	1,181	17.5	16.7
	December	421,391	6,762	1,214	17.4	15.9

recreational fishing trips from private boats (measured in angler trips, SD ±27,540) to target reef fishes off the west coast of Florida. The highest percentage of total fishing trips took place in the peninsula region (42%), followed by the panhandle (34%) and Big Bend (24%; Figure 2). Overall, artificial reefs were utilized during an estimated 46% of fishing trips (360,522 trips, SD ±18,098). Half of the estimated fishing effort on artificial reefs took place in the panhandle region (176,725 angler trips, SD ±12,012), 18% (66,436 SD ±8,133) took place in the Big Bend, and the peninsula region constituted 32% (114,820 SD ±10,314; Figure 3).

The majority of fishing trips that targeted reef fishes took place nearshore in state-managed waters, regardless of whether artificial reefs were visited (Figures 2 and 3). In the panhandle region, 77% of fishing trips that visited artificial reefs took place in state waters, followed by 69% in the peninsula and 59% in the Big Bend (Figure 3). In the panhandle, artificial reefs were visited during a majority of trips (68%), whereas anglers in other regions were less likely to report utilizing them (Figure 4).

Fishing seasons for reef fishes managed in state and federal waters off the west coast of Florida varied over the months of this study and apparently influenced artificial reef use by recreational anglers (Table 3). In the panhandle region, fishing effort on artificial reefs peaked during May through October, which coincided with open recreational harvest seasons for Red Snapper (Figure 5).

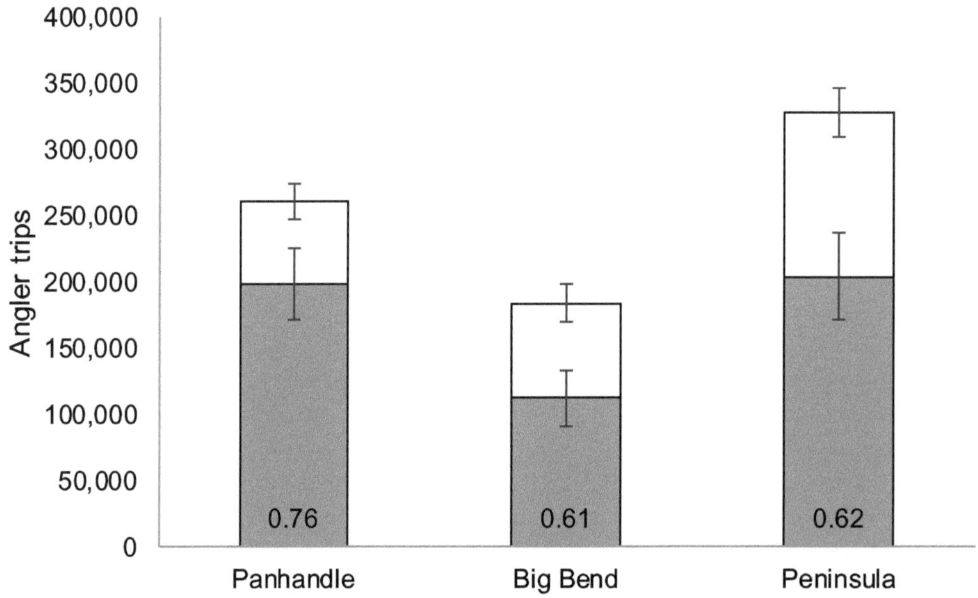

Figure 2. Total estimated recreational fishing effort for reef fishes (regardless of whether an artificial reef was visited) in each region over the 20-month study period by area fished. State (shaded bars) = trips taken in inland or state territorial seas, and exclusive economic zone (white bars) = trips taken in the exclusive economic zone. Proportion of trips that occurred in state waters are labeled.

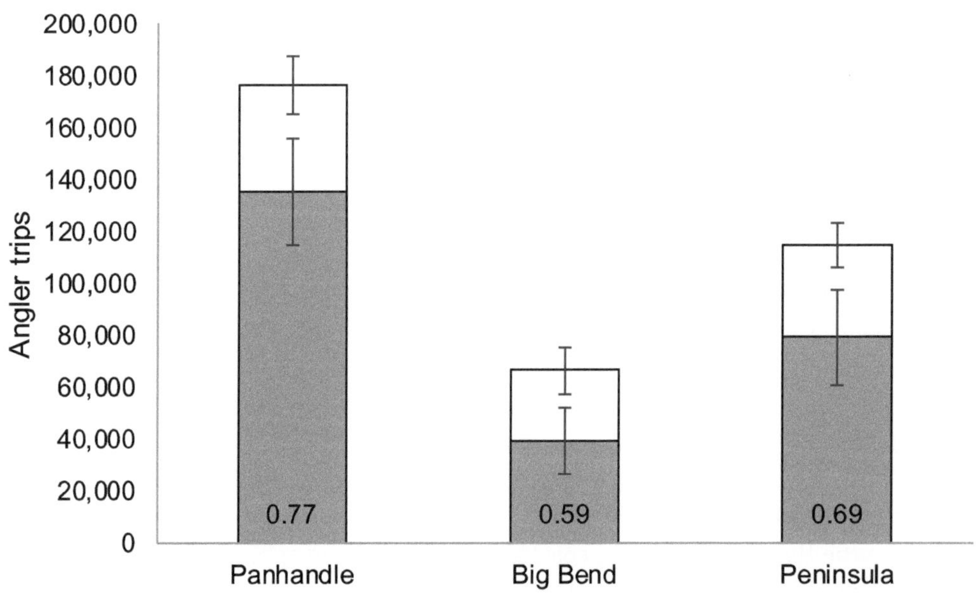

Figure 3. Estimated recreational fishing effort on artificial reefs in each region over the 20-month study period by area fished. State (shaded bars) = trips taken in inland or state territorial seas, and exclusive economic zone (white bars) = trips taken in federal waters of the exclusive economic zone. Proportion of trips that occurred in state waters are labeled.

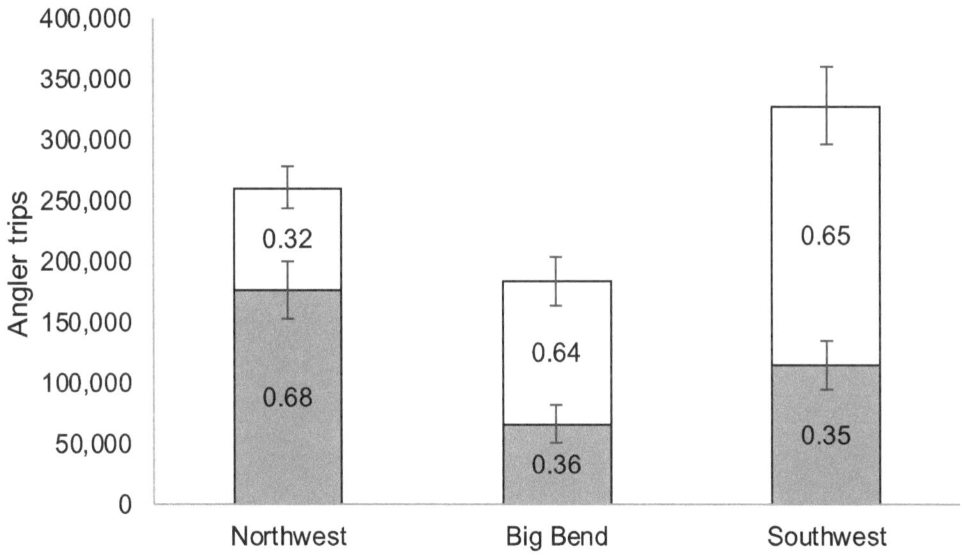

Figure 4. Estimated recreational fishing effort in each region over the 20-month study period where artificial reef habitat was reportedly used (shaded bars) or not (white bars). Relative proportions in each region are labeled.

Table 3. Months during the study period (May 2016 through December 2017, unshaded cells) when recreational harvest was open one or more days in all state waters (S), state waters adjacent to four northernmost counties in the Big Bend region (B), and the exclusive economic zone (E) for six frequently targeted reef fish species.

	Year	Jan	Feb	Mar	Apr	May	Jun	Jul	Aug	Sep	Oct	Nov	Dec
Red Snapper	2016					S	S,E	S		S	S		
	2017					S	S,E	S,E	S,E	S,E			
Gag	2016					B	S,E	S,E	S,E	S,E	S,E	S,E	S,E
	2017				B	B	S,E	S,E	S,E	S,E	S,E	S,E	S,E
Greater	2016					S,E			S	S			
Amberjack	2017	S,E	S,E	S,E	S								
Gray	2016					S,E							
Triggerfish	2017									S	S	S	
Red Grouper	2016					S,E	S,E	S,E	S,E	S,E	S,E	S,E	S,E
	2017	S,E	S,E	S,E	S,E	S,E	S,E	S,E	S,E	S,E	S,E	S,E	S,E
Vermilion	2016					S,E	S,E	S,E	S,E	S,E	S,E	S,E	S,E
Snapper	2017	S,E	S,E	S,E	S,E	S,E	S,E	S,E	S,E	S,E	S,E	S,E	S,E

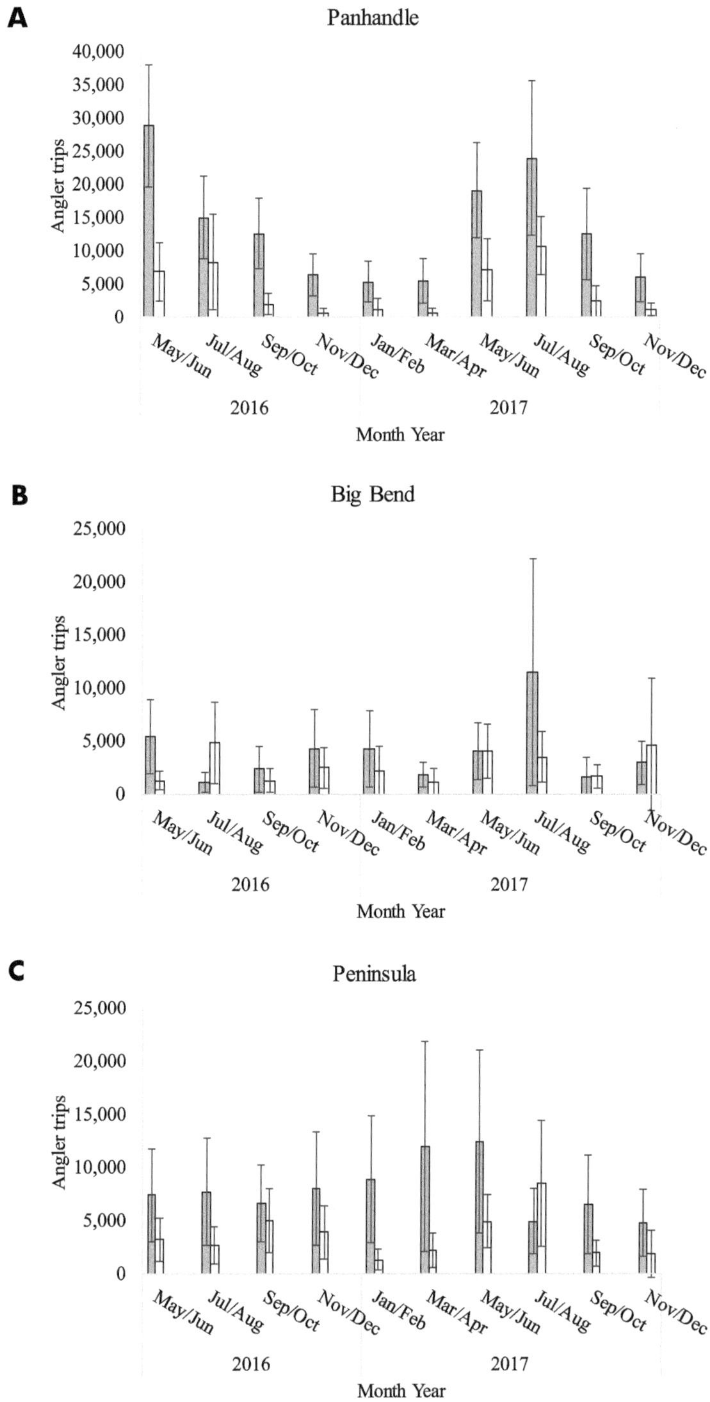

Figure 5. Bimonthly estimated fishing effort on artificial reefs in **(A)** the panhandle, **(B)** Big Bend, and **(C)** peninsula regions by area fished. State (shaded bars) = trips taken in inland or state territorial seas, and the exclusive economic zone (white bars) = trips taken in federal waters of the exclusive economic zone.

In 2016, recreational harvest of Red Snapper was permitted in federal waters during the first 11 d of June only but was more protracted during 2017 and encompassed the months of June, July, and August and the first 4 d of September. During July and August 2017, fishing effort on artificial reefs was elevated in the panhandle and Big Bend regions (Figure 5). In the peninsula region, where Red Snapper are not as abundant close to shore, fishing effort on artificial reefs also peaked in the exclusive economic zone during July and August 2017 (Figure 5).

Discussion

The results from this study indicate that a significant portion of recreational fishing effort for reef fishes along the Gulf Coast of Florida occurs at artificial reefs, particularly in the panhandle where Red Snapper is most abundant. The implications of this finding are many. Artificial reef programs have been implemented to enhance marine fisheries stocks as well as to increase recreational opportunities for anglers. Whether artificial reefs increase biomass in addition to attracting and aggregating reef-associated fishes is still debated, especially for Red Snapper in the Gulf of Mexico (Bohnsack 1989; Grossman et al. 1997; Pickering and Whitmarsh 1997; Dance et al. 2011). In their study at artificial reef sites in Florida's panhandle, Dance et al. (2011) found that Red Snapper was the dominant species across all reef types, though the majority (91%) were estimated to be below the legal size limit of 406 mm total length for recreational harvest. Karnauskas et al. (2017) also found that catch rates were 20 times higher on artificial reefs compared to natural reefs across similar depths, and age-1 and age-2 Red Snapper were the dominant age-classes. The absence of larger, older individuals across several important fishery species on artificial reefs may indicate that these habitats become less preferential with size or age or that fishing pressure is concentrated at artificial reef sites and fish are removed at higher rates (Dance et al. 2011). Thus, monitoring trends in recreational fishing effort among natural and artificial reefs is important for assessing changes in catch rates, discarding of undersized fish, and biomass of landed fish.

Potential shifts in recreational fishing effort toward more frequent usage of artificial reefs, which may also serve as attractants for some species such as Red Snapper, may have important implications for selectivity and catchability, both of which may profoundly influence the outcome of stock assessments (SEDAR 2009; Maunder et al. 2014). Selectivity is still poorly understood for reef fish stocks in the southeastern United States (Cowan 2011) and is often assumed to be constant over time in stock assessments. However, as artificial reefs have accumulated in the Gulf region, this may not be an accurate assumption. Catchability was also assumed constant until recently. In 2009, the Southeast Data, Assessment and Review (SEDAR) held a special workshop to address the influence of time-varying catchability in regional stock assessments (SEDAR 2009). Most of the discussion at the workshop focused on the influence of technological advances, such as electronic fish finders and GPS; however, the gradual accumulation of artificial reefs is an additional factor that should be considered when evaluating catchability back in time. The method we present here can be useful for monitoring future trends in recreational fishing effort with respect to artificial reefs.

A secondary component of Florida's Gulf Reef Fish Survey is a dockside intercept survey that collects information from private boat parties as they return from recreational fishing trips targeting reef fishes. These data could not be evaluated in time for this publication but will be used in the future to account for additional fishing trips taken by anglers that are not subscribed to the mail survey. Anglers interviewed in the dockside intercept survey are also asked to report whether an artificial reef was visited during the trip, and catch data collected through this survey may be used in the future to evaluate potential differences in species composition, size and age composition, and CPUE for landings and discards when anglers take advantage of artificial reefs.

Acknowledgments

The authors express appreciation to Lynne Stokes, Jean Opsomer, Jay Breidt, and Virginia Lesser for their statistical expertise during the initial design and development of the mail survey of Gulf reef fish anglers. Funding for the mail survey was provided by the National Fish and Wildlife Foundation's (NFWF) Gulf Environmental Benefit Fund.

References

Adams, C., B. Lindberg, and J. Stevely. 2006. The economic benefits associated with Florida's artificial reefs. University of Florida, Institute of Food and Agriculture Services, Florida Cooperative Extension Service, Food and Resource Economics Department, EDIS Document FE649, Gainesville.

Addis, D. T., W. F. Patterson, M. A. Dance, and G. W. Ingram. 2013. Implications of reef fish movement from unreported artificial reef sites in the northern Gulf of Mexico. Fisheries Research 147:349–358.

Bohnsack, J. A. 1989. Are high densities of fishes at artificial reefs the result of habitat limitation or behavioral preference? Bulletin of Marine Science 44:631–645.

Coleman, F. C., C. C. Koenig, and L. A. Collins. 1996. Reproductive styles of shallow-water groupers (Pisces: Serranidae) in the eastern Gulf of Mexico and the consequences of fishing spawning aggregations. Environmental biology of fishes 47:129–141.

Coleman, F. C., C. C. Koenig, K. M. Scanlon, S. Heppell, S. Heppell, and M. W. Miller. 2010. Benthic habitat modification through excavation by Red Grouper, *Epinephelus morio*, in the northeastern Gulf of Mexico. Open Fish Science Journal [online serial] 3:1–15.

Cowan, J. H. 2011. Red Snapper in the Gulf of Mexico and U.S. South Atlantic: data, doubt, and debate. Fisheries 36:319–331.

Dance, M. A., W. F. Patterson, III, and D. T. Addis. 2011. Fish community and trophic structure at artificial reef sites in the northeastern Gulf of Mexico. Bulletin of Marine Science 87:301–324.

Grossman, G. D., G. P. Jones, and W. J. Seaman, Jr. 1997. Do artificial reefs increase regional fish production? A review of existing data. Fisheries 22(4):17–23.

Huth, W., O. A. Morgan, and C. Burkart. 2014. Measuring Florida artificial reef economic benefits: a synthesis. Final Report to the Florida Fish and Wildlife Conservation Commission, Tallahassee.

Karnauskas, M., J. F. Walter, M. D. Campbell, A. G. Pollack, J. M. Drymon, and S. Powers. 2017. Red Snapper distribution on natural habitats and artificial structures in the northern Gulf of Mexico. Marine and Coastal Fisheries 9:50–67.

Kish, L. 1965. Survey sampling. J. Wiley, New York.

Koenig, C. C., F. C. Coleman, C. B. Grimes, G. R. Fitzhugh, K. M. Scanlon, C. T. Gledhill, and M. Grace. 2000. Protection of fish spawning habitat for the conservation of warm-temperate reef-fish fisheries of shelf-edge reefs of Florida. Bulletin of Marine Science 66:593–616.

Lindberg, W. J., T. K. Frazer, K. M. Portier, F. Vose, J. Loftin, D. J. Murie, D. M. Mason, B. Nagy, and M. K. Hart. 2006. Density-dependent habitat selection and performance by a large mobile reef fish. Ecological Applications 16:731–746.

Maunder, M. N., P. R. Crone, J. L. Valero, and B. X. Semmens. 2014. Selectivity: theory, estimation, and application in fishery stock assessment models. Fisheries Research 158:1–4.

Pickering, H., and D. Whitmarsh. 1997. Artificial reefs and fisheries exploitation: a review of the 'attraction versus production' debate, the influence of design and its significance for policy. Fisheries Research 31:39–59.

Powers, S. P., J. H. Grabowski, C. H. Peterson, and W. J. Lindberg. 2003. Estimating enhancement of fish production by offshore artificial reefs: uncertainty exhibited by divergent scenarios. Marine Ecology Progress Series 264:265–278.

Saul, S. E., J. F. Walter, D. J. Die, D. F. Naar, and B. T. Donahue. 2013. Modeling the spatial distribution of commercially important reef fishes on the west Florida shelf. Fisheries Research 143:12–20.

Seaman, W., Jr. 1982. Enhancement of Florida marine fisheries using artificial reefs: a review. Proceedings of the Annual Conference Southeastern Association of Fish and Wildlife Agencies 36:312–322.

SEDAR (Southeast Data, Assessment and Review). 2009. Stock assessment of Red Snapper in the Gulf of Mexico: SEDAR update assessment. SEDAR, North Charleston, South Carolina.

Simard, P., K. R. Wall, D. A. Mann, C. C. Wall, and C. D. Stallings. 2016. Quantification of boat visitation rates at artificial and natural reefs in the eastern Gulf of Mexico using acoustic recorders. PLOS (Public Library of Science) ONE [online serial] 11(8):e0160695.

Strelcheck, A. J., J. H. Cowan, and W. F. Patterson. 2007. Site fidelity, movement, and growth of Red Snapper: implications for artificial reef management. Pages 147–162 *in* W. F. Patterson, III, J. H. Cowan, Jr., G. R. Fitzhugh, and D. L. Nieland, editors. Red Snapper ecology and fisheries in the U.S. Gulf of Mexico. American Fisheries Society, Symposium 60, Bethesda, Maryland.

Switzer, T. S., T. C. MacDonald, R. H. McMichael, and S. F. Keenan. 2012. Recruitment of juvenile Gags in the eastern Gulf of Mexico and factors contributing to observed spatial and temporal patterns of estuarine occupancy. Transactions of the American Fisheries Society 141:707–719.

Tinsman, J. C., and W. H. Whitmore. 2006. Aerial flight methodology to estimate and monitor trends in fishing effort on Delaware artificial reef sites. Bulletin of Marine Science 78:167–176.

Wall, C., B. Donahue, D. Naar, and D. Mann. 2011. Spatial and temporal variability of Red Grouper holes within Steamboat Lumps Marine Reserve, Gulf of Mexico. Marine Ecology Progress Series 431:243–254.

American Fisheries Society Symposium 86:279–288, 2018

Evaluating a *Teranas* Natural Reef off Malaysia Using a Multibeam Echosounder: A Method to Support Artificial Reefs

RAZAK ZAKARIYA* AND LENNY SHARINEE SAKAI
School of Marine Environmental Sciences, Universiti Malaysia Terengganu Kuala Terengganu, Terengganu 21030, Malaysia

Abstract.—In recent years, as reported by the Malaysian Department of Fisheries, the decline in demersal fish resources is due to excessive exploitation. This indicates that management and protection of these resources and their habitat are required. *Teranas* is identified as important critical marine habitat and locally in Malay is defined as a hard rock substrate on the seabed. Teranas can appear in all shapes and sizes, depths, and locations. The combination of these various bottom features often make a teranas attractive to fish. The objective of this study is to detect a teranas so that it can be located and managed through mapping. Sonar in the form of a multibeam echosounder was used to map a teranas with a combination of bathymetry and water column features to identify the signature for teranas. The result produced a signature library to readily identify a teranas with the multibeam echosounder. An important factor in determining the signature for the teranas was the seabed structure and its associated marine life. Bathymetry identified the seabed structure, and the water column characteristics helped verify the presence of marine life. Fishing and visual surveys with scuba diving were also conducted to verify the effectiveness of the multibeam echosounder to identify the teranas. These study results can be used as an aid for planning and managing a teranas to help sustain the associated fisheries resources. Moreover, since a teranas is similar in structure and position to an artificial reef, the multibeam echosounder, when coupled with visual diver-based surveys and fishing effort, could be used to identify the extent and marine fauna associated with an artificial reef. This research will help in providing strategic location selection for artificial reef placement.

Introduction

Sandbars or shoals are natural bottom formations consisting of sand or other unconsolidated material and rise from the seabed, thus increasing the bottom profile. Offshore sandbars are formed from sedimentary deposits, typically dominated by sand or gravel (Finkl and Hobbs 2009) with bathymetric relief of a meter or greater, that provide potentially important habitat for fisheries re-

* Corresponding author: ajak@umt.edu.my

sources. Offshore sandbars with marine life are fishing humps known as *teranas*. Teranas occur locally in Malay and are characterized as sandbars with a hard substrate core on the seafloor, as defined in the Dewan Bahasa and Pustaka Dictionary. Teranas can occur in various shapes and sizes, depths, and locations. In fact, a combination of these bottom features makes them attractive to fish. The submerged sandbar serves as important habitat for schooling marine fishery resources and as a resting area and ground feeding area. Larger fish predators also frequent these areas searching for prey. Once the teranas is established, these relatively small bottom features serve as habitat that supports multiple trophic levels through colonization and ecological succession.

Thus, a teranas is a part of the marine habitat that plays an important role in the marine environment and the natural ecosystem. A disturbance to this natural system will upset the marine environment and the processes on which marine organisms depend. A loss of marine habitat will impact the food web and, concomitantly, lead to a severely altered natural system. Impacts to the ecosystem will eventually impact the humans that rely on these fish resources.

Fish resources are one of the major sources of protein in Southeast Asia. Although they are large, these fisheries resources are not inexhaustible and, therefore, require management and protection. Consequently, managing these fish resources is vital in providing resources to the human population. Reduced fishery resources as a result of a reduction in fish habitat will eventually affect the lives of humans. Similarly, a reduction in fishery resources will also affect fishermen, especially in Southeast Asia, who depend on the fishing industry for income. Ultimately, the entire region may no longer be sustainable for humans if regional fisheries are reduced. Consequently, identifying and protecting the teranas habitat is important for the future of fisheries in Asia. Moreover, since an artificial reef mimics a teranas, applying the methods used to identify a teranas will help fishery managers assess the location and relative productivity of artificial reefs deployed in Asia as well.

Methodology

The multibeam echosounder is a relatively new technology presently used in many marine associated industries for many applications, but especially for mapping the seabed (Farr 1980). The prime reason for using a multibeam echosounder is to obtain bathymetric and water column data. Here, bathymetric information can help identify the structure of the seabed to detect a teranas, and in concert with the water column data, we can verify the quality and abundance of marine life and fish density associated with the teranas. With new developments, multibeam echosounder now has the capability to record sonar returns (i.e., bottom images) throughout the entire ping for each beam, thus recording the water column data along with the seabed bathymetry. The ability to gather information from the water column, between the ocean surface and the seabed, expands the use of a multibeam echosounder to include the direct mapping of a structure, structure-associated fish and marine mammals, and organisms in the water column.

Figure 1 shows the methodology of how the research was conducted. A multibeam echosounder was used here for primary data collection as both bathymetric and water column information was directly collected and indirectly derived. The multibeam echo-

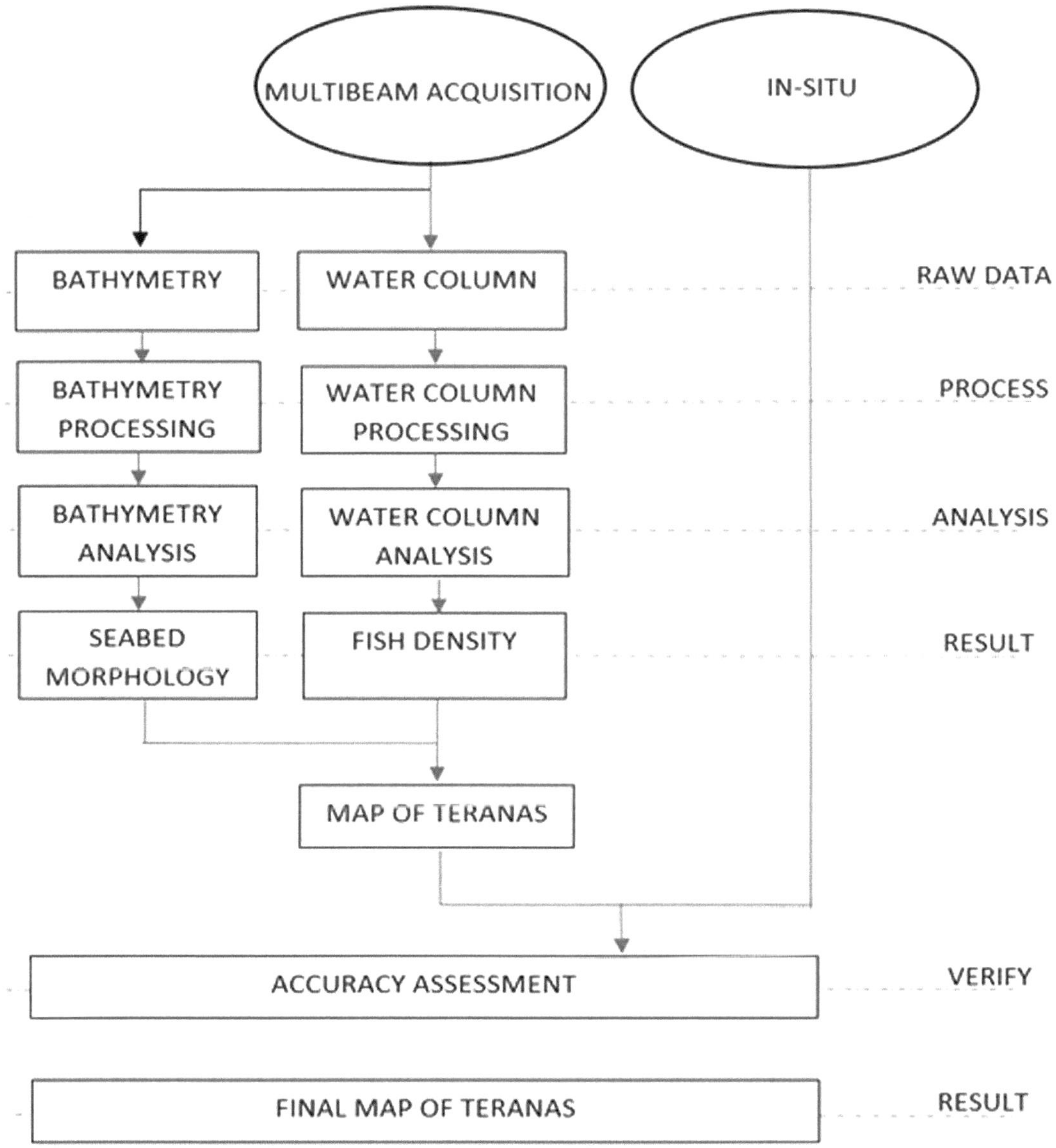

Figure 1. The method used for detecting teranas with the multibeam echosounder.

sounder used here was a R2Sonic 2020 with 2 × 2 degree beam angle resolution.

The survey plan for the region where the suspected teranas implemented a line (linear) survey with the multibeam echosounder and the line was surveyed in both directions for a total of two surveys per line. After the survey of the teranas area, we conducted a sonar survey above the teranas with several passes to gather data from the water column.

In situ ground truth data were collected for comparison and to verify the multibeam echosounder data. The in situ survey was conducted by divers using scuba and with underwater photography. Additionally, fishing was conducted in the area to verify the multibeam echo sonar data.

The multibeam echosounder data were collected in concert with QINSy navigation software (Quality Positioning Services BV [QPS], www.qps.nl) and processed with Qimera (QPS) and Fledermaus (QPS). Bathymetry data processing was done with the Qimera software package. A correction was applied to the bathymetry data to correct for tidal changes in depth due to the time differences of the surveys. Predicted tide from the nearest station, Sungai Setiu (Terengganu, Malaysia), was used. Outliers of multibeam echo-sounding data caused by errors during the data collection were removed from the analyses. Usable data were subsequently analyzed with Fledermaus to determine the bottom profile of the teranas.

The multibeam echosounder data from the water column were processed with FM Midwater (Fledermaus). The combination of the raw data and processed data were entered into the database and then converted to water column data format (i.e., .wcd file format) to enable the software to read and identify water column features from the sonar data. These features were extracted from the data and exported to point cloud data for further analysis with the main Fledermaus. A flow diagram of the data gathering, processing, and analytical procedures is depicted in Figure 1.

Results and Discussion

Seabed Morphology

A multibeam echosounder survey was conducted in the target area with a previously identified teranas. The bottom bathymetry was gridded to 0.25-m cell size to a surface. The depth varied between 17 and 18.1 m (Figure 2). Some of the features of the teranas varied from 20 cm to 1 m and were scattered on the seabed.

From the analysis and observation, the multibeam echosounder data indicated that the area is almost flat with a slight rise in the center of the teranas site in comparison with the substrate surrounding the teranas. Based on the profile of the seabed, the rise is about 0.3 m from the surrounding area. The depth surrounding the hump is –18.1 m and the highest point of the teranas is –17.07 m deep (Figure 2).

Rugosity is a measure of small-scale variations in the height or profile of a structure. The rugosity of the teranas was conducted on the profile generated by the bathymetry data. The surface was classified into three categories (i.e., course, fine, and very fine; Figure 3) based on the relative scale of rugosity. Course (highest) rugosity was found in one area. The primary feature there is higher than 1 m and is a large boulder or rock. The area surrounding this peak is mainly flat and smooth with a lower rugosity. Fine (intermediate) rugosity, when the large boulder or rock was ignored, was a more frequent rugosity than course rugosity and was in a region with small boulders. Fine rugosity only occurred on the hump or central part of the teranas. The surrounding seabed had low rugosity, and the seabed where we found fine rugosity was smoother than the central part of the teranas. Very fine (lowest) rugosity analysis indicated that the teranas surface was more rugose than the surrounding area. This indicated that the teranas has a different seabed structure than the region surrounding the teranas. It also shows that the teranas surface is rougher than the surrounding area, which has lower rugosity. This analysis helped define the boundary or limits of the teranas.

Surface slope analysis (Figure 4) indicated that the teranas identified area was less

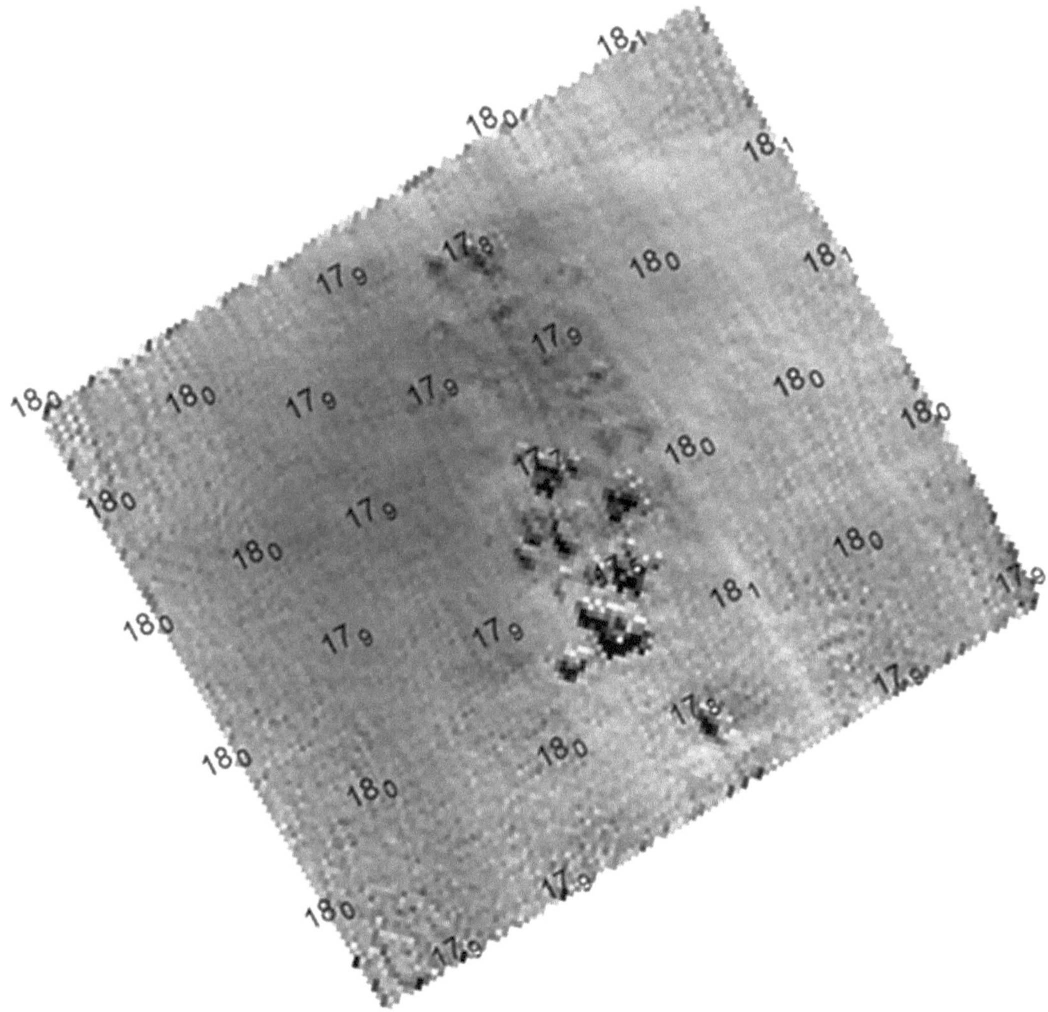

Figure 2. The bathymetry surface gridded to 0.25 m with sounding depth values indicating the water depth on the grid. Numbers on the grid indicate depth (m).

regular than the surrounding area. The presence of the boulders were responsible for a steep slope at the site. The large boulder provided the steepest slope, indicating that the boulder was sizable and prominent. The smaller boulders, with its smaller features, offered a lower slope.

It can be concluded from the bathymetry analysis that the teranas has a different feature and structure from the surrounding and that it can be readily identified.

Multibeam Water Column

Based on the analysis of water column data obtained from the multibeam echosounder, a fish school was detected above the identified teranas area. The fish school was not only detected above the teranas, but over the surrounding area as well. To confirm this finding, survey run 3 (Figure 5) was conducted at a different time in the water column. As a result, a fish school was detected at the same location for all three runs, indicating that the

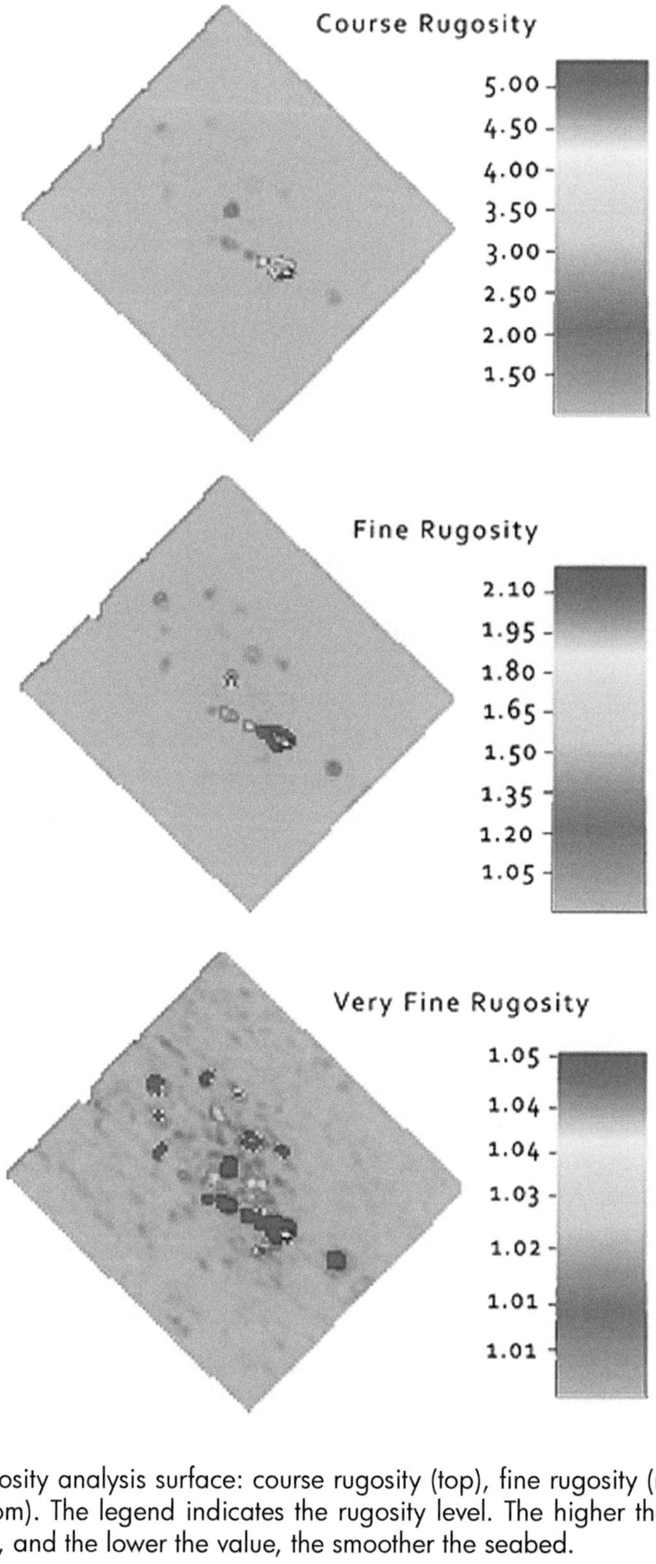

Figure 3. Rugosity analysis surface: course rugosity (top), fine rugosity (middle), and very fine rugosity (bottom). The legend indicates the rugosity level. The higher the value, the more rugose the seabed, and the lower the value, the smoother the seabed.

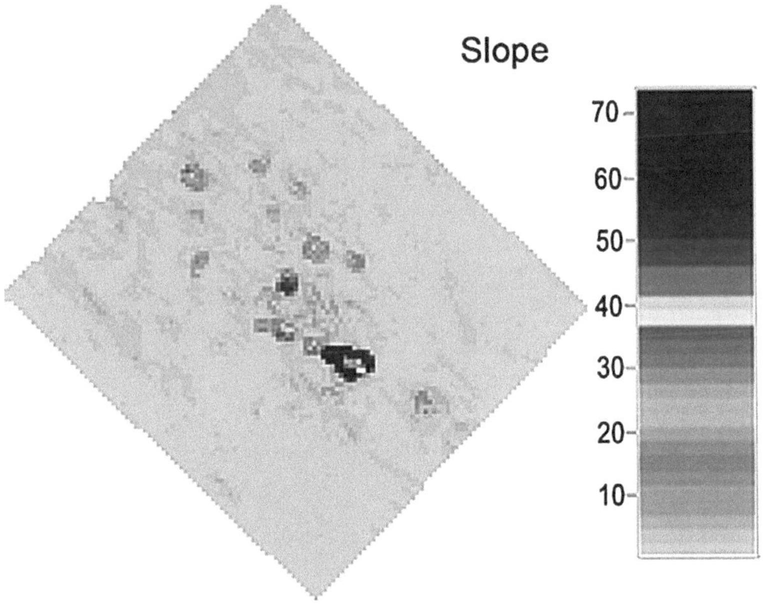

Figure 4. Slope analysis surface. The legend indicates the degree of slope. Light shading represents less-steep slope and the darker shading indicates steeper slopes.

fish school remained above the teranas. The fish school surrounded the top of teranas and occupied about 74.9 m^2 of the teranas surface. Based on the analysis of a single run, the school covered an area of 24 × 13 m in a dense group. The center of the fish school was up to 2.4 m off the seabed.

Based on the combined analysis of the bathymetry and the water column from the multibeam echosounder, the teranas boundary can be defined. The rugosity analysis differentiated the seabed material; thus, the terrain has influenced the seabed material. Moreover, the multibeam echosounder analysis for the water column indicated that fish remained above the teranas. The additional backscatter from the multibeam can help identify the seabed sediment type, and this will be included in the next sampling method.

From this study, the particular area of the teranas was determined based on the location of the fish school and the seabed structure and features. The result here is that the location of the teranas, as well as its shape and boundary, was determined (Figure 6). Using information from the multibeam echosounder for the water column, we noted that fish coverage of the teranas was 42% (run 1), 24.4% (run 2) and 24.4% (run 3). The total area covered by fish was 80.84% of the teranas.

Underwater Photos and Observation

Divers (using scuba) surveyed the area to verify the data gathered using the multibeam echosounder. Divers noted that the fish school indeed occurred only above the teranas (Figure 7). Measurements were taken to verify the width of the area based on the fishes' clustering spot. These fishes were mostly demersal, but pelagic fishes were also observed. During the underwater surveys by divers, fishers were seen circling above the teranas. Some fish were observed feeding on prey items on the seabed bottom and on the

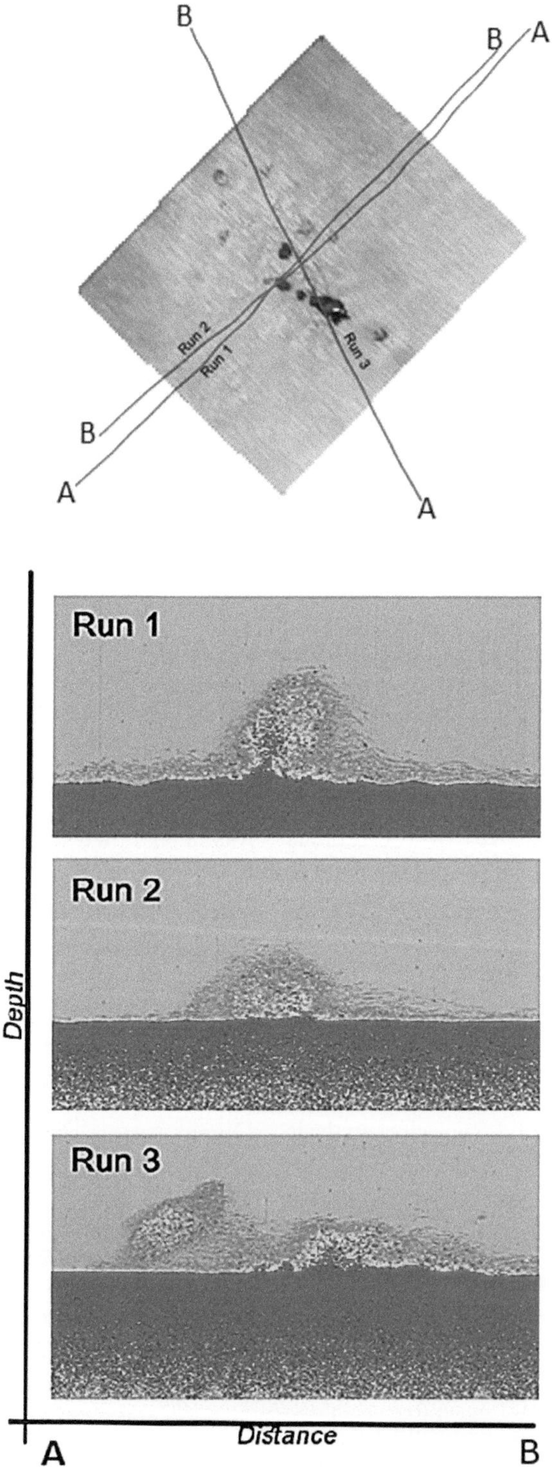

Figure 5. (Top) Cross-section profile of a water column with a fish school (dark cluster). (Bottom) Seabed with the fish school.

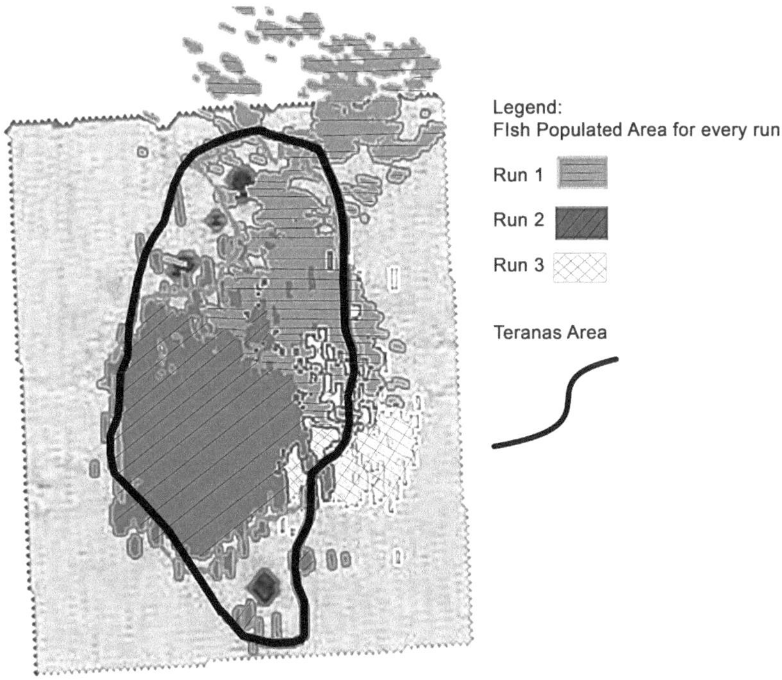

Figure 6. Teranas area determined by bottom morphology (boundary line). Run 1, run 2, and run 3 indicate the fish school detected in the water column by the multibeam echo sounder.

boulders. The divers verified that the boulders, as detected by the multibeam echosounder, had been identified correctly. The seabed of the teranas, as indicated via the multibeam echo sounder, was also correctly identified as rugose and more irregular than the surrounding substrate. On this teranas, the different nature of seabed surface was observed.

Conclusions

This study demonstrates the value of the multibeam echosounder to detect teranas areas. The multibeam echosounder provided information on the morphology of the teranas, which significantly improved our understanding of the teranas in term its characteristics and extent. When using the multibeam echosounder, it is important to compare the results with another survey method. It is especially important to compare survey results with satellite and aerial remote-sensing information. Concomitantly, diver surveys and fish sampling provide comparative information, but these may be limited by high turbidity and strong currents. Multibeam echosounder is fast and cost-effective and is especially useful when evaluating large areas for coverage, mapping, and fish density.

This study confirms how important the teranas is as habitat for fish. Thus, it is important to identify teranas and protect them, especially from bottom trawlers. With the multibeam echosounder technology, teranas can be identified, mapped, and protected in a cost-effective manner. Artificial reef researchers would do well to adopt the regular use of the multibeam echosounder in their efforts to document the extent and relative productivity of existing artificial reefs and artificial habitats.

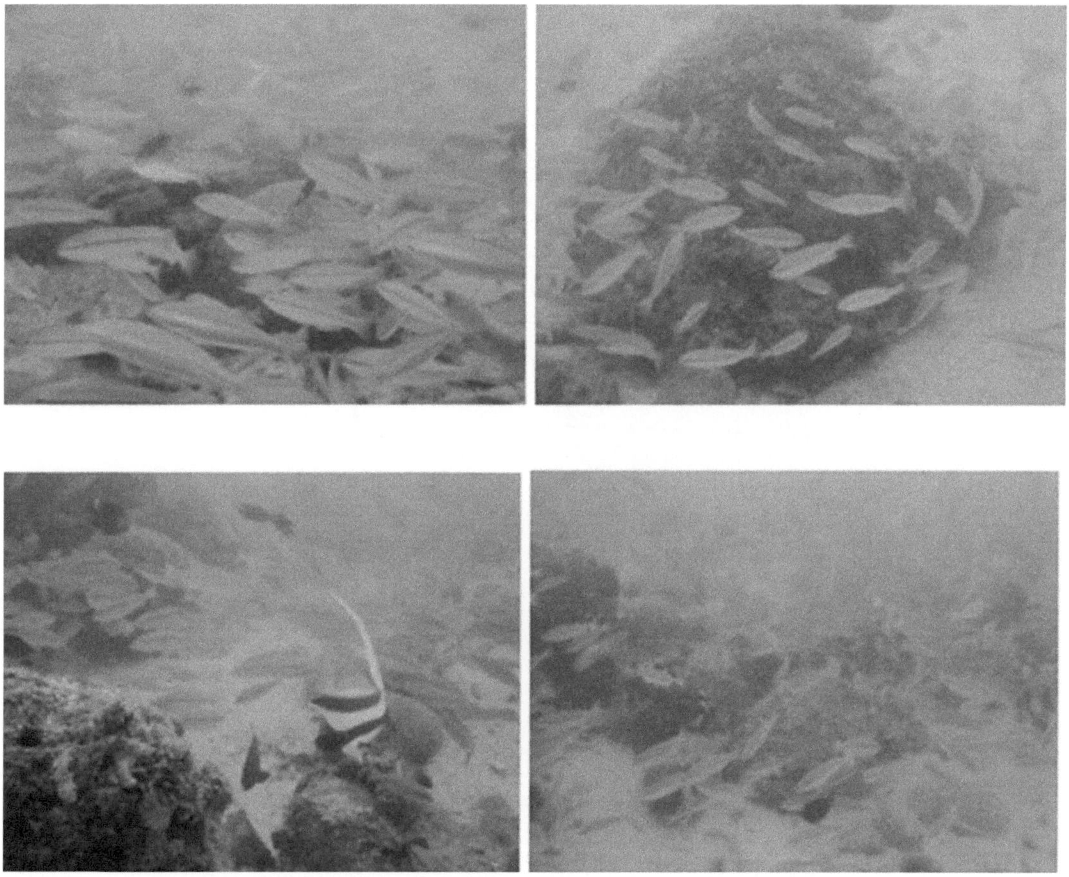

Figure 7. In situ photos taken by scuba divers at the teranas location. A school of fish was present, as indicated by the water column multibeam echosounder.

References

Farr, H. K. 1980. Multi-beam bathymetric sonar: sea bream and hydro chart. Marine Geodesy 4:77–93.

Finkl, C. W., and C. H. Hobbs, III. 2009. Mining sand on the continental shelf of the Atlantic and Gulf coasts of the U.S. Marine Georesources and Geotechnology 27:230–253.

American Fisheries Society Symposium 86:289–313, 2018

Research and Development on the Design and Construction of Artificial Reefs in Malaysia (1975–2017)

Ahmad Ali*
Southeast Asian Fisheries Development Center
Marine Fishery Resources Development and Management Department
Chendering, Terengganu 21080, Malaysia

Nur Iskandar Tajudin and Rafezi Hazizi
Department of Fisheries Malaysia
Aras 1-6, Blok Menara 4G2, Putrajaya 62628, Malaysia

Abstract.—Research and development (R&D) on the design and construction of artificial reefs in Malaysia was initiated by researchers from the Fisheries Research Institute, Department of Fisheries Malaysia (DoFM) based in Penang in 1975. Initially, R&D only focused on simple designs and construction using discarded tires, wooden fishing boats, polyvinyl chloride (PVC) pipe, reinforced concrete, and ceramic sewerage pipe. These reefs were used in fisheries management to maximize resource enhancement, resource conservation, and habitat rehabilitation and mitigation, as well as being an aid to alleviate the problem of depleting fish resources in coastal waters. However, in 2006, the DoFM made a major policy change in the design and construction materials of artificial reefs in the country. The focus was to construct large, heavy, and robust artificial reefs using durable materials such as reinforced concrete and steel to deter the encroachment of destructive fishing gears, especially trawlers, along the coastal and in marine protected areas. These new artificial reefs addressed several issues, such as fish behavior, target species, physical oceanography, and marine engineering, in their design. The reinforced concrete artificial reefs were built according to the British Standard 8110 under the supervision of officers from the Engineering Division and Research Division of the DoFM. Presently, the DoFM has designed 21 types of artificial reefs that have been deployed at 130 selected locations in Malaysia. This paper reviews the progress of R&D on the design and construction of artificial reefs using tires, wooden fishing boats, reinforced concrete, PVC pipe, ceramic materials, and steel in Malaysia between 1975 and 2017.

* Corresponding author: aaseafdec@seafdec.org.my

Introduction

Artificial reef development is part of a government program in Malaysia that contributes to the increase in fisheries resources, prevents encroachment of trawlers, reduces conflict between commercial and traditional fishers, and increases opportunities for small-scale fishers to improve their income from fishing (Jothy 1982). According to Wong (1991), artificial reef deployment was practiced by traditional fishermen along the east coast of Peninsular Malaysia since the early 1900s using derelict wooden boats and other wood-based materials. These reefs provided only short-term benefits since the wood was attacked and destroyed by Teredo worms (phylum Mollusca: family Teredinidae) as well as by other natural forces including strong currents and wave turbulence.

According to Ali (2015), the first government-sponsored artificial reefs in Malaysia were deployed in the 1970s by the Royal Australian Air Force, based at Butterworth, in Penang, in conjunction with the Kedah Fisheries Office. These reefs were composed of discarded tires, and the reef was deployed near Songsong Island, in Kedah, to enhance recreational activities such as diving and fishing. The Australian government supported this deployment during the 1970s as it was actively involved in the construction of artificial reefs to annually dispose of millions discarded tires (Sanders 1974).

Research and development of artificial reefs in Malaysia began in 1975 based on the research findings that coastal fisheries resources in peninsular Malaysia were in a continuous decline (Jothy 1982; Wong 1991; S. A. L. Mohammed Shaari, paper presented at the Malaysian Society of Marine Sciences seminar on management of marine resources, 1985). In 1965, trawling surveys conducted by the Department of Fisheries Malaysia (DoFM) using RV *Selayang* in the depth range of 10–50 m yielded an average catch rate of 167 kg/h of commercial and trash fish (Pathansali et al. 1967). Five years later, in 1970, a second fisheries survey recorded that the average catch rate was reduced to 131 kg/h (Mohammed Shaari et al. 1974). A drastic decline in landings was recorded in 1974 when only 92 kg/h was recorded (Mohammed Shaari and Leong 1976). The decrease in average catch rate from 167 kg/h in 1965 to only 92 kg/h in 1974 (a decline of about 55%) was an indication of a serious depletion in coastal fisheries resources. Construction and deployment of artificial reefs was one of the management measures taken by the DoFM. The deployment of artificial reefs also would serve as a hindrance to trawlers in traditional fishing areas. The result of the artificial reef deployments was a reduction in conflict between traditional fishers and trawlers (Jothy 1982).

Methods

This paper is based on the knowledge and experiences of the authors through their studies in Malaysia since 1992. It is also based on the accumulated information from published literature that includes journals, books, conference proceedings, unpublished technical papers, and technical reports on the research and development (R&D) of the design and construction artificial reefs in Malaysia implemented by the DoFM from 1975 to 2017 using tires, wooden boats, reinforced concrete, ceramic sewerage pipe, polyvinyl chloride (PVC) pipe, and steel. The chronology of these artificial reef efforts, along with an indication of materials used, first deployment sites, and sources of initial information, is shown in Table 1.

Table 1. Chronology of artificial reef construction in Malaysia by the Department of Fisheries Malaysia, 1975–2017.

Duration	Material	First deployment site and state	Source of initial information
1975–1995	Tires	Payar Island, Kedah	Wong (1991)
1984–2012	Derelict and confiscated fishing boats	Kapas Island, Terengganu	Jothy (1986)
1986–2017	Reinforced concrete	Payar Island, Kedah	Latun and Mohamed Pauzi (1991)
1990–1992	Polyvinyl chloride pipe	Payar Island, Kedah	Latun and Ismail (1994)
1992–1993	Ceramic pipe	Redang Island, Terengganu	Sukarno et al. (1994)
2014	Steel	Jerlun, Kedah	Nor Azman and Mohd Saki (2017)

Design and Construction of Artificial Reefs Using Tires, Wooden Boats, Reinforced Concrete, Ceramic Pipe, Polyvinyl Chloride Pipe, and Steel

Tires (1975–1995).—In 1975, the R&D effort was inclined to try new technology with regard to the design and construction of artificial reefs based on previous efforts conducted in other countries, especially efforts in the United States and Australia. Initially, only tires were used in artificial reef construction in Malaysia because of indications in the published literature of the advantages and successes of tires over other materials. Having only limited experience and prior knowledge, all artificial reef activities were conducted on a trial-and-error basis. None of the researchers from DoFM had any formal training regarding the design and construction of artificial reefs, and they had no training in marine engineering as well. At that time, available information and knowledge, especially with regard to marine engineering and oceanography, were limited. Prior to this effort, the government had neither introduced nor implemented any policies on artificial reef development and management and had not provided guidelines for material selection, designs, and construction (Wong 1991).

The first paper describing artificial reefs in Malaysia, as well as the design and construction of tire artificial reefs, was written in 1978 by Alexander A. Jothy, a research officer attached at the Fisheries Research Institute (FRI) in Penang (Jothy 1982). Based on references listed in that paper, the original idea and knowledge base for the construction of artificial reefs came chiefly from previous efforts in Australia, the United States, and Japan. He referred to papers written by Ahr (1974) on site selection, Carlisle et al. (1964) on ecology, Stone (1972) on using waste materials for artificial reefs and fishery management, Sanders (1974) on the development and progress of artificial reefs in Australia, Ogawa (1973) on the biological aspects of artificial reefs, and Ino (1974) on the history of artificial reefs in Japan. The idea of using tires to construct artificial reefs in Malaysia was based on Sanders (1974). He described the success when using tires as artificial reefs in Australia and highlighted that the tires could provide an excellent substrate for the attachment of encrusting organisms.

He also mentioned that the use of discarded tires for artificial reefs could also reduce the disposal problem of the more than 2 million discarded tires available annually in Australia. According to Jothy (1982), Malaysia also faced a similar problem with discarded tires that could be relieved by using tires for artificial reefs.

Jothy (1982) indicated situations elsewhere where it had been shown that tires were good artificial reef materials. The arguments were that tires are readily available, are relatively inexpensive to assemble into modules, and are handled with ease by scuba divers during their underwater deployment. However, similar programs in other countries experienced problems when using discarded tires for artificial reefs. Nevertheless, the absence of hurricanes and typhoons in Malaysia permitted the use of unballasted tires without any major problems of reef instability or loss of the structural integrity of the tire modules (Wong 1991). Besides discarded tires, rejected (factory imperfect) tires were also provided gratis by Goodyear, the largest tire and rubber company in the world at that time. Wong (1991) reported that the Goodyear Company in Malaysia also provided hundreds of thousands of tires for artificial reefs projects along the west coast of Peninsular Malaysia.

Early on, a tire module consisted of three to five tires (Jothy 1982; Wong 1991; Sukarno et al. 1994). To assist in the deployment, two circular holes (6–8 cm diameter) or two rectangular openings (75 cm^2) were made in the periphery (tread wall) of the tires to enable entrapped air to escape. The tires were then lashed together with polyethylene ropes to form a small tetrahedral modules shape (Figure 1). The original idea for this design was not specified, but the design was similar to that used in Virginia, USA (Meier et al. 1986). The tire units could be assembled in various forms (cylindrical, triangular, pyramidal, tetrahedral, chain-links, etc.; Jothy 1982).

Figure 1. Tire artificial reefs showing an early 1975 design module consisting of three to four or four to five tires.

Each module was then tied together by divers with a 1.6-cm-diameter polypropylene strap. About 100–150 modules consisting of a total of 300–500 tires were transported to the selected site by fishing boats or FRI's research vessel (Wong 1991).

According to Jothy (1982), the modules were lowered to the seabed by an anchor rope. The sunken modules were then arranged on the substratum and lashed together by divers, thereby increasing the stability and mass of the modules as a set (sensu Grove and Sonu 1983). Wong (1991) indicated that the modules were placed on the seabed by sliding them down a weighted rope tied to an existing module to prevent the modules from separating. The Fisheries Research Institute and private divers then tied these modules together with other adjoining modules to limit the possibility of the modules becoming scattered. Similar deployment protocols were also used in the United States and Australia, as described by Sanders (1974) and Meier et al. (1986). As a result, the modules had irregular height profiles, but none exceeded 1 m. Coordinates (latitude and longitude) of the reef sites were recorded and coded according to type of materials used and deployment date.

The first deployment activity, led by FRI researchers, was at two locations in the northern part of the Straits of Malacca near the Telur and Payar islands in the state of Kedah on May 6 and October 11, respectively, in 1975. Basic site-selection surveys to collect sediment samples were conducted by FRI researchers. A total of 6,329 discarded tires were deployed at Telur Island, and 5,229 tires at Payar Island (Jothy 1986).

Initially, the plan was to deploy between 5,000 and 6,000 tires at each site, and depending upon the degree of success of the reefs, these sites could be expanded further to include additional tires. Each site was expected to cover an area of 557–929 m^2 of the seabed, with a height not exceeding 1.2 m (Jothy 1982). In July 1976, a total of 10,193 tires were deployed near Aman Island (a small island near Penang Island) to become the third artificial reef site in Malaysia. The project received positive response from local fishermen in terms of increasing their daily catch rate and the catch of preferred fish species such as high-quality grouper (Jothy 1986). The purpose of the tire reefs deployed at Telur and Payar islands was for research and resource enhancement (conservation) while the artificial reefs near Aman Island were deployed to aggregate fish to increase fishing efficiency. The progression of fish assemblage development (i.e., colonization and succession) of the artificial reef projects near Telur, Payar, and Aman islands were filmed by the FRI's divers. Schools of juvenile snappers and groupers and the dense growth of encrusting, fouling, and sessile organism on the reefs were recorded. This information encouraged the DoFM to deploy artificial reefs at other sites in other Malaysian states.

In 1978, the project to deploy artificial reefs was extended to the states of Perak and Johor, situated in the middle and southern parts, respectively, of the Straits of Malacca. The first tire reef for the east coast of peninsular Malaysia was deployed in 1979 near Ekor Tebu Island in the state of Terengganu (Wong 1991). The artificial reefs program was expanded to Sabah and Sarawak (in Borneo) during the Fourth Malaysia Plan (1981–1985). The first tire artificial reef project in Sabah and Sarawak began in 1982 using 2,834 tires and 2,096 tires, respectively (Jothy 1986).

In 1984, the Minister of Agriculture launched a national program to collect 1.4

million tires for artificial reefs. The objective was to deploy 28,000 tires at every artificial reefs site in Malaysia (Jothy 1986). The program was scheduled to utilize 2 million tires to establish 65 large-scale tire reef sites by the end of 1990. Each site was to be comprised of 50,000 tires; thus, 1.5 million tires would be needed by the end of 1990. The construction of large-scale artificial reefs, with up to 50,000 tires per site, was one of the strategies recommended as part of the Fisheries Act 1985 (Act 317). These reefs were deployed in concert with other management measures by DoFM policymakers to promote the recovery of coastal fisheries resources in coastal areas. Under this accelerated program of artificial reef development, in 1986, the number of tires used for each module was increased from 3 to 4 tires per module to 42 tires per module. The tires were lashed into a pyramid shape using 42 tires with a resultant height of about 2 m

Figure 2. Tire artificial reefs in 1986 showing tires lashed into a pyramid using 42 tires with a resultant height of about 2 m.

(Figure 2). The sea floor occupied by a single pyramid was about 3 m^2. By the end of 1987, the number of tire-reef sites increased to 76. The highest number of tire deployed at one site was 63,523 tires (Wong 1991). By the end of 1995, the number of artificial reef sites that employed tires remained maintained at 76 and the total number of tires deployed was 3,145,856. Peninsular Malaysia had the highest number of sites with 58, followed by 10 sites at Sarawak and 8 at Sabah.

In 1996, the use of tires in artificial reef deployments significantly dropped and, subsequently, was totally forbidden by the DoFM due to the claim by many involved that automobile tires leached toxins into the marine environment (Ali et al. 2013b). In large storms, tire arrays moved considerable distances and eventually broke apart. The individual tires then became undersea hazards, impacting benthic habitats for many years (Figure 3). Additionally, in terms of providing substrate for colonizing marine life, tires were the least satisfactory material for artificial reefs as they often became covered by filamentous algae. The tire surface was far less attractive to marine life than materials such as concrete or rock, whose properties more closely resemble the materials that form most natural reefs (Alevizon 1990).

Several acoustic surveys were conducted by the DoFM in collaboration with the Universiti Malaysia Terengganu (University of Malaysia, Terengganu) using side-scan sonar and a sub-bottom profiler as well as visual observations by divers to verify reef coordi-

Figure 3. Loose tires strewn on seafloor and buried by scouring after the constraining rope broke.

nates and map the location of tire artificial reefs. Ali et al. (2008) reported that all tire reefs deployed on the sea floor in the coastal waters of Terengganu along the east coast of peninsular Malaysia were destroyed by strong currents. Most of the tires comprising these reefs had been submerged on the sea bottom. The ropes, used to lash the tires together, broke, and individual tires then became buried into the seafloor by scouring. Some artificial reefs composed of tires at sites located between the islands are still in good condition but have only become colonized with a few marine organisms and have attracted assemblages of small fish. Similar surveys conducted by Ali et al. (2013a), Azizi et al. (2014), and Azizi et al. (2016) along the west coast of peninsular Malaysia indicated that all tire reefs deployed on soft-mud seabeds sank into the substrate. These reefs did not function as hard-substrate artificial reefs for a considerable period of time, but functioned only as short-term fish aggregating devices.

Derelict and confiscated fishing boats (1984–2012).—In Malaysia, the deployment of fishing boats as artificial reefs took place on an ad hoc basis at predetermined sites. This occurred only when boats became available as reef materials, as these structures would otherwise become an environment nuisance along the shore or at jetties. Hence, there were no optimal sizes established for boats being used as artificial reefs (Jothy 1986). The first boat deployed as an artificial reef in Malaysia was near Kapas Island in Terengganu in 1984. The project later expanded to the states of Kedah, Kelantan, Perak, and Selangor. Prior to deployment, the boats were cleaned of all debris and loose pieces, which would otherwise result in flotsam upon sinking of the boat. All water and fuel tanks were opened or punctured to prevent the entrapment air, thus hindering sinking of the boat. The engine remained on the boat as a ballast to facilitate deployment (Wong 1991). According to Ali et al. (2008), the DoFM sank more than 1,000 fishing boats and Vietnamese refugee boats since 1984 in the state of Terengganu's coastal waters. In Malaysia, more than 2,000 confiscated wooden fishing boats, as well as refugee wooden boats, were deployed as artificial reefs until 2015 (Ali 2015). Ali et al. (2008) reported that all boat reefs deployed on the sea floor in the coastal waters of Terengganu were destroyed by strong currents during the monsoon season and become an environmental nuisance along the Terengganu shore. Visual observations by divers generally found that only the engines remained attached to the main part of the boats, with little or no marine fouling, sessile, or encrusting organism growth on the hull (Figure 4). Sukarno et al. (1994) reported that wooden fishing boats only remained as a functional reef structure for less than 4 years. Based on scientific evidence reported by DoFM researchers and others, the use of confiscated wooden fishing boats significantly decreased and then was banned in some states by 2012. This action was taken in accordance with the London Convention and Protocol/United Nations Environment Programme 2008 guidelines for the placement of artificial reefs (London Convention and Protocol/UNEP 2008).

Reinforced concrete (1986–2017).—The first concrete reef in Malaysia was deployed in 1986 near Payar Island in Kedah. Concrete is one of the most widely used materials in the world for the fabrication of reef modules (Grove et al. 1991). The basic plan for using concrete in artificial reef construction follows previous efforts in Japan (Jothy 1986). Two basic structural components (i.e., con-

Figure 4. Photographic record by divers showing the engine attached to the main part of the boat. The boat lacks marine fouling, sessile, and encrusting organisms.

crete pipes, 0.6 m long × 0.6 m diameter, and concrete drainage culverts, 0.6 × 0.6 m), were initially used in 1986 (Wong 1991). A pyramid was constructed by 20 modules (pipes or culverts) of concrete arranged on a hardwood platform and secured with steel cables. The resulting concrete pyramid was about 1.2 m long × 2.4 m wide × 2.4 m high and weighed about 1.5 and 2 metric tons (mt) for concrete pipes (Figure 5) and culvert pyramids, respectively. The development of concrete artificial reefs changed when the purpose for deployment was reoriented toward attracting and enhancing commercially important marine species through directed artificial reef design. Two target species were lobsters and squids. A reef designed specifically for lobsters was deployed near Redang Island in 1990, and a reef oriented to attracting squids was deployed near Jambu Bongkok Beach in Terengganu in 1992 (Che Omar and Sukarno 1994; Sukarno et al. 1994)

The first artificial reef modules for lobster included six blocks with two openings stacked on top of each other and facing opposite directions (Figure 6). The lower part of the module consisted of four blocks and the upper part of the module was composed of another two blocks. The lower portion of the module was placed on a concrete slab while the upper part was stacked with another concrete slab. Another design fashioned of concrete was the igloo type. The igloo structure had two arches and a rounded hole that only had one opening to serve as a compartment in which lobsters could hide. There was no other opening on the side, back or top of the module.

The concrete squid reef was designed to attract squid, especially pharaoh cuttlefish *Sepia pharaonis*, to aggregate, mate, and lay their eggs. These concrete-table artificial reefs measured 1 m long × 1 m wide × 0.72 m high, made up of reinforced concrete (grade

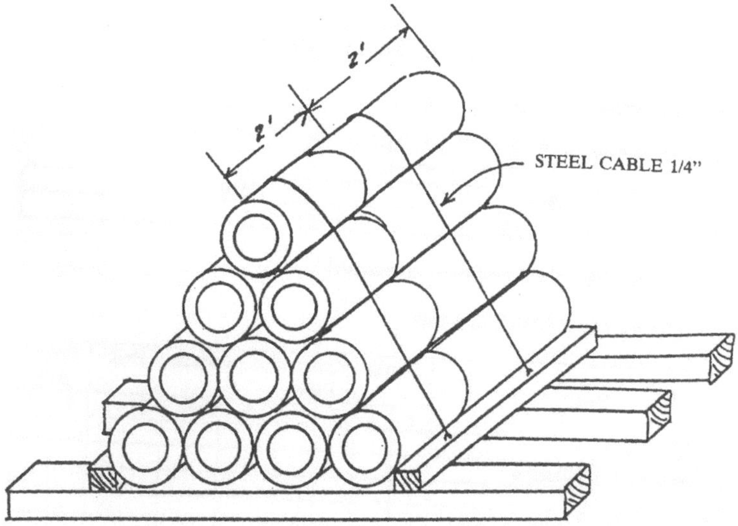

Figure 5. Early 1986 artificial reef design using concrete pipe measuring 0.6 m long × 0.6 m diameter.

Figure 6. The first concrete artificial reef for lobster constructed in Malaysia in 1990; the lower part of the module was a combination of four blocks and the upper part was composed of two additional blocks.

30) that weighed about 250 kg/module. However, after monsoon season, almost all of these structures deployed within the 170-m^2 area were found scattered on the sea floor, destroyed by illegal trawlers (Che Omar and Sukarno 1994). These squid artificial reefs were not built according to marine engineering guidelines, and some of the reefs broke apart less than 1 year after deployment.

In 2000–2001, the Southeast Asian Fisheries Development Center (SEAFDEC) sponsored short- term R&D on the design and construction of artificial reefs and fish aggregating devices (FADs). These efforts mostly focused on the improved design and construction of traditional FADs and artificial reefs for recreational fishing. Newly designed FADs used plastic strips to aggregate pelagic fishes and had rounded, hard-plastic buoys as floats. These midwater features coupled with 3.2-mt concrete artificial reefs to serve as anchors to attract demersal fish species were produced in 2002 (Figure 7). Since 2001, FADs have served as popular sites for recre-

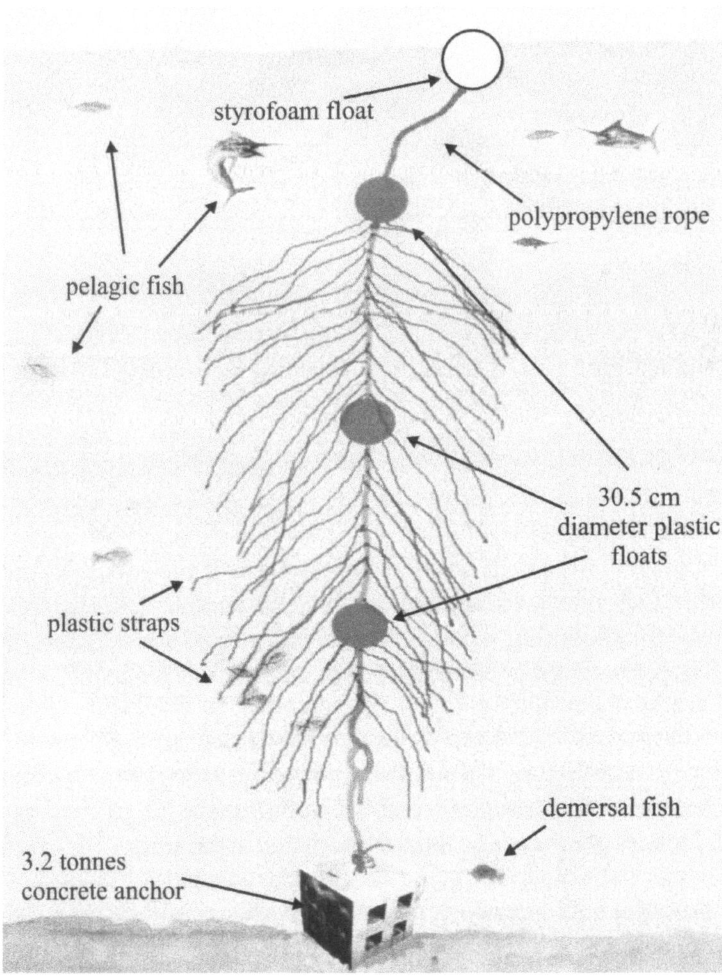

Figure 7. Fish aggregating devices designed in 2002 using plastic strips and round, hard plastic buoys (floats) to aggregate pelagic fishes with 3.2 metric tons of concrete as a base to attract demersal fish species.

ational anglers and divers because the reefs provide convenient sites with concentrations of pelagic and demersal fishes. Moreover, the diversity and abundance of flora and fauna attracted to these structures is similar to that found on coral reefs (Ali et al. 2004). The project was initially considered a success, but later, the reefs were found to be ineffective in aggregating pelagic fish species. Consequently, most of the sites were abandoned. Acoustic surveys and visual observations by divers reported by Ali et al. (2008) indicated that the plastic strips became heavily fouled with sessile and encrusting organisms. This caused the plastic strips and plastic floats to sink to the seabed. However, the concrete anchors remained firmly emplaced on the seabed and continued to function as intended (Ali et al. 2004).

The experience and knowledge gained since 1986 on the design and construction of artificial reefs using reinforced concrete helped improve the planning and management of the artificial reef program in Malaysia beginning in 2006. After 2006, the artificial reef program was implemented in all Malaysian coastal waters for the first time since 1975 and involved both federal and state government funding. New objectives were developed that included the development of new artificial reef designs to deter site damage by fishers using environmentally detrimental fishing gear (e.g., trawls) on traditional fishing ground and in specific zones. The focus of R&D at this time was to design and construct large, reinforced-concrete artificial reefs for installation on both hard and soft substrates. Reinforced-concrete artificial reefs were more effective in aggregating demersal fish species and tended to remain on site for a long time if the construction process followed accepted marine construction procedures (Grove and Sonu 1985; Grove et al. 1991). The smallest artificial reefs measured 1.65 × 1.65 × 1.65 m and the largest was 3.75 m long × 3.75 m wide × 3.85 m high. These weighed from 5 to 42 mt/module.

The initial reef design provided by researchers from SEAFDEC-Marine Fishery Resources Development and Management Department (MFRDMD) was based on the researchers' knowledge of fish behavior, marine engineering, physical oceanography, and target species. The basics of reef architecture were included in the reef design so that materials were optimally used to attract fishes to the module. These features included reef volume, size of holes in the reef structure, reef height, and profile orientation to prevailing currents that provides turbulence and current shadows. Maximum reef surface area was also included in the design to facilitate colonization by sessile, fouling, and encrusting organisms. Final and detailed technical drawings of each design were produced by the Engineering Division of the DoFM.

By 2017, the DoFM had produced 21 new innovative designs for large artificial reefs. These included soft-bottom artificial reefs (11 designs), recreational artificial reefs (4 designs), cube artificial reefs (2 designs), cuboid artificial reefs (2 designs), and lobster artificial reefs (2 designs). Figures 8–18 show several designs of the large artificial reefs constructed between 2006 and 2017. The construction protocols followed the British Standard 8110 in which concrete was at least 50 mm thick, column and beam rebar included four steel rods of Y12, links used R8 at 200 mm c/c, and slab reinforcement used BRC A10 (Figure 19). Ready-mix concrete grades 30, 40, and 50 from a batch plant were used. Concrete cube tests were conducted after 7 and 28 d of construction at government and private laboratories. Curing was allowed for

Figure 8. Soft-bottom artificial reef designed in 2006 measuring 3 m long × 3 m wide × 3.6 m high and weighing 16 metric tons/module using grade 30 concrete with column and slab thickness of 20 cm.

Figure 9. Soft bottom artificial reef designed in 2007 measuring 3 m long × 3 m wide × 3.6 m high and weighing 19 metric tons/module using grade 50 concrete with column and slab thickness of 25 cm.

Figure 10. Cubic artificial reef designed in 2007 measuring 2.0 m long × 2.0 m wide × 3.0 m high and weighing 10 metric tons/module using grade 40 concrete with column and slab thickness of 25 cm.

Figure 11. Lobster artificial reef designed in 2008 measuring 1.65 × 1.65 × 1.65 m and weighing 5 metric tons/module using grade 40 concrete with column and slab thickness of 15 cm.

Figure 12. Cubic artificial reef designed in 2009 measuring 2.5 × 2.5 × 2.5 m and weighing 16 metric tons/module using grade 40 concrete with column and slab thickness of 25 cm.

Figure 13. Cubic juvenile artificial reef designed in 2010 measuring 2.5 × 2.5 × 2.5 m and weighing 20 metric tons/module using grade 40 concrete with column and slab thickness of 25 cm.

Figure 14. Soft-bottom anti-trawler artificial reef designed in 2010 measuring 3.75 m long × 3.75 m wide × 3.85 m high and weighing 32 metric tons/module using grade 50 concrete with column and slab thickness of 25 cm.

Figure 15. Soft-bottom artificial reef for juvenile fish designed in 2010 measuring 3.75 m long × 3.75 m wide × 3.85 m high and weighing 42 metric tons/module using grade 50 concrete with column and slab thickness of 25 cm.

Figure 16. Mini soft-bottom anti-trawler artificial reef (1) designed in 2013 measuring 2.5 m long × 2.5 m wide × 3.5 m high and weighing 17 metric tons/module using grade 40 concrete with column and slab thickness of 20 cm.

Figure 17. Mini soft-bottom anti-trawler artificial reef 2 designed in 2013 measuring 3.0 m long × 3.0 m wide × 3.5 m high and weighing 19 metric tons/module using grade 40 concrete with column and slab thickness of 25 cm.

Figure 18. Mini soft-bottom anti-trawler artificial reef 3 designed in 2017 measuring 3.0 × 3.0 × 3.0 m and weighting 18 metric tons/module using grade 40 concrete with column and slab thickness of 25 cm.

at least 28 d before deployment. Modules not build according to the specifications stated in the quotation or other legal documents were not deployed (Zaidil Abdilla et al. 2010).

During installation, the free-fall deployment method was applied, using a 50–100-mt crane and a special mechanical-release device. Each module was placed on the seabed and spaced between 2 and 3 m from other modules. Soon after the deployment was completed, divers inspected the posi-

Figure 19. Construction work of concrete artificial reef were build according to the British Standard 8110.

tion of artificial reefs and video-recorded the reef for future reference (Zaidil Abdilla et al. 2010). A total of 130 new sites consisting of 21 different designs of concrete artificial reefs were installed by the DoFM during 2006–2017. Each site consisted of 12–130 modules. The success of large artificial reefs deployed along the east coast of peninsular Malaysia, designed especially to protect coastal habitat and enhance fisheries resources, was reported by Ali et al. (2013b).

Polyvinyl chloride pipe 1990–1992.—Polyvinyl chloride pipes were used for constructing artificial reefs in 1990 as a pilot project near Lembu Island in Kedah, based on the success of this material for artificial reefs in the United States (Raja Mohammad Noordin et al. 1994). Initially, the PVC reefs were constructed with 5-cm-diameter PVC pipes. The pipes were assembled into a rectangular frame. The base of each module was filled with cement for added weight (Abdul Razak and Ismail 1994).

During the Sixth Malaysia Plan (1991–1995), a large-scale reef using 10-cm-diameter PVC pipe was deployed near Perhentian Island in 1991 and funded by the Asia Development Bank. According to Raja Mohammad Noordin et al. (1994), 100 "X"-shaped modules measuring 2 m high and 2 m wide were deployed at a depth of 18 m (Figure 20). The "X"-shaped design was constructed in which each arm measured 0.9 m long. Two diagonal arms were joined together to form both the top and the bottom parts of the module. The top and bottom components were connected vertically by 12 pipes that were 1.8 m high. All pipes were connected via a T-joint and glued together using solvent cement. The bottom part of each module had a special hole in the center to attach an anchor. The structures covered about 1,860

Figure 20. The "X"-shaped polyvinyl chloride (PVC) artificial reef design in 1991 using 10-cm-diameter PVC pipe.

m^2 of seafloor and were designed for a sea ranching project similar to a sea ranch in Japan (Grove and Sonu 1985). All pipes were ventilated with holes measuring 5 cm in diameter to serve as entrances into the pipes for juvenile fish.

At the deployed site, the modules were deployed in a larger-shaped "E." The centers of each module were 2.4 m apart. Each bar of the "E" arrangement had two modules placed in two parallel lines, and between these lines, a single line of modules is placed in a "plus sign" arrangement. Since it weighed little, no heavy machinery such as a crane was needed for deployment. However, the structures were of insufficient strength and destroyed within 1 year of deployment (Raja Mohammad Noordin et al. 1994). Since PVC is unsuitable for constructing artificial reefs, especially in the open sea where the material can be exposed to various fishing activities, the DoFM ceased using this material after 1992. Research conducted by Ali et al. (2008) using side-scan sonar could not detect the PVC-constructed artificial reefs at the site.

Ceramic (1992–1993).—A ceramic artificial reef was first deployed at Redang Island in 1992. Two sizes of domestic ceramic cylindrical sewerage pipe was the main material used for the construction of the three- layer artificial reef module to provide habitat for lobsters. Each layer in the module consisted of two cylinder pipes (Figure 21). Two large ceramic pipes (0.4 m diameter) were used for

Figure 21. Artificial reefs composed of ceramic pipe deployed in 1992 using ceramic cylindrical sewerage pipe for the construction of each of three layers to target lobster.

the base. The middle and top layers consisted of smaller (0.25 m diameter) cylinder pipes. A flat concrete slab was placed on top of the top layer of pipes. All pipes had only one opening and were placed so that both openings faced the same direction. The first and third layers faced the same direction while the second layer was perpendicular to the top and bottom layers (Sukarno et al. 1994). This artificial reef did not achieved its objectives as lobsters were never found in or near the modules during a series of monitoring activities conducted by DoFM researchers. Nearly 100% of the module had become buried in the sea floor soon after deployment. Due to this result, the DoFM stopped using this material for constructing artificial reefs since 1993.

Steel (2014).—Large artificial reefs using steel were deployed in the states of Kedah and Terengganu in 2014. Material used was the kind of steel typically used in marine environments because of its greater resistance to pitting and corrosion. The design for artificial reefs deployed on the soft-mud seabed in Kedah was similar to the concrete reefs designed for soft bottoms, but it measured 3.5 × 3.5 × 6.125 m and 13 mt in weight (Figure 22). Diameter and thickness of the main structures were 20 and 0.81 cm, respectively. A total of 55 modules were deployed at two sites within five nautical miles (ap-

Figure 22. Steel artificial reef design in 2014 measuring 3.5 m long × 3.5 m wide × 6.1 m high and weighing 13 metric tons/module using steel diameter and thickness of 20 cm and 8.1 mm, respectively.

proximately 9.3 km) in the coastal waters of Kedah. Another smaller reef was deployed on a hard-sand seabed in the coastal waters of Terengganu and measured 3 × 2 × 2 m and 3.5 mt in weight. A total of 15 modules were deployed at two sites in Terengganu. Visual observations by divers in 2017 indicated that all modules were firmly emplaced on the seabed. Artificial reefs structures deployed on the soft-mud seabed (muddy) in Kedah were found to have sunk about 1.5 m into the substrate. Artificial reefs deployed on firm sand were stable and no scouring occurred. However, due to high logistic costs required for transportation and construction (approximately three times higher than using reinforced concrete) and considering the immediate economic returns, artificial reefs made of steel could were not justified.

Discussion

Based on findings by several DoFM researchers, only newly designed and large artificial reefs deployed after 2006 using reinforced concrete and steel successfully functioned to attract and aggregate fish. These same reefs also deterred illegal trawling. These statements are made in light of the results observed for earlier artificial reef designs constructed of tires, wooden fishing boats, ceramic pipes, and PVC pipes. Fauzi (2010) reported that large artificial reefs composed of concrete deployed on firm-sand seabeds in the coastal waters of the east coast Peninsular Malaysia since the year 2007 remained emplaced on the seabed. Additionally, these reefs had become nursery areas and breeding grounds for multiple fish species, lobsters, and crabs. These artificial reefs also were a hindrance to illegal trawlers as the nets and cod ends of trawlers were often found entangled on the artificial reefs (Ali et al. 2013b). The larger-sized artificial reefs are superior to smaller reefs in attracting more marine flora and fauna. The larger reefs are excellent in performing the dual functions of creating new habitats and deterring the encroachment of trawlers into inshore areas.

In terms of stability, large concrete artificial reefs and steel artificial reefs deployed on soft-mud seabeds along the west coast of Peninsular Malaysia sank into the seabed and become less effective as a reef. Ali et al. (2013a), Azizi et al. (2014), and Azizi et al. (2016) reported that at least 1 year after being deployed, the lower 1.5 m of the reefs became embedded in the soft-mud substrate along the west coast of Peninsular Malaysia. Nevertheless, the modules remained upright and did not overturn. It is predicted that these structures will maintain at their present height and not sink further into the bottom since the reefs were presently resting on a hard bottom layer. This assumes that the reefs will not be subjected to additional sedimentation in the area. All artificial reefs deployed on hard sand were stable and no scouring occurred. The lower part of the artificial reefs were designed to allow the flow of bottom currents above the module base. This design offered little resistance across the module base.

Conclusion

The design, construction, and suitability of deployment sites are of great importance to the success or failure of artificial reefs. Coastal resource enhancement will only achieve its objectives if appropriate habitat is created. At the initial stage of development, most artificial reefs were not designed, constructed, or deployed according to conventional marine engineering construction guidelines. Most early structures were destroyed by nature. In some cases, during the initial stages of deployment, some reefs submerged into the seabed in less than 1 year due to site instability. As expected, most of the early designs were not effective as artificial habitats for demersal fish species, but they did function as short-term FADs.

While implementing artificial reefs programs in Malaysia from 1975 to 2017, researchers and managers gathered valuable experiences that will guide future habitat enhancement efforts. Materials such as tires, confiscated wooden fishing boats, and PVC pipes, although initially presumed to be suitable materials, have all demonstrated a lack of suitability for artificial reef construction. Reinforced concrete of good design, size, and strength will hopefully prove to be the best material to use in artificial reef construction. Concrete reefs offer strength and design flexibility and can be formed in a size that is most suitable for creating new habitat. Additionally, concrete reefs resist environmental conditions and can also withstand the onslaught of illegal trawler activities. Department of Fisheries Malaysia researchers and engineers continue to seek new designs for artificial reefs that will closely imitate natural reefs.

Acknowledgments

The authors express their gratitude to the Director-General of Fisheries Malaysia, the Chief of SEAFDEC/MFRDMD, and the Director of the Engineering Division of DoFM for their endless support for this project.

References

Abdul Razak, L., and I. Ismail. 1994. An analysis of the efficacy of three artificial reefs designs, at Lembu Island, Kedah. Pages 410–419 *in* Proceedings fisheries research conference, 4–6 October 1993. Department of Fisheries Malaysia,Kuala Lumpur.

Ahr, W. M. 1974. Geological consideration for artificial reefs site location. Pages 31–33 *in* L. Calunga and R. Stone, editors. Proceedings on an international conference on artificial reefs. Texas A & M University (TAMU) Sea Grant College, TAMU-SG-74-103, College Station.

Alevizon W. 1990. Statement on tires as artificial reefs for Malaysia by Dr. Alevizon, a scientific expert on artificial reefs. Available: www.reefball.com/map/malaysiatires/Statement%20on%20Tires%20as%20Artificial%20Reefs%20For%20Malaysia%20by%20Dr.htm. (April 2018).

Ali, A. 2015. Assessment on the development and implementation of artificial reefs in Malaysia. Doctoral dissertation. University of Malaysia–Terengganu, Kuala Terengganu. (In Malay.)

Ali A, A. Azizi, A. Shukri, M. Fathy Kameel, and L. Roslan. 2013a. Detecting and mapping of artificial reefs using side scan sonar and sub bottom profiler, volume 1. Southeast Asian Fisheries Development Center (SEAFDEC)-Marine Fisheries Resources Development and Management Department (MFRDMD), Kuala Terengganu, Malaysia.

Ali, A., M. I. Mahyam, A. R. Solahuddin, and I. Jamaludin. 2004. A guide to make and set durable artificial reef fish aggregating devices (ARFAD's) for coastal areas. Southeast Asian Fisheries Development Center (SEAFDEC)-Marine Fisheries Resources Development and Management Department (MFRDMD), SEAFDEC-MFRDMD/SP/6, Kuala Terengganu, Malaysia.

Ali, A., M. I. Mahyam, I. Zaidnuddin, A. Mohamed Pauzi, K. Y. Ku Kassim, and Z. Nor Azman. 2008. Status of artificial reefs in Terengganu waters based on images analysis using side-scan sonar. Southeast Asian Fisheries Development Center (SEAFDEC)-Marine Fisheries Resources Development and Management Department (MFRDMD), Kuala Terengganu, Malaysia. (In Malay.)

Ali, A, A. Mohamed Pauzi, H. Rafezi, M. Abdul Halim, and R. H. Raja Bidin. 2013b. Protecting coastal habitats and enhancing fisheries resources using big size artificial reef in the east coast of peninsular Malaysia. Malaysia Journal of Science 32(SCS Special Issue):19–36.

Azizi, A, A. Ali, A. Shukri, M. F. Fathy Kameel, and L. Roslan. 2014. Detecting and mapping of artificial reefs using side-scan sonar and sub bottom profiler, volume 2. Southeast Asian Fisheries Development Center (SEAFDEC)-Marine Fisheries Resources Development and Management Department (MFRDMD), Kuala Terengganu, Malaysia.

Azizi, A, A. Ali, A. Shukri, M. F. Fathy Kameel, and L. Roslan. 2016. Mapping of artificial reefs and fish aggregating devices in kedah waters. Southeast Asian Fisheries Development Center (SEAFDEC)-Marine Fisheries Resources Development and Management Department (MFRDMD), SEAFDE-MFRDMD/SP/30, Kuala Terengganu, Malaysia. (In Malay.)

Carlisle, J. G., Jr., C. H. Turner, and E. E. Ebert. 1964. Artificial habitat in the marine environment. California Department of Fish and Game Fish Bulletin 124.

Che Omar, M. H., and W. Sukarno. 1994. Early observation of artificial reefs for squid in Terengganu waters (In Malay.) Pages 404–409 *in* Proceedings fisheries research conference, 4–6 October 1993, Kuala Terengganu, Malaysia. Department of Fisheries Malaysia, Kuala Terengganu. (In Malay.)

Fauzi, A. R. 2010. Status report of artificial reefs program in Malaysia. Pages 28–54 *in* A. Ali, A. Mohamed Pauzi, A. R. Fauzi, and A. Osamu, editors. Proceedings of workshop on artificial reefs for the enhancement of fishery resources. SEAFDEC/FRA Joint Program Regarding Artificial Reefs for the Enhancement of Fishery Resources, 4th August 2009, Putrajaya, Malaysia. Department of Fisheries Malaysia, Putrajaya.

Grove, R. S., M. Nakamura, and C. J. Sonu.1991. Design and engineering of manufactured habitats for fisheries enhancement. Pages 109–152 *in* W. Seaman, Jr. and L. M. Sprague, editors. Artificial habitats for marine and freshwater fisheries. Academic Press, San Diego, California.

Grove, R. S., and C. J. Sonu. 1983. Review of Japanese fisheries reef technology. Southern California Edison Company, Report 83-RD-137, Rosemead, California.

Grove, R. S., and C. J. Sonu. 1985. Fishing reef planning in Japan. Pages 187–251 *in* L. Calunga and R. Stone, editors. Proceedings on an international conference on artificial reefs. Texas A&M University (TAMU) Sea Grant College, TAMU-SG-74-103, College Station.

Ino, T. 1974. Historical review of artificial reefs activities in Japan. Pages 21–23 *in* L. Calunga and R. Stone, editors. Proceedings on an international conference on artificial reefs. Texas A & M University (TAMU) Sea Grant College, TAMU-SG-74-103, College Station.

Jothy, A. A. 1982. A step towards alleviating the problem of declining fish resources in Malaysian coastal waters. Pages 7–82 *in* K. S. Ong and A. A. Jothy, editors. Our seas in perspective. Malaysia Society of Marine Sciences, Serdang.

Jothy, A. A. 1986. Artificial reef development in Malaysia. Pages 5–16 *in* W. T. Meng, O. J. Eong, and K. S. Ong, editors. Proceedings of the ninth annual seminar towards greater productivity of our coastal waters. Malaysia Society of Marine Sciences, Serdang.

Latun, A. R., and I. Ismail. 1994. An analysis of the efficacy of three artificial reef designs at Lembu, Island Kedah. Pages 410–419 *in* Proceedings fisheries research conference, 4–6 October 1993, Kuala Terengganu, Malaysia. Department of Fisheries Malaysia, Putrajaya.

Latun, A. R., and A. Mohammed Pauzi. 1991. Artificial reefs in Malaysia: a country review paper. Pages 423–435 *in* Proceedings symposium on artificial reefs and fish aggregating devices as tools for the management and enhancement of marine fishery resources, 14–17 May 1990, Colombo, Sri Lanka. Food and Agriculture Organization of the United Nations, Regional Office for Asia and the Pacific, RAPA Report 1991/1, Bangkok.

London Convention and Protocol/UNEP (United Nations Environment Programme). 2009. London Convention and Protokol/UNEP guidelines for the placement of artificial reef. International Maritime Organization, London.

Meier, H. M., J. R. Martin, L. D. Feigenbaum, and M. Bell. 1986. Artificial reefs in Virginia old beginnings and new directions. Pages 337–347 *in* F. M. D'Itri, editor. Artificial reefs: marine and freshwater applications. Lewis Publisher, Chelsea, Michigan.

Mohammed Shaari, S. A. L., and C. H. Leong. 1976. Demersal fish resources in Malaysian waters, 11. Third trawl survey off the west coast of peninsular Malaysia (16 November–11 December 1974). Ministry of Agriculture and Rural Development, Fisheries Bulletin 13, Kuala Lumpur, Malaysia.

Mohammed Shaari, S. A. L., G. Rauch, K. S. Ong, and S. P. Tan. 1974. Demersal fish resources in Malaysian waters, 2. Trawl survey of the coastal waters off the west coast of peninsular Malaysia (12 December 1970–22 January 1971). Ministry of Agriculture and Rural Development, Fisheries Bulletin 3, Kuala Lumpur, Malaysia.

Nor Azman, Z., and N. Mohd Saki. 2017. Steel artificial reefs. Fisheries Bulletin, Department of Fisheries Malaysia 100:3–4. (In Malay.)

Ogawa, R. 1973. Various biological question regarding artificial reefs. Ocean Age 3:21–30.

Pathansali, D., K. S. Ong, S. A. L. Mohammed Shaari, and J. Carvalho. 1967. Preliminary results of trawling investigation off Penang. Proceedings of the Indo-Pacific Fisheries Council 12:181–201, Bangkok.

Raja Mohammad Noordin, R. O., E. K. Cheah, W. Sukarno, A. M. H. Mutalib, H. Mubarak, R. H. Raja Bidin, and C. H. Che Omar. 1994. Design and construction of artificial reefs in Malaysia. Bulletin of Marine Science 55:1050–1161.

Sanders, M. J. 1974. Artificial reefs in Australia. Pages 84–90 *in* L. Calunga and R. Stone, editors. Proceedings on an international conference on artificial reefs. Texas A & M University (TAMU) Sea Grant College, TAMU-SG-74-103, College Station.

Stone, R. B. 1972. Artificial reefs of waste material for habitat improvement. Marine Pollution Bulletin 3(2):27–28.

Sukarno, W., R. O. Raja Mohammad Noordin, M. H. Che Omar, and M. N. Rosdi. 1994. Artificial reefs in Malaysia. Department of Fisheries Malaysia, Putrajaya. (In Malay.)

Wong, F. H. 1991. Artificial reefs development and management in Malaysia. Pages 392–422 *in* Proceedings symposium on artificial reefs and fish aggregating devices as tools for the management and enhancement of marine fishery resources, Colombo, Sri Lanka, 14–17 May 1990. Food and Agriculture Organization of the United Nations, Regional Office for Asia and the Pacific (RAPA), RAPA Report 1991/11, Bangkok.

Zaidil Abdilla, A. S., M. D. Safari, A. Ali, M. A. Mohd Ridzuan, and M. Abdul Rauf. 2010. Design, construction and deployment of artificial reefs in Malaysia. Pages 80–108 *in* A. Ali, A. Mohamed Pauzi, A. R. Fauzi, and A. Osamu, editors. Proceedings of workshop on artificial reefs for the enhancement of fishery resources. Department of Fisheries Malaysia, Putrajaya.

American Fisheries Society Symposium 86:315–321, 2018

An Alternative Approach to Designing Artificial Reefs to Manage Marine Fisheries

Stephen A. Bortone*
Osprey Aquatic Sciences, LLC
47 West Shore Road, Windham, New Hampshire 03087, USA

Abstract.—Most artificial reef deployments to date have been directed toward fish assemblage enhancement, with only a few that are fisheries- or species-specific. The ostensible rationale for this approach is that when the artificial reef fish assemblage increases in a given area, the component target species will also increase. This premise also implies that the target species is characteristically reef- or structure-associated. Targeted species may not be niche-specific with regard to their association with reef attributes; however, trophic groups may be niche generalists for a suite of reef attributes. Artificial reef designs could be directed toward trophic groups and not individual species or the entire assemblage. This would allow more efficient optimization of artificial reef designs directed toward generalized life history features common among species within a trophic group. The delineation of species according to reef niche should include attributes based on trophic features that will allow optimal management through trophic-group-oriented artificial reef deployments. If successful, trophic groups within an assemblage could be more efficiently managed to allow the full integration of artificial reefs into fisheries management on a worldwide basis. Thus, the application of artificial reef design principles could become universal for targeted trophic groups.

Introduction

Bortone (2015) reviewed the history of the International Conference(s) on Artificial Reefs and Related Habitats (CARAH). Each conference (held nearly every 3 or 4 years since 1974) has brought together authoritative scientists and environmental managers to highlight the latest information on research, extent, purpose, and new developments regarding artificial reefs. While each conference, naturally, has had an emphasis on artificial reefs in the region where the conference was being held, the overall effect of the 11 conferences held, to date, has been to bring together information from around the world with regard to artificial reef deployments. Although there are many reasons cited for deploying these reefs (e.g., habitat protection and mitigation, creating waves for surfing and recreational dive/snorkeling sites, etc.; Seaman and Jensen 2000), the most often cited reason was to support fisheries management. Interestingly, Bortone (2011)

* Corresponding author: steve.bortone@gmail.com

noted that while there are literally thousands of artificial reefs deployed in hundreds of programs, there are few actual examples of artificial reefs being formally incorporated into fisheries management as part of a fisheries management strategy.

Nevertheless, there are a few notable examples where artificial reefs have been regularly and successfully employed to manage fisheries (e.g., Clarke et al. 2002; Kim et al. 2011). However, most environmental resource management programs make only ancillary use of artificial reefs in their fishery management plans (see Shipp and Bortone 2009) and do not make use of the structures prescriptively, as is the case for other fishery management tactics.

Bortone (2011) offered several reasons as to why this is the situation. For example, in the United States, most artificial reef programs are under the auspices of state, county, and local governments, with most reefs being deployed to enhance recreational activities. Owing to the nature of artificial reefs in the marine environment (i.e., research is expensive, sample sizes are often too small, data variance is high, etc.; Bortone 2006), research results are inconsistent, unpredictable, and not easily comparable to other studies. These factors and others have not allowed natural resource managers the luxury of incorporating artificial reefs into management plans as most management plans require reliable prediction of actions to affect specific outcomes.

First Steps in Designing Artificial Reefs to Manage Marine Fisheries

Clearly, there is a problem that deters resource managers from incorporating artificial reefs into marine fisheries management plans. A large part of the problem can only be overcome by directed research that specifically identifies the factors responsible for variability in the data used to establish reefs as part of a management strategy. Another part of the problem, however, lies in the objectives being established for the purpose of the reef. Lindberg and Relini (2000) argued for establishing well-defined plans with clearly defined objectives in the initial stages of planning an artificial reef project.

The present perspectives on artificial reefs have approached fishery management objectives in two seemingly opposing ways. The first objective has been in the direction of improving or establishing the overall reef structure in an area, the assumption being that once a reef fish assemblage becomes established, there will be an increase in reef-fish, assemblage-associated species, some of which are the desired target species. There are a plethora of studies designed (or ostensibly designed) to increase the overall assemblage, as if establishing the assemblage is a sufficient objective for the reef design and deployment. Researchers usually evaluate assemblages before and after reef deployment and assume (whether due to attraction or production) that demonstrating an increase in the assemblage parameters (i.e., species diversity, species richness, biomass, etc.) will be sufficient to justify further deployments for management purposes.

The alternative approach, to date, has been to deploy a reef for the express purpose of improving population parameters (i.e., abundance, size, or growth rate) of a target species. The premise is that it is possible to identify the species' characteristics that the features or attributes of a reef can improve. Subsequently, reef designs that provide more of the reef habitat attributes preferred by the target species will help the species overcome structurally associated habitat features deemed to be limiting factors to the desired species' characteristics. As a result, the target species can overcome the limiting factors (as long as they are structural habitat factors) and facilitate an increase in its population parameters to the point that the fishery improves from a management perspective.

Both avenues (assemblage improvement and target species enhancement) are admirable approaches, but neither have been particularly successful to date, especially when they are being considered for inclusion as a management option. The assemblage improvement approach has limited application and probability of success in fishery management because some (or many) assumptions regarding the target species' management are voided. In the assemblage improvement model for reef design and deployment, much of the reef material deployed only indirectly pertains to the target species. Thus, the assemblage improvement approach is economically and energetically inefficient. Moreover, there is little information that indicates building a reef directed to improve an assemblage will have a level of certainty sufficient to attract (or produce) the desired species among the dozens of potential assemblage members. This is especially true when one considers the potential lottery effect of reef colonization that occurs through recruitment (sensu Sale 1978). The lottery effect implies that, owing to nearly random factors or events, it will be difficult to predict the assemblage of organisms (both in terms of species composition and relative abundance) that find their way to a structure through dispersal processes.

The target species enhancement approach assumes that it is possible to identify the specific habitat features that are limiting the target species (see Bortone, in press). Consequently, managers would then deploy a reef with features or attributes that would allow the target species to overcome the limitations on their life history characteristics (population abundance, growth, etc.). While this may seem a desirable approach, problems lie in the ability of researchers to accurately identify and quantify the reef features that can be manipulated to quantifiably improve population attributes of the target species in a predictable (and repeatable) way. Moreover, the application of such study results may be temporally and/or spatially limited in their effective application. Unless the target species is especially well adapted (i.e., preadapted) to some attribute of the reef that is being provided via the reef deployment, there is little evidence that such an exercise would do anything but lead to a never-ending series of site-specific studies (to the benefit of graduate students everywhere!), the results of which may only have application to the immediate study site.

An Alternative When Designing Artificial Reefs to Manage Marine Fisheries

Briefly, artificial reefs could be designed (shape, size, composition, structural features, etc.) and deployed (location, depth, substrate, proximity) to attract and retain (and possibly increase productivity through growth and reproduction) fishes in a predefined guild or group that comprises the assemblage rather than individual species or the entire community. Figure 1 presents a hierarchy of biological organization where guilds are positioned. Note that "assemblage" here represents the portions of the natural community that have become associated with artificial reefs in deference to the term "community," which is a co-evolved, deterministic part of the ecosystem (after Bohnsack et al. 1991).

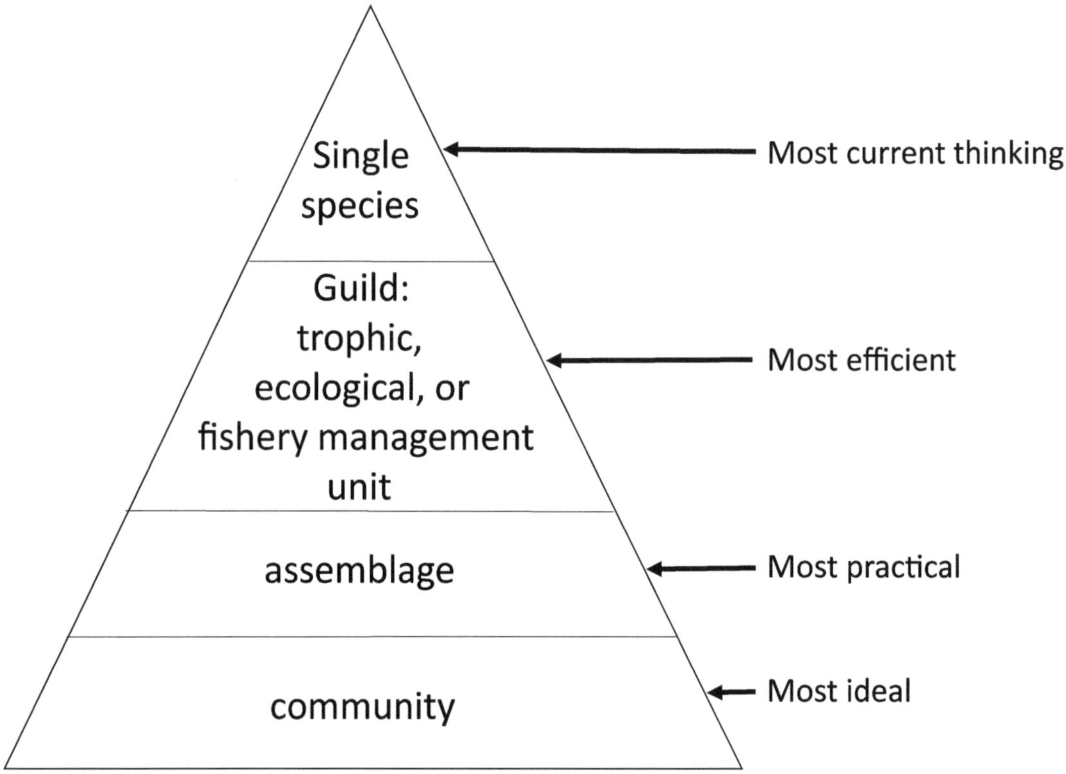

Figure 1. Hierarchical management diagram indicating the various levels of biological organization targeted by artificial reef deployments.

Nelson and Bortone (1996) examined the food habits of fishes feeding on artificial reefs in the northern Gulf of Mexico and were able to group (cluster) fishes by habitat affinity based on stomach contents. This study indicated that while the niche concept implies that a species is theoretically unique for many of its ecological attributes, most fishes in a reef fish assemblage can be assigned to a few habitat types/trophic groups based on feeding preferences (Table 1). Additionally, Farmer et al. (2016), when streamlining a management approach to stock complexes in the Gulf of Mexico, were able to group fishes in the reef complex based on various life history features and fishery attributes. The grouping criteria examined were life-history parameters as well as landings regarding presence–absence of reef fish in the commercial longline and vertical line fisheries; the observer program; and headboat, recreational, and bottom-longline fisheries. While the fishes could be grouped together differently based on the kinds attributes used for grouping, the exercise demonstrated the potential that this approach has in simplifying efforts to design and deploy artificial reefs in fishery management. This is because fish were often caught together, either as a result of the method used to capture them, their habitat affinities, or both.

Recognizing the similar habitat requirements within each identified group allows design plans for artificial reefs to more directly meet the needs of more than one target species. There are several advantages to this approach in reef design. First, because of the uneven availability of species likely

Table 1. Feeding guilds based on a cluster analysis of stomach contents (after Nelson and Bortone 1996).

Feeding guilds	Species
Reef-associated open-water feeders	Vermilion Snapper *Rhomboplites aurorubens* Greater Amberjack *Seriola dumerili*
Water column picker	Cocoa Damselfish *Stegastes variabilis*
Upper structure predators	Tomtate *Haemulon aurolineatum* White Grunt *Haemulon plumierii* Pigfish *Orthopristis chrysoptera*
Upper structure pickers	Spottail Pinfish *Diplodus holbrookii* Pinfish *Lagodon rhomboides*
Ambush predators	Gulf Toadfish *Opsanus beta*
Lower structure pickers	Gray Triggerfish *Balistes capriscus* Striped Burrfish *Chilomycterus schoepfi* Slippery Dick *Halichoeres bivittatus*
Lower structure crustacean predators	Bank Sea Bass *Centropristis ocyurus* Cubbyu *Pareques umbrosus* Belted Sandfish *Serranus subligarius* Whitespotted Soapfish *Rypticus maculatus*

to be attracted or aggregated to a structure, it is probable that the structure will provide limiting factors to more than a single species. Second, as indicated previously, determining the species-specific limiting factors may require considerable effort to be realized. A more generalized approach to a broader audience of target species that occur within a guild will likely allow a least some species to overcome limiting factors. Last, by purposefully targeting groups by providing habitat features through artificial reef deployments, more can be gleaned from learning of multiple simultaneous successes and failures of the plethora of target species associated with the structure.

As Farmer et al. (2016) indicated, species assignments to groups can be based on many factors. Reef designers should consider practical reasons for developing a structure to benefit the most efficient reef designs that meet the most important goals of the management plan. The approach of designing for ecological, fishery, or life history groups/guilds has other advantages. It aligns nicely with more recent advocacy for ecosystem-based management (e.g., Trochta et al. 2018). Thus, targeting functional groups within an assemblage jump-starts progress toward developing an ecosystem approach toward environmental management. Regardless of the rationale for the assemblage partitioning (e.g., trophic group, life history similarities, microhabitat preferences) the group inevitably will become a fishery management unit.

Broader Applications

Suggested here is a concept that the species components of an artificial-reef-associated fish assemblage can be grouped into fishery management units and identified according to the features that a designed artificial reef provides. While species-specific management often requires clearly identifying a species' needs, here a more generalized reef design should appeal to a group of species with similar but not necessarily identical requirements. These groups were partially identified by Nelson and Bortone (1996) for the northern Gulf of Mexico, but the generalized group requirements are likely to be worldwide. This is because broad habitat niches occur nearly worldwide, with many species already adapted for generalized conditions. Hence, identification of successful, group-targeted designs in one region should have application to other regions. Consequently, an era of efficiency in artificial reef design should occur with much borrowing and exchanging of plans occurring to facilitate the regular and prescribed incorporation of artificial reefs into fishery management plans.

References

Bohnsack, J. A., D. L. Johnson, and R. F. Ambrose. 1991. Ecology of artificial reef habitats and fishes. Pages 61–107 *in* W. Seaman, Jr. and L. M. Sprague, editors. Artificial habitats for marine and freshwater fisheries. Academic Press, San Diego, California.

Bortone, S. A. In press. Artificial reefs in the future management of Red Snapper. *In* S. Szedlmayer and S. A. Bortone, editors. Red Snapper biology in a changing world. CRC Press, Boca Raton, Florida.

Bortone, S. A. 2006. A perspective of artificial reef research: the past, present, and future. Bulletin of Marine Science 37:11–39.

Bortone, S. A. 2011. A pathway to resolving an old dilemma, lack of artificial reefs in fisheries management. Pages 311–321 *in* S. A. Bortone, F. P. Brandini, G. Fabi, and S. Otake, editors. Artificial reefs in fisheries management. CRC Press, Boca Raton, Florida.

Bortone, S. A. 2015. CARAH (International Conference on Artificial Reefs and Related Aquatic Habitats): an historical perspective of accomplishments. Journal of Applied Ichthyology 31(Supplement 2):3–14.

Clarke, S., A. Leung Wai-Yin, Y. M. Mak, R. Kennish, and N. Haggan. 2002. Consultation with local fishers on the Hong Kong artificial reefs initiative. ICES Journal of Marine Science 59:S171–S177.

Farmer, N. A., R. P. Malinowski, M. F. McGovern, and P. J. Rubec. 2016. Stock complexes for fisheries management in the Gulf of Mexico. Marine and Coastal Fisheries: Dynamics, Management, and Ecosystem Science 8:177–201.

Kim, C. G., S. I. Lee, H. K. Cha, J. H. Yang, and Y. S. Sun. 2011. A case study of artificial reefs in fisheries management: enhancement of Sandfish, *Arctoscopus japonicus*, by artificial reefs in the eastern waters of Korea. Pages 111–124 *in* S. A. Bortone, F. P. Brandini, G. Fabi, and S. Otake, editors. Artificial reefs in fisheries management. CRC Press, Boca Raton, Florida.

Lindberg, W. J., and G. Relini. 2000. Integrating evaluation into reef project planning. Pages 195–235 *in* W. Seaman, Jr., editor. Artificial reef evaluation with application to natural marine habitats. CRC Press, Boca Raton, Florida.

Nelson, B. D., and S. A. Bortone. 1996. Feeding guilds among artificial-reef fishes in the northern Gulf of Mexico. Gulf of Mexico Science 1996(2):66–80.

Sale, P. F. 1978. Coexistence of coral reef fishes—a lottery for living space. Environmental Biology of Fishes 3(1):85–102.

Seaman, W., Jr., and A. Jensen. 2000. Purposes and practices of artificial reef evaluation. Pages 1–9 *in* W. Seaman, Jr., editor. Artificial reef evaluation with application to natural marine habitats. CRC Press, Boca Raton, Florida.

Shipp, R. L., and S. A. Bortone. 2009. A prospective of the importance of artificial habitat on the management of Red Snapper in the Gulf of Mexico. Reviews in Fisheries Science 17:41–47.

Trochta, J. R., M. Pons, M. B. Rudd, M. Krigbaum, A. Tanz, and R. Hiborn. 2018. Ecosystem-based fisheries management: perceptions of definitions, implementations, and aspirations. PLOS (Public Library of Science) ONE [online serial] 13(1):e0190467.

Index

F

H

K

L

M

N

Q

S

T